SOILS

AND THEIR USE IN

WALES

COUNTY COVERAGE IN THE REGIONAL BULLETINS

Bulletin No.10 (Northern England)
Cleveland, Cumbria, Durham, Humberside, Northumberland, North Yorkshire, Tyne and Wear, South Yorkshire, West Yorkshire.

Bulletin No.11 (Wales)
Clwyd, Dyfed, Gwent, Gwynedd, Mid-Glamorgan, Powys, South Glamorgan, West Glamorgan.

Bulletin No.12 (Midland and Western England)
Cheshire, Derbyshire, Greater Manchester, Hereford and Worcester, Lancashire, Leicestershire, Merseyside, Nottinghamshire, Shropshire, Staffordshire, West Midlands, Warwickshire.

Bulletin No.13 (Eastern England)
Bedfordshire, Cambridgeshire, Essex, Hertfordshire, Lincolnshire, Norfolk, Northamptonshire, Suffolk.

Bulletin No.14 (South West England)
Avon, Cornwall, Devon, Dorset, Gloucestershire, Somerset, Wiltshire.

Bulletin No.15 (South East England)
Berkshire, Buckinghamshire, East Sussex, Hampshire, Isle of Wight, Kent, Oxfordshire, Surrey, West Sussex.

Plate 1. Denbigh 1 association near Talybont, Dyfed. After several weeks of dry weather, pale patches in grass fields indicate shallower soils. The Dovey estuary and Snowdonia are in the background.

SOIL SURVEY OF ENGLAND AND WALES
BULLETIN No. 11

SOILS

AND THEIR USE IN

WALES

By
C.C. Rudeforth, R. Hartnup, J.W. Lea, T.R.E. Thompson,
and P.S. Wright

HARPENDEN
1984

ISBN 0 7084 0295 X

Series Editor	R.D. Green
Art Editor	E.M. Thomson
Line Illustrators	M.J. Williamson, M.S. Skeggs, Elaine Avis, Susan Harrop, Susan Nicolson and D.V. Hogan

Text set in Times Roman
Tables set in Univers

Typeset by Method Ltd, Woodford Green, Essex
Printed by Whitstable Litho Ltd, Whitstable, Kent

PREFACE

The Survey commenced a five year project in 1979 to produce a soil map of the whole of England and Wales, at a scale of 1:250,000 and to describe soil distribution and related land quality in appropriate detail. The project was commissioned by the Ministry of Agriculture, Fisheries and Food, with the primary aim of providing a systematic inventory capable of being used or interpreted for a wide range of purposes including agricultural advisory work, but also for the many facets of land use planning and national resource use. Since England and Wales is divided by the Ministry into five regions and the Principality of Wales, the commission requested six regional maps and bulletins. The maps were published in 1983 and all six bulletins in 1984.

This ambitious project is the first complete account of our soils since the county series initiated by the Board of Agriculture in the late 18th and early 19th centuries, in which there were introductory contributions on soils and soil distribution, some including a schematic county soil map.

The soil map was constructed on the firm base of earlier detailed mapping that extended over a period of 40 years, during which time a fifth of the country was completed at 1:63,360 and 1:25,000 scales. These surveys have been updated and incorporated, so credit for the map can be extended to many scientists who were former members of staff.

The books and map provide valuable information both for those who wish to learn about our soils and those with responsibility to protect and use them effectively.

In 1983 the maps attracted the British Cartographic Society's John Bartholomew Award for excellence in small scale thematic cartography, an honour for which much credit goes to Ellis Thomson and his cartographic staff.

The project was managed by J.M. Hodgson, who with A.J. Thomasson edited this Bulletin. The soils of Wales were mapped by R.Hartnup, J.W. Lea, T.R.E. Thompson, and P.S. Wright under C.C. Rudeforth as Regional Officer.

It has been my privilege to direct the work of the Survey in this exciting period and particularly to be associated with the dedicated team that has contributed professional skill, vast energy and enthusiasm to the project.

3rd April 1984   *D. Mackney*
Rothamsted Experimental Station,   *Head of the Soil Survey of*
Harpenden, Herts   *England and Wales*

ACKNOWLEDGEMENTS

The text was prepared mainly by regional staff with contributions from staff in other regions. A.J. Thomasson is to be particularly acknowledged for new ideas which have advanced understanding of interactions between soil and climate affecting crop performance and management.

The introductory text of Chapter 4 was written by F.J. Sparkes of the Land and Water Service of the Ministry of Agriculture, Fisheries and Food. The survey is indebted to Mr Sparkes, his colleagues especially J.W. Astill, also to R.K. Fry and staff of the National College of Agricultural Engineering, Silsoe, for friendly advice and cooperation during the compilation of the table in Chapter 4. Their help is much appreciated.

We wish to thank also Miss A.E. Tarran, A.D. Hughes, Edryd Jones, H.W. Roberts and R.J. Skinner of ADAS for their guidance and cooperation in the compilation of Chapter 5 and one of the included tables. Their participation was most helpful. For help with the text and tables on Forestry in Chapter 5 we are much indebted to Forestry Commission Staff, particularly A.J. Grayson, D.G. Pyatt and R.E. Crowther. The Forestry Commission willingly allowed access to detailed soil maps of their forests in Wales.

M. Ayles, R.H. Marles, A. Woodcock and J. Stancombe of the Meteorological Office supplied the data used to compile Figures 8, 9, 10 and 11 for droughtiness, workability and crop suitability assessments. Their help and advice are gratefully acknowledged. The calculations to prepare the operational data sets could not have been completed without the efforts of Mrs C. Thomas and colleagues in the Rothamsted Computer Department.

The text was word processed by Mrs J.M. Cooles, Mrs C.M. Gosney, Mrs H. Roberts, Mrs J. Robinson, Mrs J.Y. Shuttleworth and Mrs C.M. Scott. The willingness and speed with which they learned and adopted new techniques enabled the simultaneous preparation and compilation of this and its five companion volumes. Without their skills, and development work on word processing for automatic typesetting by R.D. Green, the Bulletins could not have been prepared so quickly and efficiently. Mrs C.M. Gosney had the onerous responsibility for the final marking up and compilation of the word processor discs from which this book was printed directly. Dr Rudeforth especially wishes to thank Mrs S. Waldron for secretarial help and assistance with indexing.

The data used in the preparation of Figures 12, 13, 14 and 15 in Chapter 1 were derived from the parish summaries published in the 1980 agricultural census of the Ministry of Agriculture, Fisheries and Food, using the CAMAPGB computer mapping program prepared by Messrs B. Fletcher and J. Hotson. The survey is grateful for this help and for permission from MAFF to publish the data and details from the same census used in Table 3 and Fig.16.

P.J. Loveland is responsible for the analytical data in Appendix 2.

The survey could only be made with the agreement of farmers and landowners and we are grateful for this continuing friendly cooperation.

CONTENTS

4 LAND DRAINAGE   261

5 AGRICULTURAL AND FORESTRY INTERPRETATIONS   271

LISTS OF ILLUSTRATIONS

TABLES

INTRODUCTION

The mapping programme on which the 1:250,000 National Map and this book are based started in April 1979. Up to that time approximately one fifth of the country had been surveyed at scales of 1:25,000 or 1:63,360 (Fig.1). Between 1979 and March 1982 the field staff covered the previously unsurveyed ground, recording observations by small pits or borings to a depth of 1 metre at an average frequency of 250 per 100 km². Concurrently they described soil profiles in greater detail on a 5 × 5 kilometre grid for a systematic National Soil Inventory, the statistical results of which will be published later. These pits were sampled for laboratory analysis. Aerial photographs, geological maps and other sources were used to help position soil boundaries. The details of survey procedure varied depending on previous knowledge of the country and on the nature of the terrain. In many districts sample farms were surveyed, in others the distribution of the observations was less concentrated, or located on transects related to relief or to the underlying geological pattern. All the observations were recorded on computer compatible field cards for storage in the Soil Survey Information System.

The soil map and legend
The National Soil Map shows 296 geographic soil associations of which 92 occur in Wales and are listed on the key printed on the map sheet for the Principality. All 296 associations are briefly summarized in the separate legend booklet. The soil associations which the map, the legend and this book portray and describe, are identified by the most frequently occurring soil series and combinations of ancillary series. They are named after key series and are of two kinds. The first kind is named after the single most extensive constituent soil series in the various delineations of the association. These are shown in capital letters in the keys on the map sheets and in the legend. In some associations the most extensive soil series is dominant (greater than 50 per cent), and for these the main soil series are listed in the legend in order of decreasing extent. The second kind of soil association is named in small letters in the legend. In these there is no single most extensive soil series or the proportions of constituent soil series are more varied. In general, these associations have a larger number of component soil series.

The map has 67 colours representing the subgroups to which soil series identifying soil associations have affinity. The associations are further identified by number and letter codes. The numbers of the code indicate the predominant major soil group, soil group and subgroup. For example, association 654a is dominated by soils of the Hafren series which belongs to soil subgroup 6.54, the *ferric* subdivision of soil group 6.5 *Stagnopodzols*, which in turn is part of major soil group 6, *Podzolic soils* (§ 6).

This bulletin

This bulletin is designed to meet the needs of readers with a broad interest in the soils of Wales and, by the provision of keys, profile descriptions and references, to help those who require more specific information on soils.

Chapter 1 describes the geology, landscape, climate and current land use. Soil formation, classification and the terminology used throughout the subsequent chapters are explained in Chapter 2. The descriptions of the soil associations in Chapter 3 are in alphabetical order. For each extensive association there is comment concerning national distribution, followed by brief profile descriptions of the main component soil series and an account of regional variations in composition and distribution. Keys are included to help the user to identify component soils. Where relevant, details of soil water regimes, opportunities for cultivation, cropping and drought risk, complete the description. The less extensive associations are more briefly described, and for those in the uplands mainly under semi-natural vegetation there is a summary of ecological relationships. Chapter 4 deals with land drainage. Here the degree of waterlogging, the likely response to treatment, drain spacing, need for permeable back-fill and subsoiling and the suitability for moling are described for individual soil series. The keys to component soils in Chapter 3 allow this information to be applied at field level. Interpretations of the soil map for agriculture and forestry are developed in Chapter 5. Suitability of land is assessed for sustained production of selected arable crops and for intensive grassland use. The suitability of the land for early potatoes and forestry is also treated.

Use of the map, legend and bulletin

The map, legend and bulletin provide soil information at different levels of detail. At the simplest, most general level, the brief descriptions of the soil associations in the legend booklet should suffice. Reference to the legend, using the index colours and notation on the map as a guide, gives the names of the main component soil series and their soil subgroups. The legend also gives brief notes on geology, site, soil, typical cropping and land use. Users requiring more information should refer to the bulletin where, in Chapters 1 and 2, there is further background material and, in Chapter 3, the soil associations are fully described in alphabetical order. The keys to component soil series provided for the most extensive associations should enable the user to identify individual soil series with some confidence even though the keys have been simplified and generalized for lay use. Readers requiring detailed description of soil series are referred as appropriate either to earlier publications or, in some cases, to Appendix 2 in which full descriptions and analytical data are provided in mini-print.

After identifying the soil series, further reference can be made to Chapter 4 for details of suitable drainage systems and to Chapter 5 for agricultural and forestry interpretations.

The information given by the maps and legend is limited in several ways, so care is needed in its interpretation. Firstly, soil patterns in England and Wales are commonly complex and vary greatly in composition. Partly because of this the detail shown on the soil map varies. It varies too because of differences in the density of observations. The greatest detail is commonly in districts previously surveyed at larger scales. Secondly, the minimum area that can be shown on the map is 0.5 km² and because of this many soil associations include small patches of soils which, at a larger scale, would be correlated with a different map unit. Despite these inclusions, which commonly have a different land use potential and management characteristics, many associations delineate areas of land that are broadly homogeneous for interpretative purposes. Thirdly, only the main soil series are described in the legend. The associations usually consist of ten or more soil series many of which are inextensive and are briefly described in the bulletin.

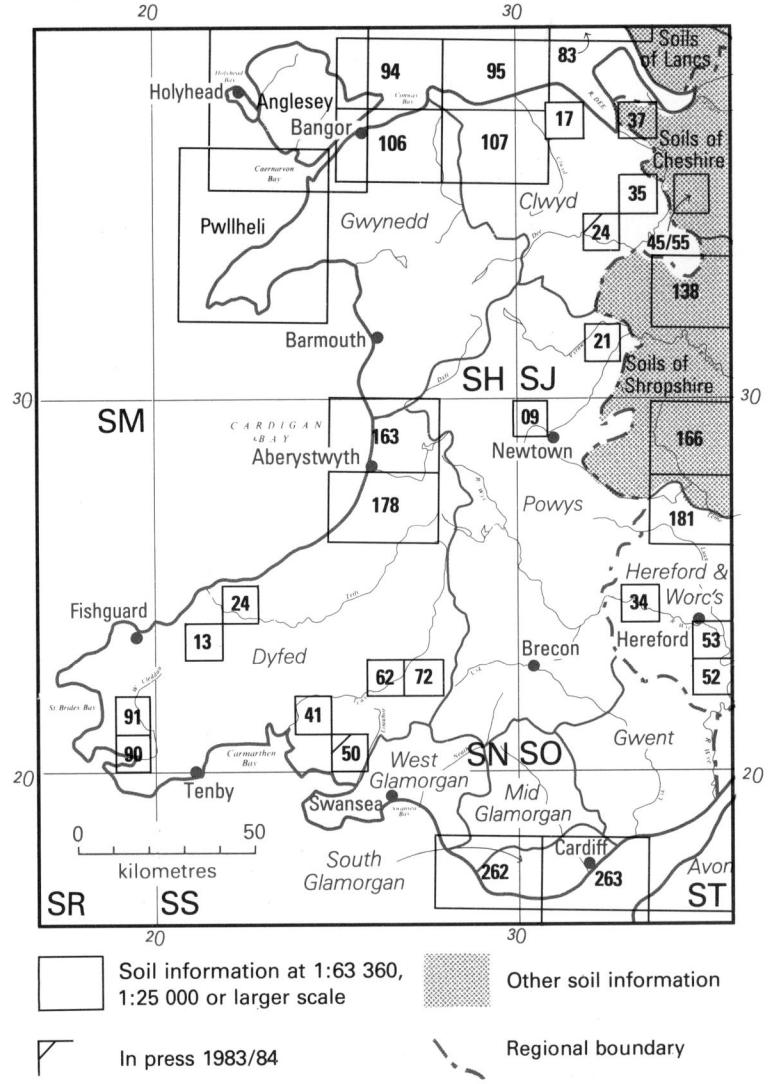

Figure 1. The region and previous surveys

1

DESCRIPTION OF THE REGION

Wales is a hilly land of contrasts bordered by England to the east and on the remaining sides by the sea. It has eight counties and covers 20,000 km^2, stretching nearly 400 km from Cardiff, the capital, to the tip of Anglesey. Weather from the Atlantic ocean makes the coastal climate moist and mild and rainfall is heavy on the hills. Nevertheless the scenically spectacular coasts are sunny and the lowlands, particularly in the more sheltered east, relatively dry.

People and their activities have been much influenced by the environment and availability of natural resources. Of the 2.8 million inhabitants, half are concentrated in the large towns and industrial valleys of south Wales with its extensive coalfield, whereas the most rural county, Powys, with few mineral resources, has a sparse population of one person to five hectares. The central hills have long provided a cultural barrier so most of the Welsh-speaking people live in the western counties of Gwynedd and Dyfed. As well as coal mining, industries include iron and steel production associated with the coalfields in the south and north-east, and slate quarrying in Snowdonia. More recently the oil industry has taken advantage of good natural anchorages as at Milford Haven to build refineries. The main rural activities are farming and forestry, although tourism and use of water resources are assuming increasing significance.

Four-fifths of Wales is used primarily for farming, and about half is under permanent or temporary grass. Rough grazings cover a quarter, and tillage, mainly in the drier lowlands, one-twentieth of the land. The grasslands are used mainly for dairying and livestock fattening in the lowlands and for sheep and beef cattle rearing in the hills. While the total areas of farmland and of tillage have decreased slightly in the last twenty years, livestock production has risen substantially (Table 4).

Forest and woodlands cover 2,350 km^2, almost double the area in 1938. Much of it is administered by the Forestry Commission. Annual production now exceeds 600,000 m^3 of felled timber. Non-productive amenity woodlands which have changed relatively little in recent years cover less than 300 km^2.

Multiple use of land is becoming common. The Forestry Commission offer picnic areas, walks and exhibitions for visitors. Many farmers earn extra income providing holiday accommodation. There are three National Parks; Snowdonia, the Pembrokeshire coast and the Brecon Beacons. Together they cover 4,110 km^2 of farmland and forest managed for the benefit of the public as well as individual owners. Other land is designated as having outstanding natural beauty or as country parks.

In the uplands, integration of seemingly conflicting uses of land has proved beneficial where planned in relation to variations of soil and site. Farming and forestry blend well, giving better access to once remote farmland along graded forest roads, the plantations providing welcome shelter from exposure. In this way, land poorly suited to agriculture, produces fair yields of timber, local employment is increased and the variety of habitats and the amenity value of the landscape improved. The same land often forms part of a catchment for reservoirs supplying water to large towns in England as well as Wales.

§ 1. GEOLOGY

Most of Wales is geologically ancient, consisting of hard Palaeozoic and in places Precambrian rocks (Fig.2). Mesozoic rocks outcrop in the south and north-east but Tertiary sediments are virtually absent. The solid rocks are widely covered by a variety of superficial deposits from a succession of Pleistocene glaciations and their associated periglacial episodes.

The lithology rather than stratigraphic age of a rock affects soil formation so lithology forms the basis of this account in which rocks are divided into igneous (including metamorphic) rocks, sedimentary rocks and drift. Narrower lithological types within these categories are described below. In Table 1 the major formations in Wales are listed in stratigraphic order with a brief description of lithology. The system and unit columns provide links with standard geological texts and maps, upon which this account is based, in particular those by Smith and George (1961), George (1970) and Institute of Geological Sciences (1977 and 1979). General relationships between the rocks in south and north Wales are summarized in geological sections (Fig.3).

Solid geology
Igneous and Metamorphic rocks
The main periods of igneous activity were during the Precambrian and Ordovician. Igneous rocks are largely confined to Gwynedd and south-west Dyfed and there are small outcrops on the Berwyn Hills and near Builth Wells. These rocks are very hard, forming characteristic craggy relief.

Acid igneous and metamorphic rocks of Precambrian age, mainly granite gneiss and allied types are extensive on Anglesey, with smaller outcrops on the Lleyn and St David's peninsulas. Ordovician acid igneous rocks are less extensive but form prominent landscape features such as Yr Eifl on Lleyn and the scarp of Cader Idris, as well as smaller outcrops in north-west Wales.

Basic and intermediate igneous rocks of Ordovician age occur as sills, dykes and volcanic plugs, representing vent-feeders of volcanic activity. In Gwynedd and Clwyd they occur as isolated outcrops throughout the mountain ranges from Cader Idris to

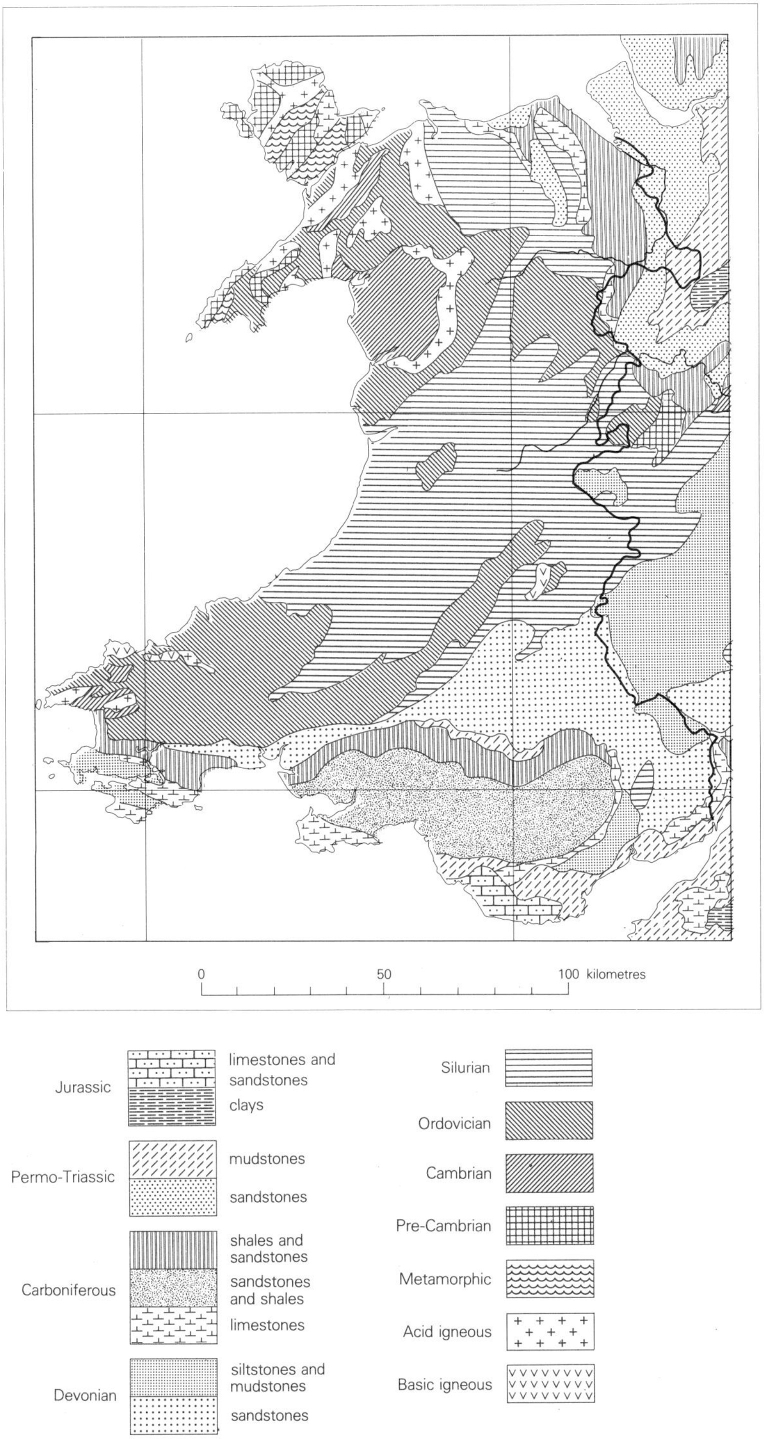

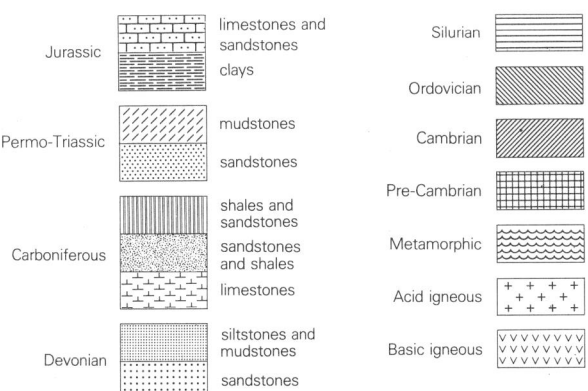

Figure 2. Solid geology

Table 1
Geological Strata and their Lithology

System	Unit	Lithology	Divisions of local importance
Jurassic	Lias	Limestone and shale	Lower Lias
Triassic (New Red Sandstone)	Penarth (Rhaetic[1])	Sandstone and shale	Quarella sandstone
	Mercia Mudstone (Keuper Marl[1])	Shale and dolomitic conglomerate, breccia	Blue Anchor Formation, (Tea Green Marl[1]); Retford Formation (Keuper Green Beds[1])
	Sherwood Sandstone (Bunter Sandstone[1])	Sandstone	Helsby and Bromsgrove sandstones (Bunter Sandstone[1]); Chester Pebble Beds (Bunter Pebble Beds[1])
Carboniferous	Coal Measures	Sandstone and shale, conglomerate	Upper, Middle, Lower Coal Measures; Pennant sandstone
	Millstone Grit	Grit, sandstone, shale, conglomerate	Basal Grit; Middle Shales; Farewell Rock Group
	Carboniferous Limestone	Limestone and shale	Lower Limestone Shales
Devonian	Upper Old Red Sandstone	Grit, sandstone, quartzite, conglomerate	Skrinkle Sandstone
	Lower Old Red Sandstone	Sandstone, siltstone, shale, conglomerate with some limestone bands	Cosheston Beds; Raglan Marl Group; St Maughan's Group; Ridgeway Conglomerate
Silurian	Ludlow	Shale, mudstone, with some sandstone and limestone	
	Wenlock	Shale, mudstone, with some sandstone and limestone	
	Llandovery	Shale, mudstone, with some sandstone and limestone	

M E S O Z O I C

P A L A E O Z O I C

P A L A E O Z O I C			
Ordovician	Bala (Caradoc and Ashgill)	Shale with grit, slate, limestone, rhyolite, lava and tuff	Mydrim Shales; Mydrim Limestone
	Llandeilo	Shale, sandstone and limestone	
	Llanvirn	Shale, grit, sandstone, rhyolitic and andesitic lava and tuff	
	Arenig	Grit, sandstone, shale, slate, with rhyolitic lava and tuff	
Cambrian		Slate, shale, sandstone and grit	Upper, Middle, Lower Cambrian; Rhinog Grits
Precambrian		Lava, tuff, sandstone, grit and shale	Benton Series
Intrusive igneous and Metamorphic		Dolerite, quartz diorite, gabbro granite, schist and gneiss	Johnston Series

[1] Old terminology

Penmaenmawr and also on the Berwyn Hills. In Dyfed, dolerites form the sill-like intrusions of the Preseli Hills and gabbros outcrop between Strumble Head and St David's Head. Precambrian diorites form the Johnston ridge further south. In central Wales there are small dolerite outcrops around Builth Wells. There are later dyke swarms of Tertiary age on Anglesey and the Gwynedd mainland.

Rhyolite lavas and tuffs of Precambrian age occur around St David's and on the Bangor and Padarn ridges north and east of Caernarfon. In Ordovician times periodic vulcanism produced lavas and tuffs interbedded with sedimentary rocks. These occur in Gwynedd where rocks which are principally rhyolitic but range to basaltic are widespread in the mountain ranges that sweep in a half circle from Cader Idris, through the Aran, Arenig and Moelwyn hills into Snowdonia. Rhyolitic, with some andesitic lavas and tuffs outcrop in south-west Dyfed around Fishguard and Treffgarne, and there are smaller outcrops around Builth Wells and on the Berwyn Hills.

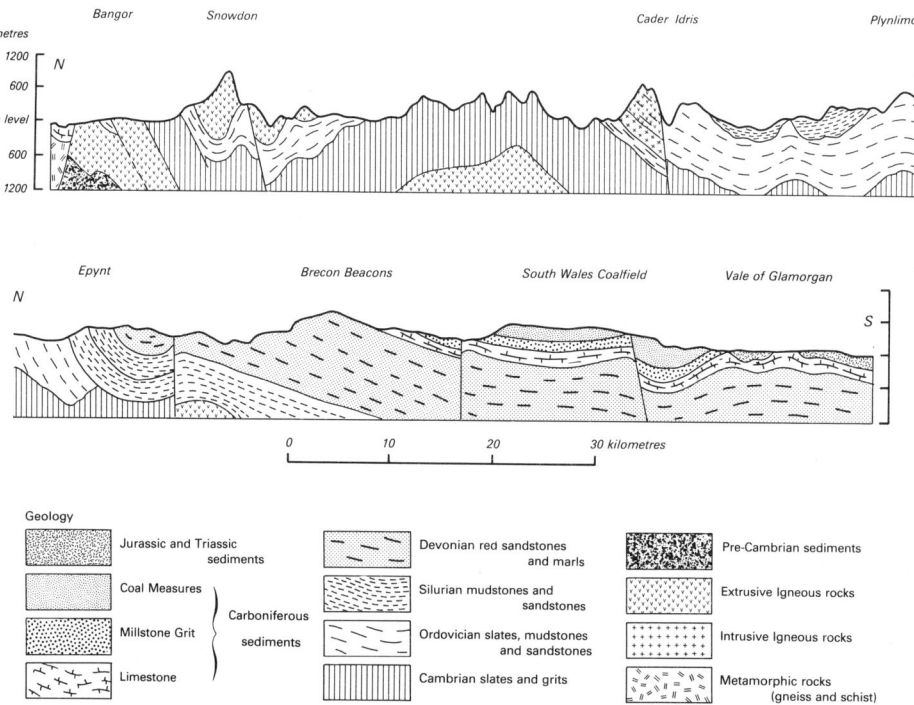

Figure 3. Geological sections across Snowdonia and the South Wales coalfield.
(From the National Atlas of Wales)

Argillaceous sedimentary rocks

Argillaceous sedimentary rocks are the most important single rock type in Wales and consist of shales, mudstones, and siltstones occurring in all geological systems. They are often interbedded with other rocks, usually sandstone, but especially in mid-Wales they form extensive uniform tracts of country. Lithology is related to depositional conditions which varied, sometimes rapidly, so there is often a sharp change from clay shales through silty shale to siltstone. These sediments weather to a relatively uniform fine earth so they are discussed together here.

Precambrian shales outcrop on Anglesey and Lleyn where they are interbedded with sandstones and volcanic rocks. In Cambrian times, shales were deposited in a transgressive sea but conditions varied and beds of sandstone and grit, representing inshore deposits form part of the succession. In north Wales, metamorphism has produced purple, green and grey slates and there are few interbedded sandstone rocks in the 'slate belt' from Bethesda to Penygroes. The main Cambrian outcrop in the Rhinog Hills is a desolate tract of rugged mountainous moorland where bands of grit come to the surface. In south-west Wales green, red and purple mudstones and shales outcrop in association with sandstones around Solva and south of Fishguard.

Grey, hard thinly bedded Ordovician and Silurian shales and mudstones are the characteristic and dominant rocks of most of central and north Wales. In places, as already described, there are small inclusions of igneous rocks. Sandstones and grits and impure limestones occur sporadically, mainly in the lower Ordovician. Some of the sediments are best described as greywackes as they contain sand-size fragments of mudstone.

Devonian shales are abundant throughout the succession that is loosely termed 'Old Red Sandstone' but they are rarely continuous, being mostly interbedded with sandstone and siltstone and occasionally limestone. Unlike other Palaeozoic sediments nearly all the Devonian rocks are red in colour. Argillaceous rocks are most common in the Lower Old Red Sandstone which forms the main part of the Welsh outcrop but their distribution is uneven, their greatest purity being in the St Maughan's and Raglan Marl groups in Gwent. A characteristic feature of the red mudstones is the frequency of calcareous deposits which range from limestone nodules to bands of limestone, commonly called cornstones.

Grey, blue and black Carboniferous shales and mudstones outcrop extensively in south Wales and to a lesser extent in Clwyd. They form part of the Millstone Grit succession and also outcrop at the base of the Carboniferous Limestone, but their main occurrence is in the Middle and Lower Coal Measures of the Welsh coalfields. These relatively soft rocks form the lowest ground of the coalfields and are generally blanketed by drift so that it is uncommon to find bedrock within the soil profile.

Triassic mudstones of the Mercia Mudstone Group outcrop only in Gwent and south Glamorgan, most extensively around Cardiff, Newport and Barry. Most widespread is a red and purplish, sometimes calcareous mudstone (formerly called Keuper Marl). It is overlain by the Blue Anchor Formation (Tea Green Marl) which consists of pale creamy green often calcareous silty mudstone. It passes to the inextensive Penarth Group (Rhaetic series) which is mainly shaly and often calcareous in south-east Wales.

Shales with limestones of the Lower Lias represent the Jurassic system from Gwent into the eastern Vale of Glamorgan. Limestones dominate further west with shales occurring only as thin partings.

Arenaceous sedimentary rocks

Sandstone, grit, quartzite and conglomerate often outcrop together producing bold relief and coarse-textured acid soils. They are of all ages except Jurassic but they are most extensive in the Devonian and Carboniferous.

Precambrian sandstones, quartzite, grits and conglomerates occur in west Lleyn and much of Anglesey along with contemporary igneous rocks. The Cambrian of north Wales includes tough, massive grits, sandstones and conglomerates mainly in the Harlech Dome where they are thick and form the highest ground. The Rhinog Grits for example are over 600 m thick at the centre of the Dome. In south-west Dyfed near St David's there are Cambrian conglomerates, grits and sandstones. These are often purple, red or green and outcrop with mudstones and shales. Grits, sandstones and conglomerates occur throughout the Ordovician and Silurian systems, though they are nearly everywhere subordinate to shales and mudstones. They are most prominent in the Lower Ordovician Arenig series which outcrops around the Harlech Dome, on Anglesey and near Carmarthen. Elsewhere arenaceous rocks of Ordovician and Silurian age are rare, and although grits and conglomerates in particular may form bold prominences they are generally thin and persistent over only short distances.

Red Devonian sandstones are mainly fine grained and grade to siltstones with some grits and conglomerates. They dominate the Lower Old Red Sandstone, particularly in the Brecon series. The outcrop from Shropshire to Carmarthen Bay is fairly uniform in lithology but around Milford Haven two local formations are intercalated in the Breconian sequence–the Ridgeway conglomerate and the Cosheston Beds of brown and grey sandstones. The Upper Old Red Sandstone, a group of sandstones, grits, quartzites and conglomerates is best developed in the Brecon Beacons where it caps the mountains. It also has a limited outcrop south of Milford Haven known as the Skrinkle Sandstone.

Arenaceous sediments in the Carboniferous system dominate the Millstone Grit and Upper Coal or Pennant Measures, but are only a minor component of the Lower and Middle Coal Measures. The Millstone Grit in north Wales has a narrow outcrop from Prestatyn to Oswestry and consists of sandstone, grit and quartzite with subordinate shale beds. In south Wales the main outcrop is north of the coalfield where three divisions are recognized. The Basal Grit is a group of grits, quartzites, sandstones and conglomerates which form a series of bold escarpments from Mynydd-y-Garreg near Kidwelly to the Blorenge near Abergavenny. Some beds are very pure with over 99 per cent quartz and have been quarried in places. These pass upwards to the Middle Shales which have subordinate sandstone bands, overlain by the Farewell Rock group of sandstones and quartzites.

Forming the high ground of the coalfield are the massive scarped Pennant sandstones. These are bluish grey when unweathered, turning rusty brown on exposure. Some beds are coarse grained or conglomeratic and contain derived pebbles of coal and ironstone. The common description of these sandstones as feldspathic (Strahan *et al.* 1907, George 1970) is not supported by recent studies in the Gwendraeth valley district (Archer 1968). Here the characteristic lithology is classed as a subgreywacke, containing about 60 per cent detrital quartz grains with fewer included lithic fragments in a clay mineral matrix. Only minor amounts of detrital feldspar and muscovite are present. In north Wales the Upper Coal Measures are less prominent with more shale beds and are predominantly red in colour.

The Triassic New Red Sandstone is only represented by the Bunter sandstones and pebble beds of north Wales in the Vale of Clwyd and the western edge of the Shropshire–Cheshire Plain. Minor intercalations of sandstone occur in the Mercia Mudstone of south Wales, and, near Bridgend, massive gritty sandstones (Quarella sandstone) form the Penarth (Rhaetic) succession.

Limestones

Thin bands of limestone, such as the Ordovician Mydrim Limestone and Devonian cornstone occur in all geological systems, but only the Carboniferous and Jurassic limestones and Triassic dolomitic conglomerates occur widely as soil parent materials.

The hard Carboniferous Limestone series forms wide outcrops in South Pembrokeshire, Gower, the Vale of Glamorgan and south-east Gwent and rims the coalfield in a narrow belt along the north and east crops. In north Wales it outcrops in Anglesey, around Llandudno, in the Vale of Clwyd and on the eastern flank of the Clwydian Range. Further south it forms the prominent escarpment of Eglwyseg Mountain at Llangollen. The limestone is a very pure deposit, yielding on solution only small amounts of residual material (Findlay 1965). Mineralization with lead, zinc, copper and iron ores is widespread. Shaly intercalations are rare except at the base of the formation and in southeast Gwent where there is an undulating terrain of limestone ridges and shale valleys.

Triassic dolomitic conglomerate and breccias of Carboniferous limestone fragments in a red matrix outcrop in the west of the Vale of Glamorgan around Cowbridge, Coity Hill and Porthcawl.

Jurassic Lower Lias limestones with subordinate shales form much of the Vale of Glamorgan. The proportion of shales is greatest in the east, whereas in the west, shale partings are very thin.

Drift geology

The term drift encompasses all Quaternary deposits from Pleistocene till, gravels and Head to recent alluvium, windblown sand and peat (Table 2). Such deposits blanket nearly the whole of Wales, modifying the relationship between solid rocks and soils. Some drift is far-travelled, for example tills from deposits flooring the Irish Sea, while some Head and scree deposits are very local, derived entirely from nearby country rock. As with

Table 2
Quaternary Formations and their Lithology

Deposit	Type	Lithology	Divisions of local importance
Peat	Basin	Reed-swamp fen-carr	
	Blanket Bog	Bog moss, cotton grass, sedge	
	Raised Bog	Bog moss	
Alluvium	River	Sandy, loamy, clayey	
	Marine	Sandy, loamy, clayey	
Aeolian sand	Dune	Calcareous and non-calcareous sand	
Head (and scree)		Stony loamy and rocky	
Glaciofluvial and River Terrace drift		Sandy and gravelly	
Till	Reddish	Stony loamy and clayey	Irish Sea Drift
	Other	Stony loamy and clayey	Welsh Drift, Wolstonian till, some Irish Sea Drift

solid rocks, the lithology of the deposits rather than their age or genesis is more important for soil studies so it is lithology that is discussed here. The broad distribution of drift is likewise more important than detailed stratigraphy.

Quaternary studies in Wales have often been based on coastal and quarry exposures, and deposits which are of significance in the interpretation of Quaternary events such as the *Patella* beach, or Minchin Hole breccia on Gower, have little or no surface exposure. This description of Welsh drift gives emphasis to deposits important as soil parent materials. Comprehensive accounts of the Quaternary period in Wales are in Bowen (1974) and Lewis (1970).

Wales has been glaciated at least three times during the Quaternary but only the last (Devensian) and the penultimate (Wolstonian) glaciations provide soil parent materials. The distribution of resulting drifts is shown in Figure 4. The Wolstonian glaciation was more extensive, covering the whole of Wales with ice originating not only from the high ground, but also from the Irish Sea, the latter ice overriding the coastal lowlands of west Wales, Gower and the Vale of Glamorgan. During the intervening Ipswichian interglacial

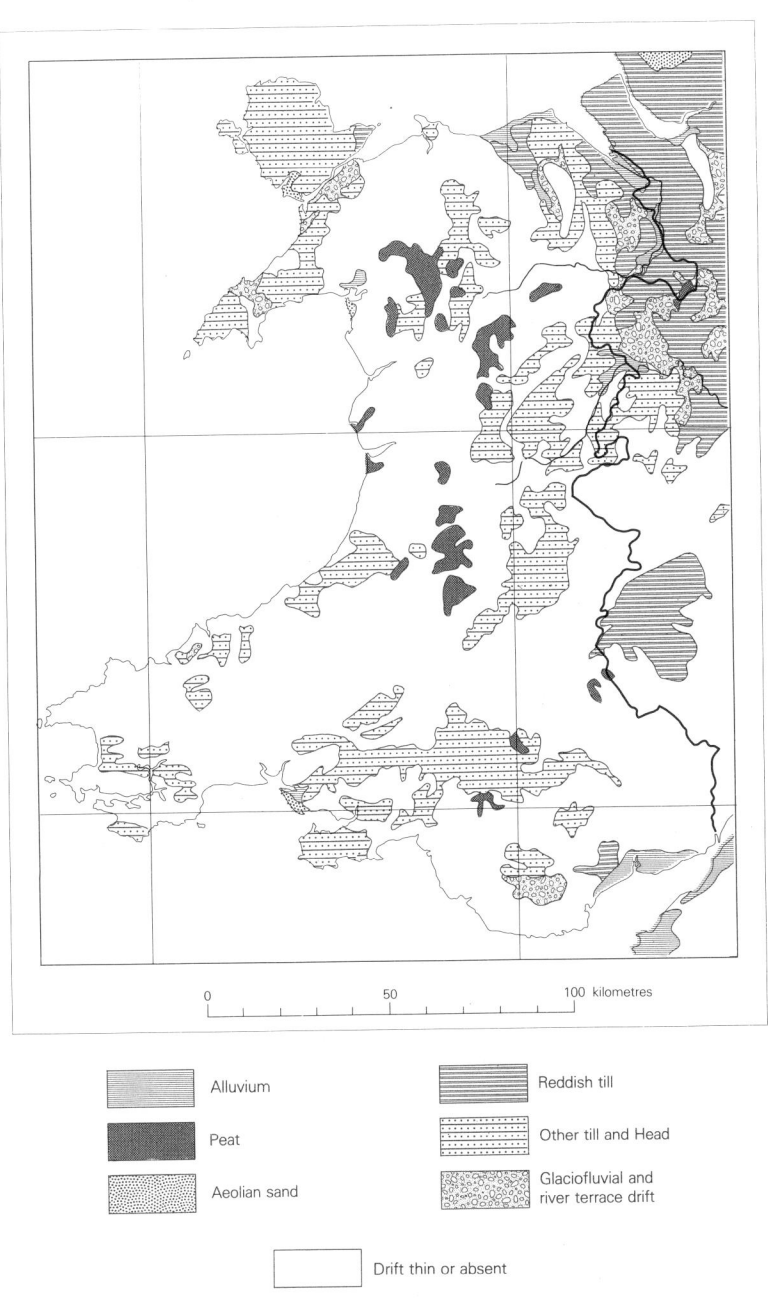

Figure 4. Drift geology

the climate was warm and temperate and reddish soil horizons over limestone may have originated at this time (§ 5, § 84). During the Devensian a large ice sheet centred on Snowdonia radiated outwards to meet Irish Sea ice which again invaded Anglesey, Lleyn, the Vale of Clwyd and the Cheshire–Shropshire plain. In the south the Irish Sea ice reached only the shore of Cardigan Bay, leaving south-west Dyfed, south Gower and the Vale of Glamorgan ice-free. In the last stage, ice retreated leaving cirques and characteristic U-shaped valleys throughout the uplands. Periglacial conditions existed beyond the ice limits and over the whole of Wales as the ice retreated, and there was mass movement and *in situ* modification of drift and solid rocks by frost action. As the climate improved during the succeeding Flandrian period the land was recolonized by trees and soil development started. In valleys, alluvial deposition started during the Flandrian and is a continuing process, while around the coasts sand dunes are still being formed.

Till

Till is material deposited by ice and is typically an unsorted deposit of rounded and sub-rounded stones in a loamy or clayey matrix. Compact lodgement till deposited between ice and bedrock is found over much of Wales in both valleys and plains. Ablation till is less compact and is carried on and in the ice and deposited as the ice melts. As the ice melted and retreated, tills and associated deposits were partly resorted by meltwater streams forming for example the 'morainic drift' marking the limit of Devensian ice in the Vale of Glamorgan. The upper layers of the till have also been at least partly reworked by periglacial processes.

'Welsh drift' deposited by ice originating in the Welsh uplands is usually distinguished from 'Irish Sea drift' although they are mixed along the contact zones. The colour and composition of till reflects the geological formations over which the ice passed and Welsh tills are often very local in character. Tills from the Lower Palaeozoic and Carboniferous shale outcrops are blue-grey or olive and clay rich. Over Pennant sandstones and Upper Old Red Sandstones tills are more sandy and also red in the latter instance. Irish Sea drift is usually red or purple and clay-rich. Calcareous where unweathered it contains shell fragments from the sea bed and far-travelled erratics from the Lake District and Scotland. However, when the ice has travelled over non-red country rock as on Anglesey and Lleyn the till loses its redness and is similar to Welsh drift apart from its content of far-travelled erratics.

Wolstonian tills occur beyond the Devensian ice limit in south-west Dyfed and Gower. They are mostly of Welsh provenance and have been reworked during subsequent episodes and mixed with glaciofluvial deposits. In the Vale of Glamorgan there is shelly drift at Ewenny and Pencoed which is mostly buried by younger drift, but Crampton (1966) traced erratics and heavy minerals indicative of Irish Sea deposits as far as east Cardiff.

Glaciofluvial deposits

Glaciofluvial deposits, the products of meltwater flowing beneath and issuing from the ice, are widespread in Wales, especially in Clwyd. Typically they consist of sorted and bedded sand and gravel, sometimes with bands of finer material. Gravel trains are found in most valleys and wider spreads occur as outwash fans or terraces such as the Wrexham delta. They often have hummocky topography with enclosed hollows or kettle holes, formed as blocks of ice melted within the gravels. In places they were reworked by rivers in the Flandrian period, producing the terraces characteristic of most Welsh valleys.

During the last stages of glaciation in Cardigan Bay, Irish Sea ice formed a barrier across the river mouths and temporary lakes developed, the largest being in the Teifi valley where laminated lacustrine clays are found from Llandyssul to Cardigan.

Periglacial deposits

Periglacial deposits formed during the permafrost and intense cold of the late Devensian period producing distinct deposits and landforms which have extensively modified the landscape. Frost shattering of rocks produced the angular scree characteristic of the present day hill slopes (Plate 2). These and other deposits were moved downslope by solifluction, a process occurring during summer when periodic thawing above the permafrost gave a viscous, mobile layer capable of gravitational movement on all but the gentlest slopes. The resulting Head is of variable character and thickness. It is commonly

R.Hartnup

Plate 2. Quarry in Lower Palaeozoic siltstones and cleaved mudstones at Mynydd Pencarreg, Dyfed. Downslope movement of rock fragmented by frost action is clearly shown.

a mixture of stones, roughly orientated downslope, in a loamy matrix. On steep hillsides it may be thin and mainly composed of angular local rock fragments, and the thickest accumulation is on valley floors where it is usually reworked till.

Involuted structures and the casts of frost polygons and wedges formed by freeze and thaw under periglacial conditions are widespread as are compact subsurface layers or fragipans (Fitzpatrick 1956).

Alluvium

Alluvium is a marine, riverine or lacustrine sedimentary deposit of Flandrian age. Most Welsh valleys are narrow, but there are extensive tracts of river alluvium several kilometres wide in the broad valleys of the Severn, Dee and Clwyd. Sorting during deposition is the main factor affecting the character of alluvium but there are differences within and between catchments which are determined by the surrounding country rocks. Alluvium from Ordovician and Silurian rocks has a larger silt content than that from Carboniferous rocks. Alluvium from Triassic and Jurassic mudstone in south Wales is invariably clayey. Likewise alluvium along rivers draining red Triassic or Devonian rocks, notably that of the Usk, is red in contrast to the brownish or greyish alluvium elsewhere in Wales.

Alluvium often rests on the glaciofluvial deposits which choked the valleys at the end of the Devensian and many alluvial soils become gravelly within the upper metre, especially in the upper and middle parts of the valleys. Low terraces are a feature of many valleys and are not always easy to distinguish from the floodplain, with which they often merge upstream. In many valleys deep stoneless alluvium is slightly raised above the active floodplain. It is possible that this is the result of accelerated soil erosion following forest clearance and ploughing by early man, and similar deposits in England have been dated to the Romano-British period (Hazelden and Jarvis 1979).

With the gradual post-glacial rise in sea level, estuarine and marine alluvium has built up in the seaward reaches of the valleys and on the coastal flats. Here interbedded bands of peat provide evidence of periods of slower eustatic rise. Most of the alluvium is silty clay loam or clay but coarser deposits occur along creek ridges. The most recent unripened alluvium beneath the present saltings is calcareous and saline but much has been reclaimed by sea walls and ditches. Where reclaimed it is now non-saline and has become largely decalcified except on the Gwent levels where most soils are calcareous within 60 cm depth.

Dune sand

The extensive dune systems of the coast are of relatively recent origin but are derived from glaciofluvial sands originally deposited beyond the present coastline or beach sand. At Kenfig Burrows in the Vale of Glamorgan archaeological evidence (Higgins 1933) suggests that dune development began in the Neolithic period. Since then stable stages have alternated with periods of dune migration, as in the mid-Iron Age and the 14th

Century. Recent afforestation has helped to stabilize the dunes, particularly at Newborough Warren, Anglesey and at Pembrey, Dyfed.

Peat
Peat deposits may be broadly divided into lowland and upland forms. Lowland peats were originally topogenous, forming in basins with a high groundwater-table but when mature they become raised and autochthonous, relying on rainfall. They are inextensive in Wales, the largest being Tregaron and Borth Bogs. There are several smaller bogs as at Crymlyn near Swansea and Llanllwch near Carmarthen. Resting on alluvium, these peats have a complex stratification going back to the late Devensian. At the base there is lacustrine organic detritus on which has developed reed-swamp and fen-carr peat overlain by a succession of *Molinia*, sedge and *Sphagnum* peats partly reflecting climatic changes.

Basin peats are also found in the uplands but blanket peat is more extensive. Most soils above 300 m O.D. have a peaty top and thick peat is widespread above 600 m O.D. Blanket peat growth requires acid soil conditions, low temperatures and adequate rainfall and is most rapid on level ground with slow surface run-off. Its development is more recent than that of basin peat. In Neolithic times man's pressure on the forest increased and there was widescale clearance, creating large areas of heath in the uplands. During this period growth of blanket peat was initiated. The basal layer of blanket peat at Glaslyn Bog near Plynlimon, gives a radiocarbon date of 4220 B.P. and the underlying soil contains pollen from species of open habitats (Taylor 1973).

By Iron Age and Roman times the proportion of tree pollen was as low as at the present day (Moore 1977), and around 2500 B.P. the cooler more humid sub-Atlantic period began. Without the protective deciduous forest cover, leaching of soils, initiated in the earlier Atlantic period, was greatly increased. The wet heathy vegetation gave a water holding organic surface layer resistant to decay. The acid products of decomposition intensified the leaching process, leading to podzolization, giving the stagnopodzols which cover the uplands today. In places where relief and climate combined to produce more waterlogged conditions, the organic surface layers developed into blanket peats and Chambers (1983) identifies a major extension of peat growth in post-Roman time. Mostly highly humified, this peat is derived from plants characteristic of the present uplands, notably *Molinia, Eriophorum,* and other sedges, mosses and the common heath species.

§ 2. RELIEF

The landscape of Wales is dominated by its mountains and high moorlands which form part of every county. More than half of the land is above 200 m O.D. and much is over 600 m O.D. (Fig.5). It is also a land of great scenic diversity. The jagged mountains of Snowdonia rising to 1085 m O.D. (Plate 3) with their cliffs, screes and lakes, contrast with the rest of upland Wales with its gentle rather monotonous moorlands, dissected by deep valleys. The high land is fringed by a succession of rounded, often steep sided, grassy hills and plateaux, in places cut by broad valleys with large, meandering rivers. The coastline is spectacular and varied with large dune systems, long sandy beaches, salt marshes and rocky cliffs.

The rocks of Wales have been affected by successive periods of earth movements during Caledonian and Hercynian, and to a smaller extent Alpine orogenies. The pre-Devonian rocks have been strongly folded, faulted and thrust during the Caledonian to form mountain chains generally aligned north-east to south-west. Devonian and Carboniferous rocks were folded more gently by the Hercynian movements that produced the

T.R.E. Thompson

Plate 3. The Llanberis Pass. View from Snowdon with Glyder Fawr to the right above scattered rock outcrops and broken screes of the Moor Gate association. Shallow soils to the left are in the Bangor association.

synclines visible today in the escarpments of the Brecon Beacons, Black Mountains and south Wales coalfield. In north Wales, Hercynian movements accentuated existing Caledonian structures such as the Clwydian anticline, while a major syncline developed within the Carboniferous rocks along the Vale of Clwyd. At the same time intense faulting produced the horst and graben structures which characterize the nearby coalfield. Mid-Tertiary Alpine movements were weak in Wales but did throw the Jurassic and Triassic strata of the Vale of Glamorgan into gentle synclines and anticlines.

Landforms

Rock structure broadly influences landform but Wales owes its upland character principally to uplift during the Tertiary and to the hardness of its rocks. Structure and rock type control landform locally but the outstanding feature of Wales is a series of dissected plateau surfaces. These are widely regarded as remnants of Tertiary and early Pleistocene erosion surfaces which cut across major differences of structure and lithology. Whether these upland surfaces are the products of sub-aerial peneplanation (Brown 1960), are marine platforms (George 1974) or are simply the result of differential erosion reflecting rock hardness and structure (Challinor 1930) is unlikely to affect soil formation as any related soft deposits have since been removed by glaciation. Brown recognized six relief regions in a useful classification later adopted by Bowen (1977), dividing Wales into

Plate 4. The scarp of Cader Idris from the north. Soils of the Manod and Brickfield 1 associations are under woodland and farmland in the foreground with the Hafren association on the plateaux, and Bangor association on the crags and screes.

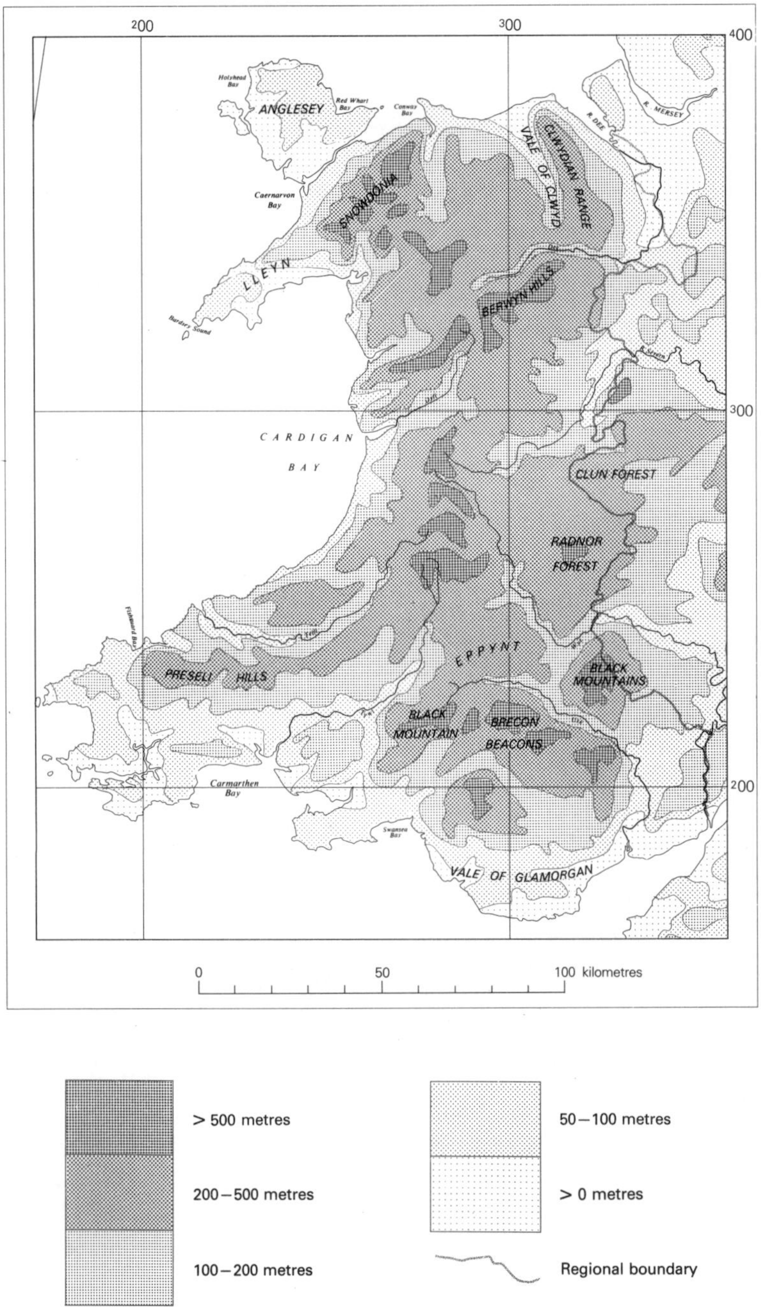

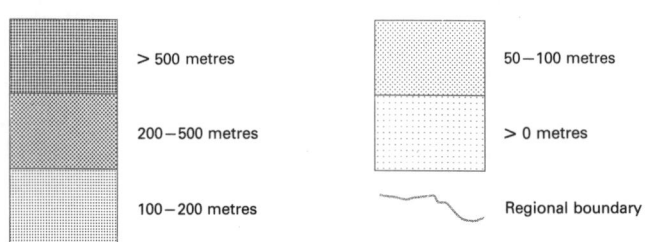

Figure 5. Relief

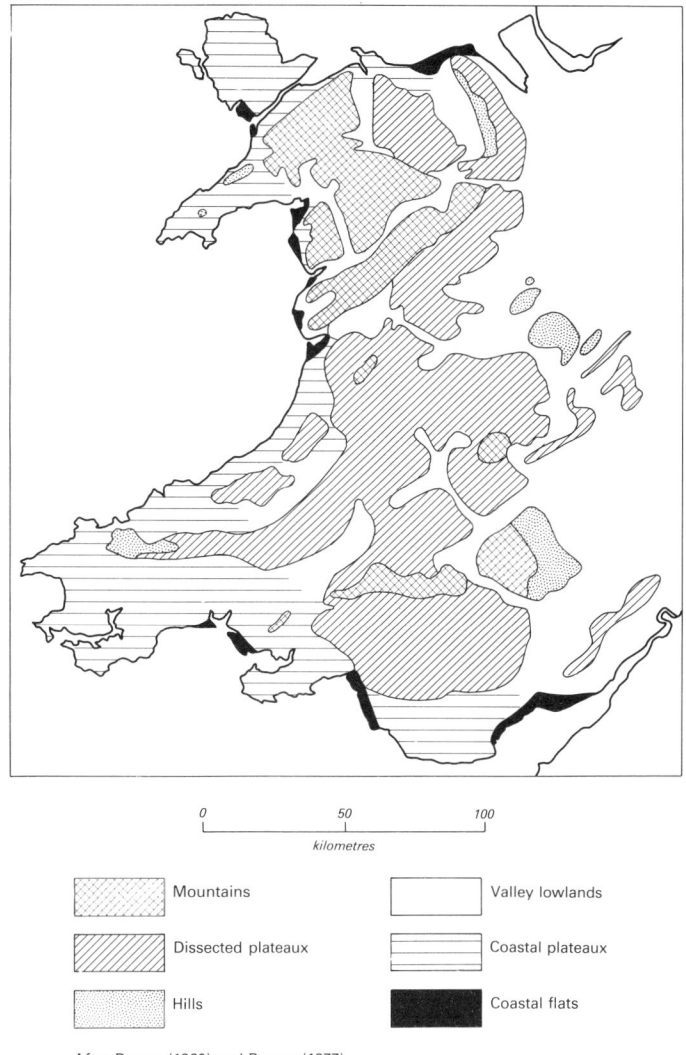

0   50   100

kilometres

Mountains   Valley lowlands

Dissected plateaux   Coastal plateaux

Hills   Coastal flats

After Brown (1960) and Bowen (1977)

Figure 6. Relief regions

mountains, dissected plateaux, hills, valley lowlands, coastal plateaux and coastal flats (Fig.6).

The mountains of Wales, higher than 600 m O.D. consist of hard Palaeozoic rocks. They are highest and most rugged in the north-west where Ordovician igneous rocks form the jagged and imposing ranges of Snowdonia, Arenig, Cader Idris (Plate 4) and the Aran range. In the Rhinog Mountains of the Harlech dome, Cambrian grits form the highest ground. In central Wales, where Plynlimon is composed of Lower Palaeozoic shales and mudstones with minor sandstone and grit bands and Radnor Forest is formed of siltstones, the mountains have smoother profiles but nevertheless stand conspicuously above the surrounding countryside. In south Wales the mountain summits form part of an escarpment marking the northern rim of the coalfield syncline where it is capped by resistant Devonian grits and conglomerates. The escarpment is most impressive on the Brecon Beacons and Black Mountains. Further west the Millstone Grit quartzites and grits form the highest ground of the Black Mountain in Dyfed.

Dissected plateaux dominate much of upland Wales forming a series of long level skylines, accordant in general elevations and unaffected by differences in rock type and structure. The landscape is relatively featureless with plateaux dissected by deep, in places broad, valleys (Plate 5). Typical plateaux include the south Wales coalfield uplands, Eppynt, the central Wales uplands, the Berwyn Hills (Plate 6) and Mynydd Hiraethog.

Areas below 600 m O.D. too dissected to be termed plateaux are described as hills. These are of small extent and include areas of geological diversity with narrow outcrops. In south-west Wales the Preselis with their numerous igneous intrusions have proved resistant to erosion producing characteristically craggy terrain (Fig.45). The mid-Wales borderland also has varied rocks while the Clwydian range is a narrow anticlinal ridge of Silurian shales.

Plate 5. Afforestation on the Manod association near Llanafan, Dyfed. The grassland in the valley is on the Denbigh 1 association.

J.W. Lea

Plate 6. The Berwyn mountains near Llangynog, Powys. Controlled heather burning has produced a striped pattern on the Crowdy 2 association.

The coastal plateaux are least affected by rock structure and lithology, being erosion surfaces planed across a range of rocks from Jurassic limestones and shales to hard Precambrian igneous rocks. The most impressive and least dissected surface is found in the Vale of Glamorgan, Gower (Plate 7), south-west Dyfed, Lleyn and Anglesey at 60 m O.D. A second surface can be traced at about 120 m O.D. at the western foot of the Snowdonia range. The remnants of a third platform at 180 m O.D. are traceable as 'monadnocks' on Anglesey, Lleyn Peninsula, in Gower and near St David's Head.

The broad valley lowlands of the Tywi, Teifi, Conwy, Clwyd, Severn, Usk and Wye coincide to a large extent with lines of geological weakness such as the Tywi anticline and the Vale of Clwyd synclinal rift. The valley floors which are in places several kilometres wide provide the most level land in otherwise hilly country.

The coastal flats formed by deposits of the Flandrian marine transgression are not extensive as much of the coast is cliffed or has strong long-shore currents, but they are found in sheltered places such as Carmarthen Bay and the estuaries of the Dee and Severn.

Drainage

The broadly radial drainage pattern of Wales is shown in Figure 5. In detail it is more complex and pays little regard to geological structure. The broad pattern certainly

Plate 7. The Gower peninsula. Arable fields in the centre are mainly on the well-drained East Keswick 1 association while the central ridge carries uncultivated podzols of the Goldstone association. Dunes in the foreground form the Sandwich association, with saltmarsh behind.

predates the Quaternary period and was probably initiated in the Tertiary when the present Welsh land mass was uplifted (George 1974). Bowen (1977, 1980) partly combines the views of George and Brown and sees the landscape as consisting of marine platforms, severely modified and often replaced by subaerial surfaces. The initial drainage lines on these surfaces were little influenced by the rocks but they have since been affected by differences in rock hardness, river capture and glacial diversion.

Glaciation

There is impressive evidence of glacial erosion in the highest and currently wettest districts where the greatest thickness of ice accumulated and where glacial and periglacial conditions persisted longest. Thus cirques occur on high ground between Snowdonia, the Brecon Beacons (Plate 8) and the Rhondda. Ice overdeepened most valleys and steepened their slopes and this has led to landslips, especially over weak strata in the south Wales coalfield. Ice-moulded features such as *roches moutonnées* and striated and smoothed rock surfaces are widespread though largely hidden by drift except in north Wales.

Till is widespread in valleys but extensive till plains exist only in the north-east fringing the Cheshire–Shropshire Plain. In the valleys it is dissected and often moved to the valley floor by solifluction. Drumlins are uncommon except on Anglesey, around Mynydd Hiraethog and near Llanfyllin, Newtown, Builth Wells and Hirwaun. Morainic ridges, often with kettle holes, show the limit of piedmont ice in south Wales from Margam to Caerleon, while ridges as at Glais in the Swansea valley, and the Vale of Clwyd indicate transient ice margins as does the morainic ridge across the Teifi valley above Tregaron which has impeded the river flow and given rise to the famous peat bogs (Plate 9).

Deglaciation produced vast volumes of meltwater which cut channels in bedrock and spread sand and gravel across the landscape as kames, kame terraces, eskers and deltas. These deposits have a range of forms but are commonly hummocky except where

Plate 8. Snow on the Brecon Beacons. The Lydcott association covers most of the ridge, with hagged peat soils of the Crowdy 2 association clearly visible on flatter summits. The soils in the valleys are in the Wenallt association.

Cambridge University Collection

Plate 9. Raised bogs at Tregaron. The three bogs are in the Crowdy 2 association which includes both amorphous and fibrous peat soils. Alluvial soils of the Conway association are adjacent to the river Teifi.

deposited as flat glaciofluvial terraces and deltas. Kettle-holes, a product of stagnant-ice wastage, are common in valley terraces throughout Wales. They also occur in proglacial lake delta terraces at Wrexham and elsewhere. Eskers are less common and mostly small and unspectacular.

Periglacial deposits mantle and tend to smooth the landscape. Solifluction moved material downslope to produce lobes and terraces especially in upland valleys (Crampton and Taylor 1967). Screes are widespread, especially in the more mountainous parts where some are unvegetated and still active. They usually skirt tors and crags of hard rock which are common in Snowdonia and the Preseli Hills. Frozen ground phenomena are dramatically displayed in the low sinuous or circular ridges plentiful in parts of mid-Wales and interpreted as the ramparts of former pingos by Watson (1971). Contemporary periglacial activity has been recorded at high altitudes in Snowdonia by Ball and Goodier (1970) where the action of ground ice is producing stone stripes and polygons.

§ 3. CLIMATE

Lying on the western fringe of Great Britain, Wales is one of the first parts to encounter the moist air masses associated with frontal systems advancing eastwards from the Atlantic. The western coasts, especially those of Anglesey and the peninsulas of Lleyn, south-west Dyfed and Gower, are exposed to westerly winds but are among the sunniest and mildest parts of Britain. The extensive hills and mountains of the interior are often cloudy, cold and exposed with high rainfall.

Local differences in climate relate mainly to altitude and proximity to the sea. With increasing height above sea level average temperatures fall and precipitation increases although there is a marked rain shadow east of the main mountain mass (Fig.7) such that the Black Mountains of Gwent have much less rain than the Brecon Beacons in whose rain shadow they lie. The highest rainfall is on the summit of Snowdon which has an average annual total of 4500 mm. The sea modifies the annual range of temperature so that coastal regions have cooler summers and milder winters than inland.

This account emphasizes those aspects of climate which are important for agriculture, forestry and soil formation. Much of upland Wales is under semi-natural vegetation and climate is the main limitation to agricultural improvement. In the foothills and lowlands interactions between climate and soils significantly affect land use and farming practice. For example, high rainfall combines with impermeable soils to cause waterlogging and restrict cultivation while drier climates interact with soils having small water reserves to cause droughtiness. Such interactions are described for specific soils in Chapter 3. The main climatic factors upon which they depend are mean temperature and moisture balance, and these are described below.

Temperature
The accumulated temperature above 0 degrees centigrade for the early half of the year (January to June) is now preferred as a measure of the heat energy available for crop growth, because of recent work on grass (Peacock 1975, 1976) and cereals (Biscoe and Gallagher 1978) which shows that leaf extension is maintained, albeit slowly, almost down to 0 degrees centigrade and many arable crops ripen early and therefore do not benefit from high temperatures later in the year. Data were calculated for 109 representative meteorological stations for the period 1959 to 1978 and average (median) values of accumulated temperature above 0 degrees centigrade, which correlate well with the length of the growing season, have been chosen to represent the broad thermal differences across the region. Figure 8 was drawn using a relationship based on the lapse of these data with altitude, latitude and longitude. The more usual summation above a threshold of 5.6 degrees centigrade for the whole year (Bendelow and Hartnup 1980) is, however, more appropriate for crops such as grass and sugar beet which respond to high summer temperatures and continue to grow well into the second half of the year provided there is enough moisture in the soil.

Only the southern coastal fringe has average accumulated temperatures higher than

1500 day degrees centigrade, while the remaining lowlands are in the range 1400 to 1500 day degrees centigrade. Temperatures are much lower in the hills (less than 1300 day degrees centigrade), and there are large areas in central and north Wales with less than 1100 day degrees centigrade. The Welsh mountains have a warmer climate than high ground further north in Britain and the arctic zone (Birse 1971) is reached only on the highest summits. Pearsall (1950) shows that summer temperatures in British mountains are similar to those in west Greenland and therefore much lower than those at comparable altitudes and latitudes in continental areas of Europe and north America. This accounts for the low tree line and the development of climatic peats on relatively low ground. Snow

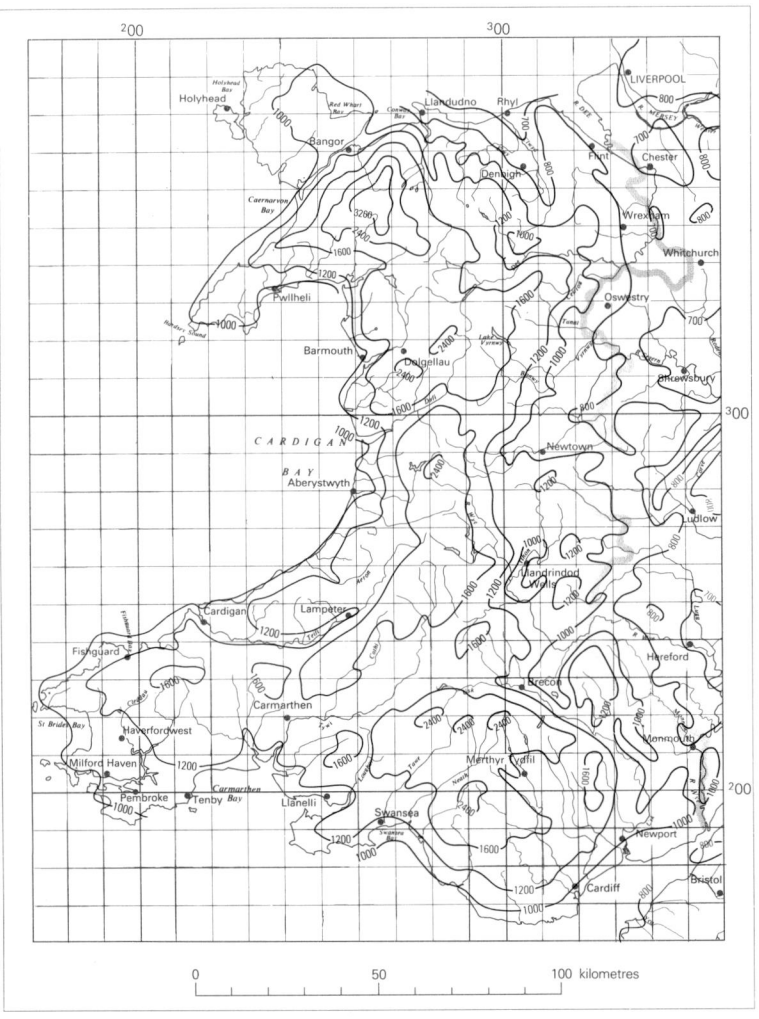

Figure 7. Average Annual Rainfall (mm), 1941—70. (Meteorological Office data)

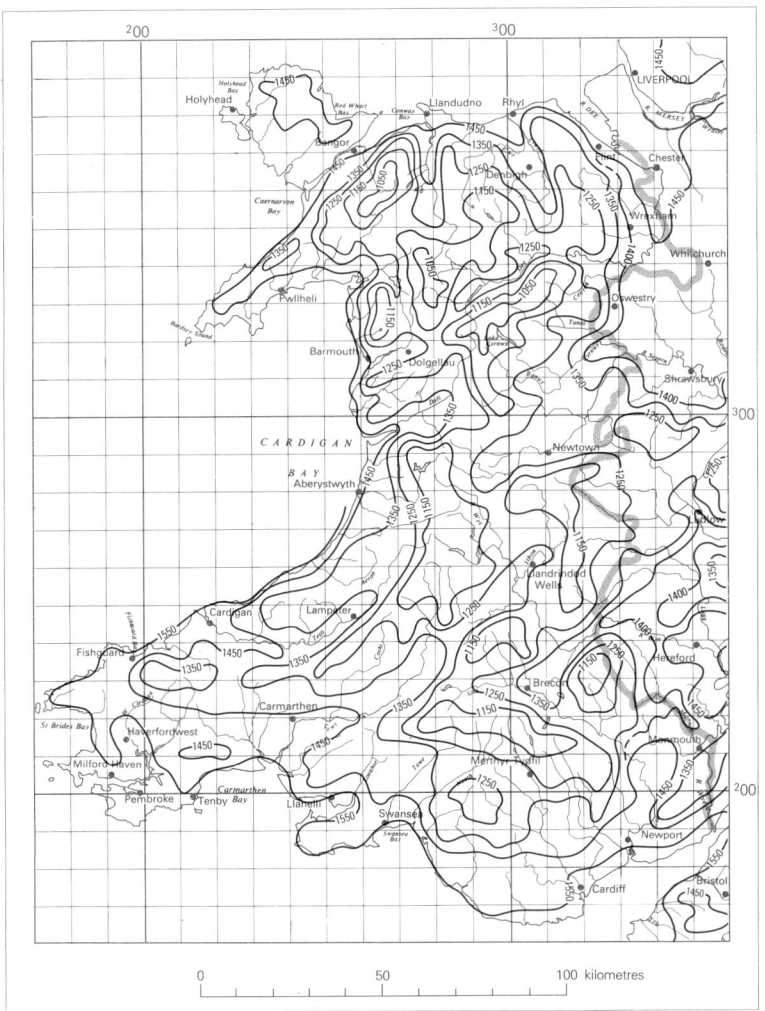

Figure 8. Median Accumulated Temperature above 0 deg C (day-degrees),January to June, 1959—78

lies on the hills for long periods in winter and drifting is common on exposed areas, causing serious problems for the stock farmer, especially if fences are sited so that sheltering animals are buried.

While summer temperatures are relatively low, the western coasts benefit from an early rise in spring temperatures which enables crops to use the lengthening daylight more fully than inland. With Cornwall, Devon and parts of Dorset these western and southern coasts are the only parts of Britain where averages of daily mean temperatures (reduced to sea level) rise above 5.5 degrees centigrade before mid-February (Taylor 1967) and, as a result, it is possible to grow early potatoes in south-west Dyfed, Lleyn and Anglesey.

Soil moisture

Potential transpiration is defined as the amount of water transpired by an actively growing short green crop, which completely shades the ground and is adequately supplied with water from soil reserves or irrigation (Penman 1948). This concept can be extended to include evaporation losses from bare moist soil in winter. Although actual transpiration from plants can be substantially different, the concept has proved valuable for assessing the effect of climate on water use by crops and seasonal variations in soil moisture conditions. Potential transpiration varies less than rainfall with annual averages ranging from 575 mm in west Dyfed to 400 mm on the mountains in Snowdonia. Distribution throughout the year is uneven with a peak in June or July and only 10 to 15

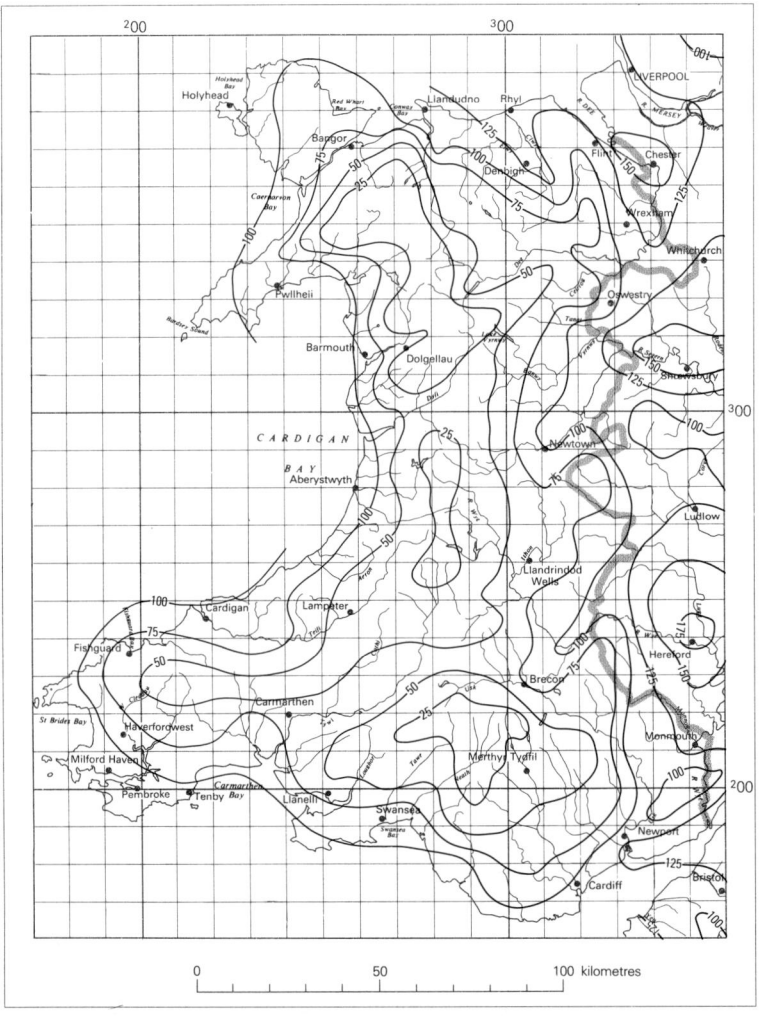

Figure 9. Mean Accumulated Maximum Potential Soil Moisture Deficit (mm), 1961—75

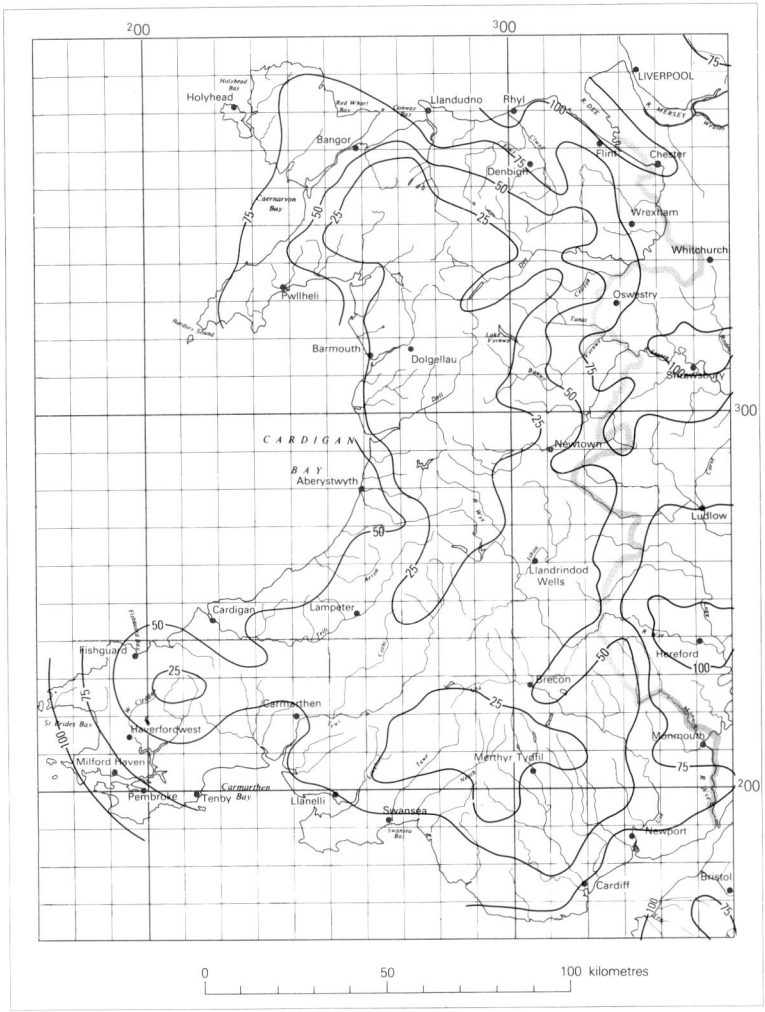

Figure 10. Mean Accumulated Potential Soil Moisture Deficit (mm) for mid-July, adjusted for Winter Wheat, 1961—75

per cent of the annual total is transpired between October and March. Year to year variation is small with standard deviations of 25 to 30 mm common in most districts.

Soil and climatic interactions are examined below using the concept of potential soil moisture deficit (P.S.M.D.) which is calculated as the difference between potential transpiration and rainfall during the growing season. Transpiration usually first exceeds rainfall during April in an average year and a soil moisture deficit develops which then accumulates, usually reaching a maximum in June or July in west Wales and a few weeks later in the east; thereafter rainfall is normally greater than potential transpiration and soil reserves are replenished during autumn. There is then a surplus of moisture which persists until the spring. Using data supplied by the Meteorological Office, monthly moisture balances were calculated for 954 stations in England and Wales for each of the years 1961 to 1975, and deficits were accumulated month by month.

Figure 9 shows the distribution of mean maximum P.S.M.D. in the growing season; over most of Wales it is less than 125 mm. The resulting moist conditions encourage fungal diseases in arable crops and only in soils with small available water capacity is droughtiness a serious problem in normal years. The maximum P.S.M.D. is appropriate for studying grass or other perennial crops. Most arable crops, however, do not fully cover the ground early in the season and some, for example cereals, stop growing before the maximum deficit has developed. Figure 10 shows the distribution of mean P.S.M.D.

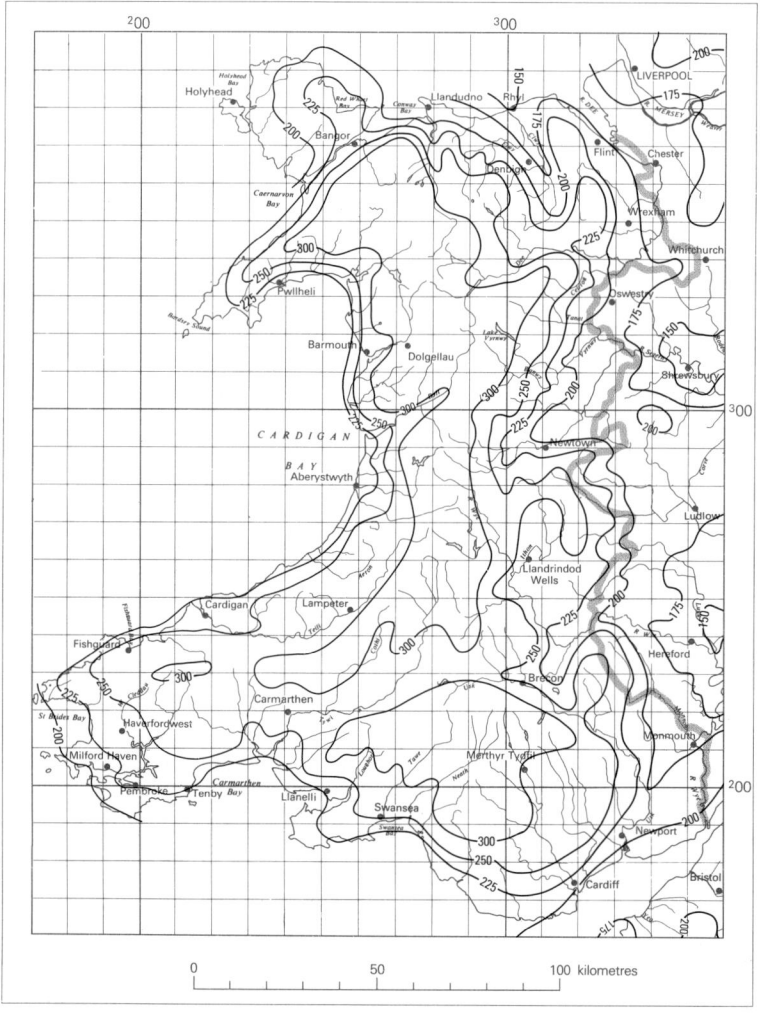

Figure 11. Median Field Capacity Period (days), 1941—70

for mid-July adjusted for the average growth pattern of winter wheat, termed M.D. (winter wheat) (§ 106). To assess site drought risks for a specific crop, the adjusted moisture deficit for that crop is subtracted from the available soil water reserves, a procedure described in detail by Thomasson (1979) and applied to extensive agricultural soils in Chapter 3. On average, mid-July deficits are smaller than the maximum P.S.M.D. and cereal crops suffer less from water stress than grass growing in the same type of soil.

Soil wetness is a more common problem in Wales than soil droughtiness. Field capacity (Chapter 2) is a useful concept for assessing the effect of climatic wetness on agriculture. It is calculated from rainfall and potential transpiration using a standard soil water abstraction model for short rooted crops (Smith 1967). The duration of field capacity determined in this way varies broadly with annual rainfall and its distribution has been plotted (Fig.11) from regression of average annual totals (1941–70) on the median dates given in Smith and Trafford (1976). Duration is shortest (less than 175 days) in the Vale of Clwyd with periods between 175 and 250 days over much of the remaining lowland, whereas field capacity lasts longer than 300 days in most of the uplands.

The field capacity period can be used as a direct measure of accessibility to land provided allowance is made for soil properties. Cultivation of well drained coarse textured soils is often possible within the field capacity period without harmful effects whereas clayey, other slowly permeable, or peaty soils are usually impassable, and remain wet for longer periods than well drained soils under the same weather conditions. The methods of assessing workability, trafficability and poaching risk from the interaction of these data with soil factors are described in Chapter 2 (§ 7) and estimates for specific soils are given in Chapter 3. The drainage status of soils and the relationship to field capacity are discussed in Chapter 4. The isopleths on Figure 11 drawn at 25 or 50 day intervals give field capacity zones which are used in Chapter 5 as a framework for assessing the suitability of soils for growing various crops including grass.

§ 4. LAND USE

The moist climate of Wales suits the growth of grass, and livestock farms are the dominant feature of the landscape. On higher ground, enclosed pasture often gives way to less productive rough grazing or is replaced by forest. While the use made of land depends largely on its inherent quality, social and economic factors also influence land use policy. Local planning regulations aim to confine new building to existing industrial and urban areas, and other controls are exercised over large rural tracts. Public interests govern the use of 4,110 km² of land in the National Parks of Snowdonia, the Brecon Beacons and the Pembrokeshire Coast where the National Park Committees maintain the landscape under powers given in 1949. Lengths of the coastline in the north-west, the Gower and the Wye Gorge are designated Areas of Outstanding Natural Beauty (676 km²) and 26 km² of Wales have been dedicated as Country Parks under the Countryside Act 1968. In these and the National Parks, changes to the landscape are often resisted. Outside these areas though there is little apparent control, farming depends heavily on state and European Community support which guide land use and farming practice by inducement.

Over the hundred years from the 1881 census, the population of Wales grew from 1.6 to 2.8 million. This resident population is swollen each year by tourists; 4.5 million stayed more than four nights in 1980. The community is served by a main rail system that has shrunk from 3,000 to 1,600 track kilometres since 1961. In contrast, the road network has grown. During the seven years from 1976, when there were 31,000 km of road, the network grew by 1,355 km (Welsh Office 1983a). Motorways have increased from 71 to 193 km in the same period. Transport accounted for a little over 2 per cent of the land area in 1970

The second Land Use Survey demonstrates the importance of grassland in Wales. Permanent or temporary grass covers half the Principality and ranges from almost two-thirds of Dyfed to a quarter of West Glamorgan. Rough grazings of heather and *Molinia* and pastures invaded by weed species such as bracken, rushes or gorse account for a third of Wales, South Glamorgan has the least with less than 5 per cent while Gwynedd has 40 per cent. Arable land represents 5 per cent of Wales while woodland occupies a tenth; West Glamorgan is remarkable in having 20 per cent woodland.

Farming

The pattern of farming (Table 3) relates to the climate, soil and the profitability of various forms of husbandry. Arable farms are few and largely confined to the drier lowland fringes, while livestock farms are far more common and take full advantage of the generally equable temperatures and adequate rainfall for grass growth. Sheep graze unenclosed land in the hills. Farming, however, is not static, and new techniques and changes of farm prices can enable the range of a crop to extend to land previously considered unsuited, or to contract from land where it once grew successfully.

The distributions of specific crops and stock are shown in Figures 12 to 15. Their relative abundance in selected representative parishes is illustrated by Figure 16. Livestock farming dominates inland Wales (Church *et al.* 1968) with dairying concentrated in the lowlands, the coast, the English borderland and major valleys. Arable farming is only significant along the south coast from Gwent to Pembroke and in isolated pockets in the north, although periodic crops of barley for feed form part of the rotation on many livestock farms. Welsh farms are small, with 70 per cent of holdings possessing less than 50 ha of crops and grassland (Welsh Office 1983b). Although there is a continuing trend towards farm amalgamation (Henderson 1977) and while many upland farms have large areas of rough grazing, Wales has only 2 per cent of the United Kingdom's holdings with more than 200 ha of crops and grassland (Aitchison 1979).

The proportion of grassland less than 5 years old (temporary grassland) varies from county to county but is 8 per cent overall. In Wales swards are on average appreciably older than in England, 24 per cent being 8 to 20 years old when ploughed in. Between the surveys of Stapledon (1936) and Green (1982) the proportions of enclosed swards containing more than 40 per cent valuable plant species increased more than tenfold. This

Table 3
Agricultural Land Use 1980

	Areas Admin. (km²)	Agricultural (km²)	(%)	Total tillage[2] (%)	Crops[1] Wheat	Barley	Sugar beet	Potatoes	Total vegetables	Fruit[3]	Grassland[1] Permanent	Young	Rough	Grazing livestock units/100 ha[4]
Clwyd	2,426	1,777	73	22	0.7	5.9	<0.1	0.3	<0.1	<0.1	60	12.7	15.4	133
Dyfed	5,767	4,587	79	21	0.1	5.0	<0.1	0.6	<0.1	<0.1	60	13.1	15.5	112
Glamorgan, Mid	1,019	507	50	13	0.3	3.2	<0.1	0.2	<0.1	<0.1	54	7.2	30.1	95
Glamorgan, South	416	248	60	29	3.1	9.9	<0.1	0.5	0.1	<0.1	62	12.3	4.8	129
Glamorgan, West	817	364	45	16	0.2	3.3	<0.1	1.1	0.5	<0.1	52	8.3	28.8	90
Gwent	1,376	847	62	25	2.3	4.4	<0.1	1.0	0.3	0.3	65	13.5	6.0	125
Gwynedd	3,867	2,922	76	12	0.1	2.7	<0.1	0.2	<0.1	<0.1	42	7.2	44.4	87
Powys	5,077	3,660	72	16	0.6	2.3	<0.1	0.1	<0.1	<0.1	56	10.7	25.7	113
Total	20,765	14,912												

[1] All categories are given as a percentage of the total area of agricultural holdings.

[2] As a percentage of the total area of agricultural holdings; tillage includes all arable crops, fallow and grassland four years old or less.

[3] Fruit includes small fruit and orchards.

[4] Apparent densities are inflated in counties with significant areas of common grazing land. Powys has about 20 per cent common land, West Glamorgan 15 per cent, Clwyd, mid-Glamorgan, Gwent and Gwynedd 10 per cent.

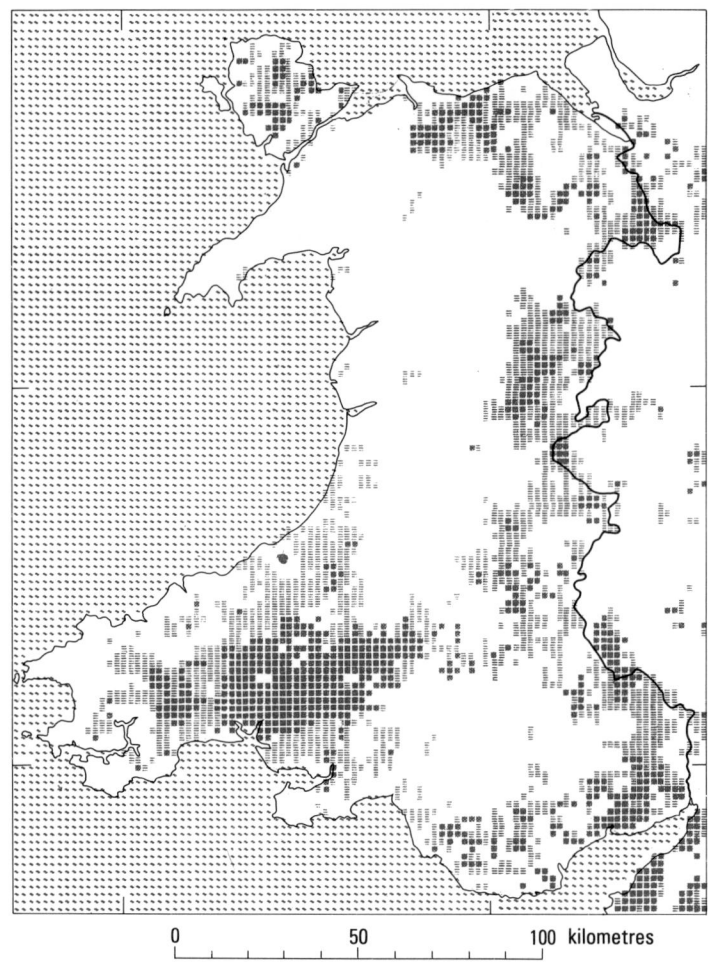

Grass >4 years old as a percentage of total area (MAFF agricultural census, parish summary 1980).

62.5–72.5 %   > 72.5 %

Figure 12. The distribution of permanent grassland (more than 4 years old) in 1980, shown as a percentage of total area

better grassland is maintained by reseeding as the better grasses and herbs die out with age. Between 1947 and 1971, improvements in productivity enabled stocking rates to increase by 30 per cent while between 1951 and 1971 grass satisfied a 33 per cent increase in the dietary requirement of livestock. Grass is conserved for winter feed either as hay or silage. During the period 1938 to 1977 hay yields from temporary grassland rose from 2.6 to 4.6 tonnes/ha and those from permanent grassland from 2.1 to 4.4 tonnes/ha. Between

1938 and 1982 silage production leapt from 31,000 to 3,161,200 tonnes, including a small amount of ensiled cereals, reflecting its suitability in the unpredictable climate of Wales. The average yield of silage in 1982 was 22.2 tonnes/ha.

The grass is used for beef, sheep and milk production. Most of the beef cattle are from dairy cows crossed with Hereford, Aberdeen Angus and European breeds that impart fast live-weight gain. Some beef produced on dairy farms is pure bred Friesian. Most beef calves are either sold the first year to farms on higher land or kept for a year and sold for

0   50   100 kilometres

Grass 4 years old or less as a percentage of crops and grass (MAFF agricultural census, parish summary 1980).

▦ 15.0—20.0 %   ▦ > 20 %

Figure 13. The distribution, 1980, of grassland (4 years old or less) shown as a percentage of crops and grassland

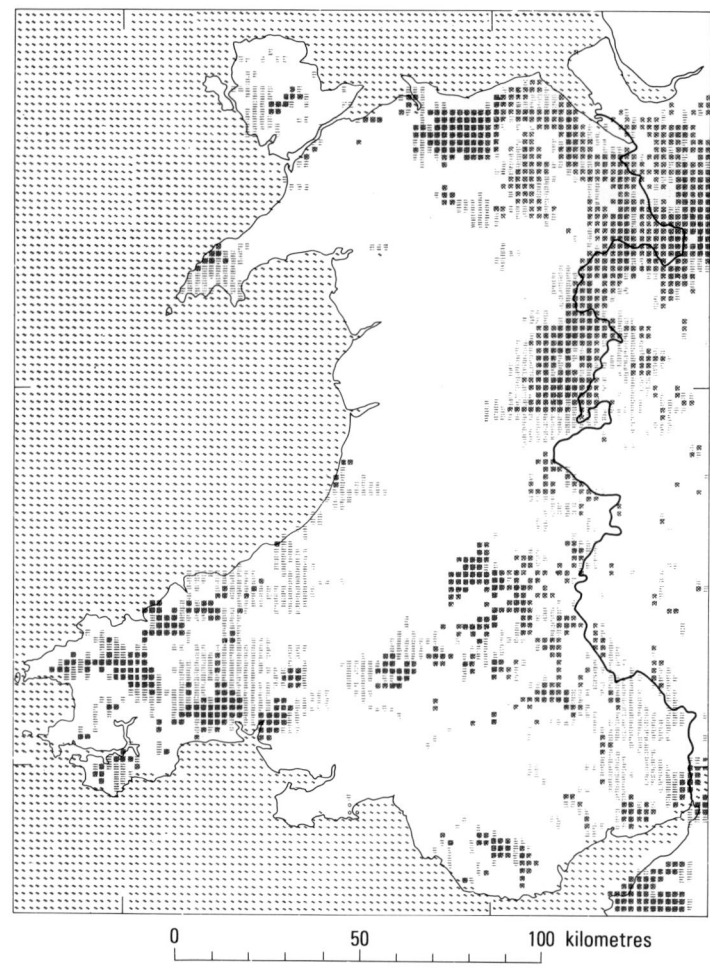

0   50   100 kilometres

Grazing livestock units per 100 hectares of crops and grass (MAFF agricultural census, parish summary 1980).

▦ 120.0—135.0   ▦ > 135.0

Figure 14. The distribution of grazing livestock units per 100 ha of crops and grassland 1980

Table 4
Livestock on Welsh agricultural holdings (millions) (Welsh Office 1983a)

	1961	1982
Cattle	1.2	1.4
Sheep and lambs	5.5	8.4
Pigs	0.2	0.1
Poultry	4.4	7.7

Tillage as a percentage of the total area (MAFF agricultural census, parish summary 1980).

26.0−36.0 %   36.0 %

Figure 15. The distribution, 1980, of tillage shown as a percentage of total area

fattening in England. Store cattle from the hills are sold after an average two and a half years for fattening in the lowlands. Farms at 200 to 350 m O.D. often have less-productive pastures so concentrate either on rearing pure-bred beef cattle or store cattle, some stock is fattened on the best land. Above 350 m O.D., beef enterprises are mostly based on store cattle, sometimes in conjunction with a lowland home farm on which the cattle are reared and fattened. The number slaughtered in Wales has dropped since 1961 but the total including dairy cattle has increased (Table 4).

Parish	Size (km²)	Soil Associations	Comment
Capel Curig	96	Bangor (30%) Hafren (20%) Wilcocks (15%) Moretonhampstead (15%)	2% commonland; 12% forestry
Llanuwchllyn	116	Hafren (25%) Brickfield 1 (20%) Crowdy 2 (20%) Bangor (20%)	3% commonland; 10% forestry
Llanfor	130	Brickfield 1 (30%) Hafren (25%) Manod (20%) Crowdy 2 (15%)	7% commonland
Llansadwrn	30	Manod (50%) Cegin (40%) Wilcocks (10%)	2% commonland; 3% forestry
Clydey	34	Manod (75%) Parc (15%) Cegin (5%)	1% commonland; 15% forestry
Aberffraw	31	East Keswick 1 Brickfield 2	7% commonland
Manorbier	14	East Keswick 3 (55%) Milford (35%) Neath (10%)	No commonland
Penmark	14	Ston Easton (90%) Denchworth (10%)	Partly urban and includes airfield; 8% forestry
Llanddewi	8	East Keswick 1 (70%) Brickfield 2 (30%)	No commonland

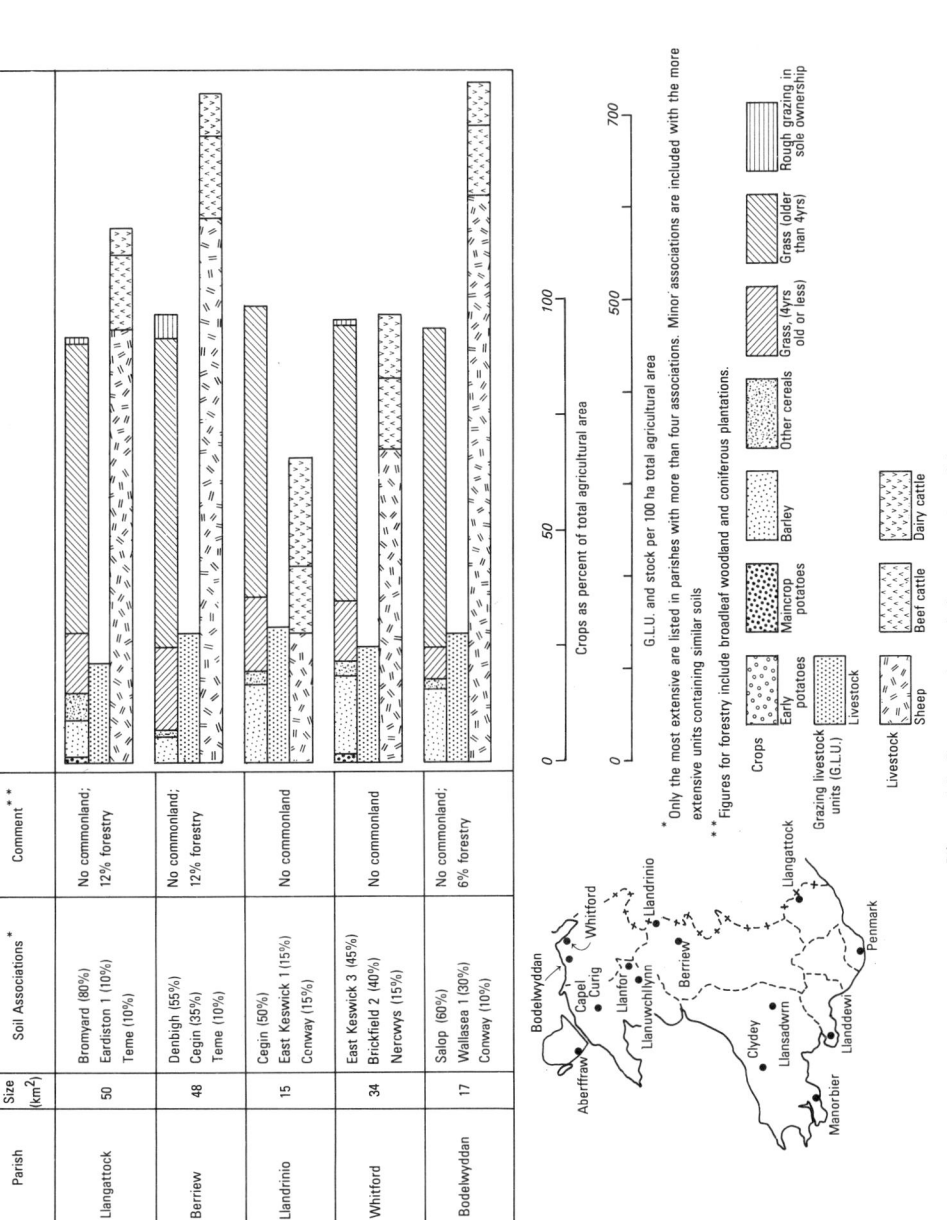

Figure 16. Crops and stock in selected parishes

Milk production is concentrated in the lowlands (Fig.14). Since the mid-19th century milk has been sent to London by train from the farms of the renowned Tywi valley. Nearly all milk is now taken by tanker to creameries where it is pasteurized and used in a range of dairy products. This has concentrated dairying on larger units capable of justifying tanker collections. Most herds are Friesian and the average dairy farm has two to three milkers per hectare producing 4,460 litres of milk per cow per year.

Sheep in Wales can be separated into mountain flocks and the larger more prolific lowland breeds. Ewe lambs from the mountain flocks are sent to lowland pastures for their first winter while the castrated lambs are sold fat or sent for fattening in the lowlands. While mountain ewes seldom twin, lowland flocks are normally crossbred with heavier breeds making them more productive. The Welsh flock has risen appreciably since 1961 (Table 4) but the number slaughtered has varied with an overall increase to 1.1 million in the years up to 1982.

Some indication of the productivity of grassland and rough grazing in Wales is given by the map of grazing livestock units (Fig.14). Density is greatest in parts of Clwyd, Powys and Dyfed. Grazing livestock units are a measure of grazing intensity and incorporate all grazing livestock.

Barley is grown in the lowlands of the south coast, west Dyfed, Anglesey, north-east Clwyd and the Welsh borderland. It is favoured in moist western Britain because of its earlier maturity and ripe heads that dry relatively quickly after rain. In a few parishes, barley accounts for 20 per cent of total agricultural land but the overall figure for Wales is only 2 per cent. More or less continuous arable production is restricted to the most workable soils of the lowlands. Despite soil conditions that favour autumn sowing and the general United Kingdom trend away from spring barley, encouraged by improved winter crop protection and new varieties, spring barley continues to be the main cereal in Wales. Although some is grown for malting, most is sown to meet on-farm demands for straw and stock feed, and need not be of prime quality. Since 1961 the Welsh harvest has risen from 64,000 to 260,000 tonnes (Fig.17). The present area is 55,000 ha and average yields have also grown (Table 5).

Table 5
Crop yields in Wales (tonnes per ha) (Welsh Office 1982)

	1961	1982
Wheat	3.3	6.2
Barley	3.1	4.6
Oats	2.6	4.0
Potatoes (early)	16.9	22.8
Potatoes (main)		38.6
Turnips and swedes	44.4	54.9
Mangolds and fodder beet	55.9	84.7

Wheat is grown mainly in parts of Gwent and South Glamorgan, north-east Powys and lowland Clwyd. The crop is sown in the autumn but harvested after barley, so is less suited to districts with an early return to field capacity (Fig.11, § 3). Market conditions in the early 1980s have favoured it and there have been strong inducements to extend its range. The Welsh harvest in 1982 was 50,000 tonnes. Estimated average yields have risen significantly (Table 5) and with improved crop protection and husbandry, yields of 10 tonnes/ha are becoming increasingly common.

Oats are distributed patchily throughout the lowlands but are also grown in the upper Wye valley and on the Lleyn Peninsula. The crop grows well on acid soils but is not well-suited to combine harvesting because the grain is shed too readily from the ripe head. Mainly for this reason it has been supplanted by barley and the harvest has dropped from 115,000 tonnes in 1961 to 25,000 tonnes in 1982. Most oats go into stock feed but the best are used for cereal and health foods. From a peak of 112,500 ha in 1948 the area planted had fallen to 5,900 ha in 1982 but average yields per hectare have increased (Table 5).

Rye is grown sporadically throughout Wales mostly as a forage crop. Sown in autumn, it provides an early bite for livestock, and fields are then normally left until a following root crop.

Turnips and swedes are grown in south Powys and parts of Clwyd. Many crops are

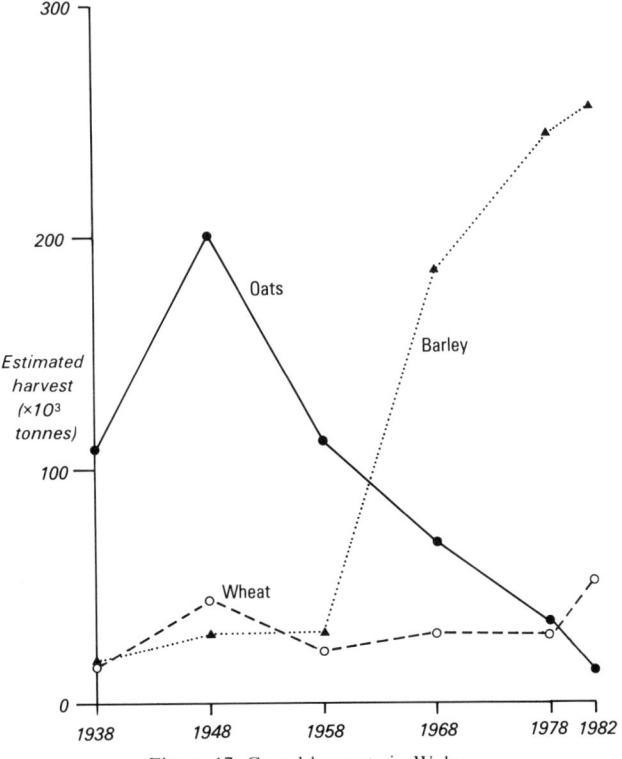

Figure 17. Cereal harvests in Wales.

drilled into killed grassland or cereal stubble. A small amount of sugar beet is grown in north-east Powys, the Maelor and in the Vale of Clwyd, mostly on freely draining soils. The roots, harvested late, are processed at Allscott in Shropshire.

Early potatoes are a traditional high-value crop in the mildest parts of Dyfed, the Gower, Lleyn and south-east Anglesey. About 3,000 ha are planted in favoured fields from mid-February but yields are erratic, depending greatly on soil conditions at planting. A wet season delays planting and the week or fortnight advantage over drier but colder districts of Britain is lost. Currently, growers aim to harvest in early June following the Channel Island crop. Irrigation increases yields dramatically and eases harvesting. On most of the soils, early potatoes are only grown about one year in four and are followed by a grass ley to counteract damage to soil structure. On coarse textured soils associated with igneous rocks they are grown in consecutive years. The two most common varieties are Home Guard and Red Craigs Royal. The crop is harvested to optimize profit usually at around 18 tonnes/ha.

Main crop potatoes are grown throughout the lowlands especially near centres of population, mainly to diversify cropping and increase profit in good years. The area planted has fallen from 4,300 ha in 1961 to 3,000 ha in 1982 but yields have doubled to about 38 tonnes/ha (Table 5).

Maize is restricted to parts of Gwent, Dyfed, South Glamorgan and Clwyd. It is either fed green to stock or ensiled. The soil remains bare throughout the year and erodes from slopes unless the crop is drilled along the contours. Market gardening has diminished. The climate is not ideal as there is too little sunshine and warmth. Recent increases in heating costs and imports from Europe have put many glasshouse growers out of business. Those who remain are close to ready markets along the south and north coasts or rely on 'pick-your-own' trade, much of it from holiday makers and people from nearby towns.

A constant theme in present-day Welsh agriculture is increased efficiency and productivity. While partly due to improved husbandry, better breeds and varieties, and ever more effective crop protection and control, much of the increase comes from using artificial fertilizers. Following increases in the use of phosphorus and potassium fertilizers in the fifties and early sixties, rates of application have remained roughly constant (MAFF 1981). The steep rise in phosphorus fertilizer costs, combined with the enforced replacement of basic slag by alternative forms of phosphorus, reduced consumption in the early seventies. In contrast, nitrogen fertilizer use has increased consistently such that overall rates in Wales in 1981 were almost ten times those of 1957. In early years most was as compound fertilizer but since 1960 straight nitrogen fertilizers have been increasingly important (Church and Lewis 1977). Rates of application are three times higher on the most intensive dairy farms than on mainly stockrearing farms. For silage, rates are twice as high on dairy and arable units. Nitrogen use in the uplands has not kept pace with the more intensively grazed lowlands where on some farms grass yields can approach 20 tonnes dry weight/ha. These trends are reflected in the figures for average national consumption, which also show that nitrogen use in Wales is considerably less than in England.

To demonstrate regional variations in cropping within Wales, Figure 16 illustrates the agricultural statistics of 14 parishes. These have been chosen for their homogeneous climate, topography and soils and are spread throughout Wales. The relative proportions of the most extensive soil association in each parish are listed. Because the agricultural statistics include rough grazing with sole rights only, the extent of commonland, most of which is grazed, is listed separately. MAFF agricultural statistics for a parish are based on returns from farmers resident there who own or rent land both inside and outside that parish; data from holdings in the parish owned by absentee farmers is excluded.

Upland parishes tend to be mostly lightly stocked. Capel Curig with the largest proportion of rough grazing supports only 40 grazing livestock units per 100 ha, most of them sheep. In the more productive lowlands where more nitrogen fertilizer is used, Llandrinio with a concentration of dairying carries nearly 150 units per 100 ha. Parishes such as Aberffraw and Berriew support more than 100 units per 100 ha with large sheep flocks as well as beef and dairy cattle. Penmark with a large area of Ston Easton soils has the largest arable acreage and is the only parish in Figure 16 in which wheat, as opposed to barley, features prominently. Manorbier on the south coast shows the greatest variety in cropping, but even here permanent grassland is the single most extensive crop.

Forestry

Forests have for many centuries contributed significantly to the resources of Wales. Place names like coed (forest), derw (oakwoods), bedw (birchwoods) and onn (ash trees), bear witness to much more extensive former woodlands. In medieval times they provided refuge and sustenance for local communities and as feudal power increased with improved communications, large tracts were subject to laws designed to conserve the forests (Linnard 1982). By the 14th century, the woods were put to a wide variety of uses. They supplied large timber for buildings and ships, oak bark for tanning, windthrown trees and dead wood for construction and fuel, and yielded charcoal for castle fuel, smelting and burning lime. Trees provided foliage (especially elm) for cattle fodder and mast for fattening swine. Herbage was grazed by livestock. Fruits, berries, nuts, honey and wax were other benefits along with wood ash, rich in potassium for fertilizer and used for soap making and dyestuffs.

Despite continued forest clearance, it was not until the 17th century that a few estate owners began planting for the future. In the 19th century some private estates employed professional foresters who for the first time began to systematize tree measurements and, by studying productivity, laid the foundation of modern scientific forestry. Only since 1919, with the formation of the Forestry Commission have substantial areas of hill land been planted (Plate 10), and grants provided to encourage private landowners to replant.

In 1981 there were 44 forests in Wales owned by the Commission (Forestry Commission 1982) producing about a tenth of Britain's timber need. Today the forest

T.R.E. Thompson

Plate 10. Abandoned farmhouse, Machno valley, Gwynedd. Economic changes have led to the afforestation of uplands with fast-growing conifers.

products include building materials, furniture, telegraph poles, pit props, fencing, chipboard and paper pulp. The forests also provide rural employment and can enhance the scenery and recreational value of the countryside.

Conifers dominate most plantations since they grow relatively rapidly on the rough hill land less favoured for agriculture, but some 50,000 slower growing broadleaved hardwoods requiring better land are planted each year, principally beech, oak, ash and sycamore. Of the conifers some five million Sitka spruce and one quarter-million Lodgepole pine are planted each year on exposed uplands, while many hundreds of thousands of larches, Norway spruce, Western hemlock and firs are set in more sheltered positions. Most of the trees are raised from seed in nurseries near Pembroke, Cardiff and Caernarfon where several million seedlings are grown at any one time.

Mineral extraction

Extractive industries have moved large amounts of material, especially in the mines of north and south Wales. The main activity came with the industrial revolution of the last century, but small scale working goes back to Roman times.

Slate quarrying is largely confined to Snowdonia. Rock waste rather than the quarries dominates the landscape of the 'slate belt'. Here 12 to 20 tonnes of rock are quarried to

P.S. Wright

Plate 11. East Pit Colliery, Gwaun-Cae-Gurwen. Wilcocks 1 association is on the lower land with the Withnell 1 association on distant hillsides. The Gelligaer association covers the summit plateau on the skyline.

obtain 1 tonne of finished slate. Spoil heaps form vast man-made screes, fringed by lobes and fans of outspreading debris.

In south and north-east Wales, coal mining is the main extractive industry. By the end of the last century most accessible seams were being worked and annual production has since declined to about 12 million tonnes. The conical waste tips left a great mark on the landscape (Plate 11) but many are now being flattened and reclaimed for other uses. About one-fifth of present day coal mining is by opencast methods which produce less waste, as overburden in replaced and the land restored (§ 33).

Large reserves of iron ore in the Carboniferous rocks were also mined in the last century and a string of mining and smelting townships grew up, notably between Blaenavon and Dowlais. Other metal ores including those of copper and nickel were processed in the lower Swansea valley leaving much of the land derelict for many years although there have recently been determined efforts at reclamation. Lead and zinc were worked in mid- and north Wales in hundreds of scattered mines until the early twentieth century, and their spoil heaps are still visible.

Plate 12. Limestone quarrying near Carew Newton, Dyfed. Limestone beneath deep drift was once transported by water to many parts of West Wales. Tidal creeks and saltmarshes border the main waterways.

Other extractive industries having a considerable impact on the landscape include quarrying for limestone (Plate 12), sandstone, hard igneous rocks and sands and gravels. Some 13.6 million tonnes of limestone were taken in 1980 for roadstone or construction. At present 2 per cent of extracted limestone is used in agriculture to fertilize 6 per cent of crops and grass in any one year. Sandstone totalled 1.5 million tonnes, land-won sand and gravel 2.5 million tonnes, and igneous rock 1.8 million tonnes. Derelict land (Plate 13) covered 7,200 ha in the 1960s, but 3,326 ha were reclaimed with grants of £65m during the ten-year period up to 1980 (Welsh Office 1983a).

P.S. Wright

Plate 13. View down Ebbw Vale. The Withnell 1 association covers the steep valley sides.

2

BACKGROUND TO THE SOILS

§ 5. SOIL FORMATION

Soil is the upper layer of the earth's mantle and is the product of several complex interacting processes, which are most intense near the surface. The *soil profile* is a vertical column of soil as seen in the exposed face of a pit. It usually consists of layers known as *horizons* roughly parallel to the ground surface passing down to relatively unaltered material, which may represent the *parent material* in which the soil above has formed, although upper horizons are frequently formed in a different, often more recent, material. The term parent material is used here to indicate the principal source of weatherable material. Soil profiles are essentially three-dimensional and range in depth from a few centimetres to several metres, but only the upper 1.5 metres are used for soil classification purposes in England and Wales (§ 6).

Under a cover of vegetation, plant remains are added to the soil where they are more or less decomposed and incorporated with the mineral material. Part of the organic matter decays, releasing nutrients for plant growth, but part is transformed to dark coloured humus which decomposes at a slower rate. In base-rich, well-aerated soils, rapid decomposition and faunal activity, particularly that of earthworms, achieve intimate mixing of mineral and organic components. In strongly acid soils where there is little mixing by fauna, breakdown is retarded so that partly decomposed plant litter accumulates on the mineral soil surface. Where waterlogging or low summer temperatures severely inhibit decomposition, thick accumulations of organic matter or peat develop at the surface. The practices of ploughing and cultivation mix crop residues into the soil, where they are further incorporated by microbial activity. Most regularly cultivated soils have less organic matter than those under semi-natural vegetation or grassland.

Soils further develop through the accumulation of unconsolidated mineral grains and rock fragments by *weathering*, and the horizons become differentiated by physical and chemical as well as biological processes. Physical breakdown by the action of freezing and thawing, or wetting and drying, into progressively finer particles renders rock more accessible to chemical weathering. The chief agent is rainwater containing dissolved oxygen, carbon dioxide and acids derived from decomposing plant residues and, in some districts, from atmospheric pollution. Rainwater moves through the soil and dissolves and decomposes minerals. This releases ions, especially the basic cations (bases) calcium, magnesium, sodium and potassium, into the soil solution. Thus carbonates are gradually dissolved; micas and other minerals in sedimentary rock become hydrated; ferrous and sulphide ions are oxidized and primary silicate minerals such as feldspar are subject to hydrolysis, whereby cations and some silica pass into solution. The basic cations released by weathering, and nitrate ions derived largely from organic matter or by fixation from the air, provide nutrients for plants, but *leaching* (Fig.18) tends to move them downwards beyond the root zone or laterally through the soil by seepage. The extent of leaching, which is related to annual rainfall and soil texture, is indicated by the acidity of most soils. In agricultural practice, leaching losses into drainage water are balanced by liming and fertilizer dressings, which also promote incorporation of organic residues. Relatively insoluble weathering residues such as clay minerals, hydrated ferric oxides, quartz and other resistant minerals inherited from the parent rock, are left in the soil profile after leaching. Some soluble cations, particularly calcium, may be deposited in lower horizons as coatings or nodules of carbonate.

In many soils dispersed clay is carried in suspension by water percolating to lower levels (Fig.18). The clay is redeposited as skins or coats on the surface of structural aggregates or in pores and around stones when films of water dry out. The resulting clay-enriched subsoils, characteristic of many leached soils in lowland England, are described as *argillic* horizons. Clay translocation under past climatic regimes gives *paleo-argillic* subsoil horizons which are recognized by their micromorphology and reddish colour.

Intense leaching and translocation can lead to *podzolization* (Fig.18). In this process an acid humus layer accumulates at the surface and releases organic acids. These form complexes with iron and aluminium compounds that readily migrate downwards in percolating water. A depleted, bleached subsurface horizon is thus left above horizons that have been correspondingly enriched by precipitation of humus and/or iron and aluminium.

Waterlogging leads to the reduction, mobilization and removal or redeposition of iron compounds and produces distinctive soil horizons. The reduction of ferric iron to more mobile, colourless or grey ferrous iron complexes, by micro-organisms or by products of decomposing organic matter, is known as *gleying*. Gleyed soils, in which a slowly permeable subsoil impedes surface water drainage, are called stagnogley soils or, if organic matter is accumulating in the surface horizons, stagnohumic gley soils. Where more permeable subsoils are affected by groundwater, ground-water gley soils are formed. In both kinds of soil periodic waterlogging allows intermittent or local aeration, with consequent re-oxidation, and subsoils become mottled in grey, yellowish and ochreous colours, often accompanied by black ferri-manganiferous nodules or concentrations. Persistently waterlogged soils are usually wholly grey or bluish grey. Well aerated soils retain the browns, reds or yellows of oxidized ferric iron compounds.

In uplands with high rainfall where soils have peaty surface horizons and are strongly

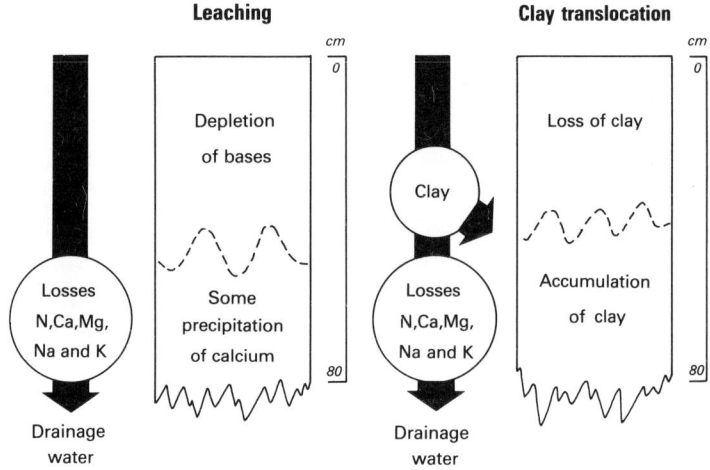

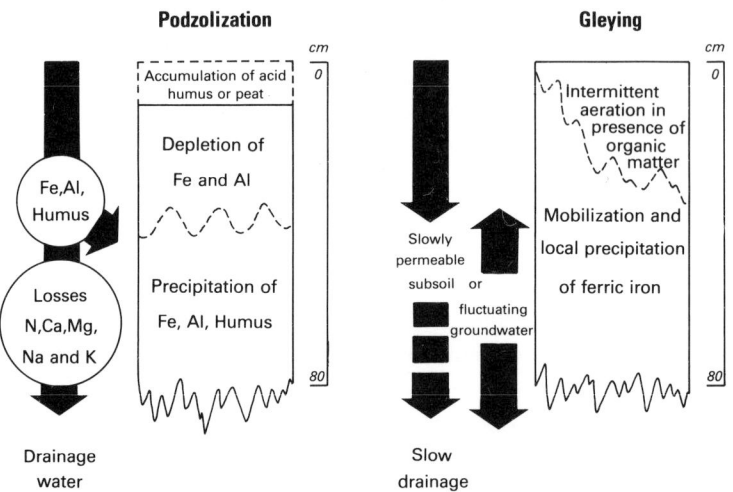

Figure 18. Soil forming processes and the movement of soil water.

leached and podzolized, a thin hard and often continuous ironpan may develop. It usually forms where descending soil water reaches less-weathered iron-rich parent material below seasonally waterlogged upper horizons. The soils thus show features of both gleying and podzolization, their name, stagnopodzols, emphasizing that the gleying results from impedence of water movement in the upper layers. Further distinctions are made between soils of this kind with ironpan development, and those with more diffuse ochreous iron-rich subsurface horizons.

In most soils, particles of sand, silt and clay adhere together in aggregates known as *peds*, and, influenced by the size and distribution of spaces between these, the soil is loose or relatively compact. The degree of this structural development is strongly influenced by texture. Sands are commonly structureless, that is, their individual grains do not combine as peds, whereas ^lays swell and shrink during wetting and drying cycles to produce well formed blocks or prisms. However, in fluid muds and organic materials that have remained saturated with water since deposition, structure develops only after drainage allows irreversible loss of part of the water and gives the soil a firmer consistence. This process is known as *ripening*. The actions of roots, soil fauna and micro-organisms assist structural development, particularly near the soil surface, increasing porosity and the flow of air and water which are necessary for good crop production.

The soil forming environment

The characteristics of the soil at any particular place depend on five main groups of factors. These are, the physical and chemical constitution of the parent material; past and present climate; relief and hydrology; the length of time during which soil forming processes have been acting; and the ecosystem, including the extensive modifying effects of man's activities.

The chief properties of parent materials that influence soil formation are the permeability, base content, hardness, grain size and the mineralogy of their weathering products. For example, the widespread grey Palaeozoic rocks are dominantly argillaceous, consisting of relatively hard shales, mudstones, siltstones and slates which weather slowly to give stony silty or loamy soils in which leaching and to a lesser extent podzolization are the main soil forming processes. Otherwise similar Devonian siltstones are red in colour from the presence of haematite and soils on these materials inherit their reddish hue. Hard sandstones locally give very shallow ranker soils on steep slopes, interspersed with outcrops of rock, but, more usually, weather to deeper leached or podzolized loamy soils. Acid igneous and metamorphic rocks are mostly hard and often give podzolized soils, but basic igneous rocks in Snowdonia weather readily to give less-leached brown earths. Base-rich soils on limestones are comparatively rare in Wales, partly because the rocks are covered by thin drift, as over the Lower Lias limestones in the Vale of Glamorgan, and partly because strong leaching on hard upland limestones gives very shallow non-calcareous ranker soils.

Most Palaeozoic shales and mudstones have a micaceous or mixed clay mineralogy, and their clayey soils lack marked swelling and shrinking properties, but Jurassic shales in the Vale of Glamorgan are smectitic and give wet swelling clay soils (Avery and Bullock 1977).

In Wales, as in most glaciated regions, many soils are developed in drift. The drift is often of local provenance, scarcely distinguishable in composition from underlying country rock. Compact fragipans, probably formed under a periglacial climate, impede water percolation and often cause gleying.

The broad relationships of climate and soils and the effects of slope and elevation are

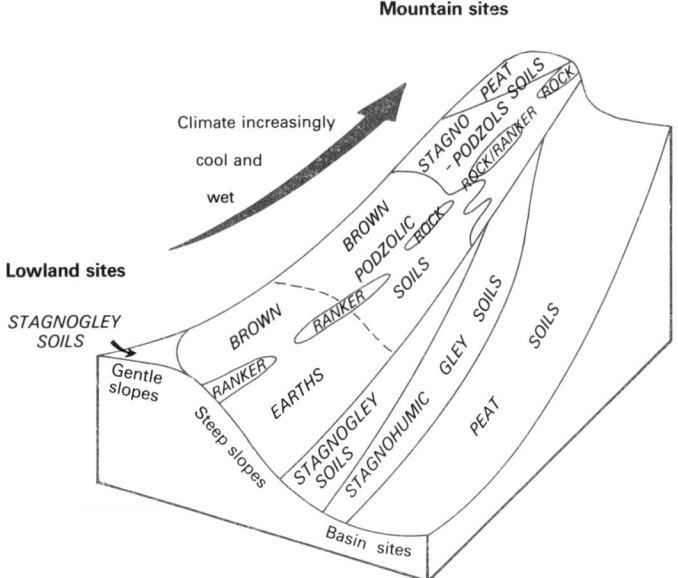

Figure 19. Relationships of soils, climate and relief in Wales

shown schematically in Figure 19. Soils in the wet uplands, are intensely leached and strongly acid. Podzolization influences most of the soils, and brown podzolic soils are most extensive in Wales. These have accumulations of ochreous sesquioxides of iron and aluminium in the subsoil without a bleached horizon above. They occupy both gently sloping and steep land usually at middle elevations above the lowland zone where less leached brown earths are typical. At higher elevations brown podzolic soils are succeeded by stagnopodzols. The soils are waterlogged for longer on high ground where more rain, lower summer temperatures and lower rates of evapotranspiration retard organic matter decomposition and encourage its accumulation at the surface. Thus podzolized and gleyed upland soils have peaty surface horizons (stagnopodzols and stagnohumic gley soils), and on level summits blanket peat is extensive. Similar acid peat soils are found in wet basin sites.

In Wales, however, stagnohumic gley soils are not widespread on summits but are usually in fine-textured till on footslopes and in basins. On the highest ground intensive frost action rearranges stones into stripes and polygons and keeps unvegetated scree slopes active (Ball and Goodier 1968).

On the coastal plateaux and valley lowlands leaching is less marked so podzols are confined to the coarsest most base-deficient materials. Soils with argillic horizons are developed on medium-textured drifts and solid rocks, mainly in districts where the mean accumulated maximum potential soil moisture deficit exceeds 100 mm (Fig.9, § 3), though the transition to non-argillic soils is gradual.

Relief and slope shape also influence hydrology and soil water regime, giving catenary sequences as in mid-Wales (Rudeforth 1967 and 1970). On undulating ground with slowly permeable parent materials, surface waterlogging causes gleying on flat ground, but soils on the slopes are drier as most rainwater runs off to lower ground producing the pattern of soil distribution illustrated in Figure 20. In sands and other permeable materials water penetrates to the subsoil, leaving relatively high ground well drained, but soils on lower land are affected to some degree by groundwater. On slopes, water moving through permeable material often reappears as springs at the junction with less permeable strata and the soil thereabouts is enriched in bases by flushing. Soils affected by groundwater and those affected by surface-water are not always as distinct from one another as Figure 20 suggests, since permeability of the parent material can change with slope and many soils possess features of both.

The length of time a soil remains undisturbed by erosion or deposition is important in

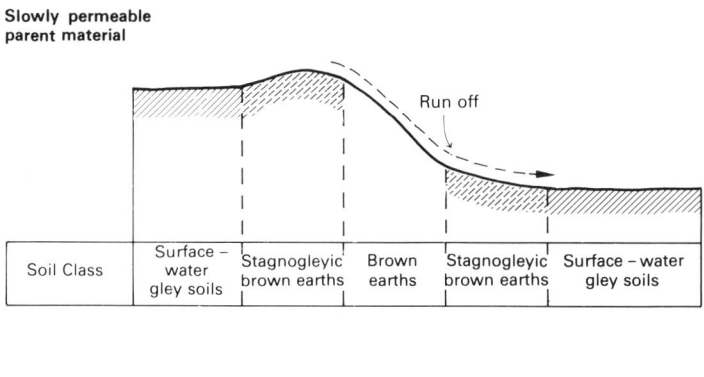

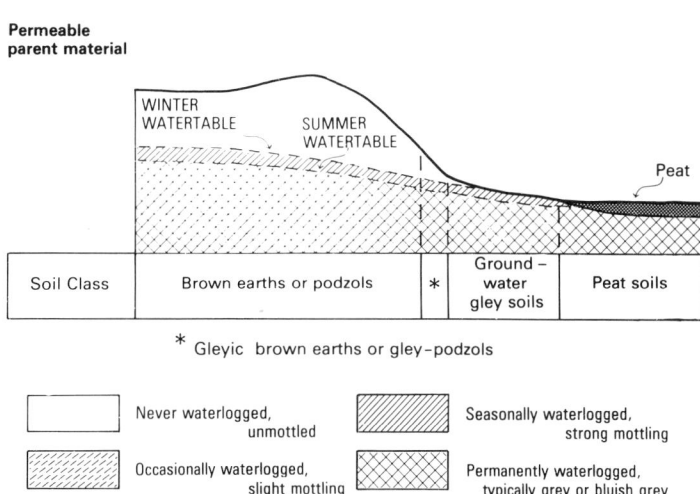

Figure 20. Relationships of relief, hydrology and soils

its evolution. Glaciation has removed the former soil cover from large areas of Wales and blanketed others with thick drift so soil formation began on new surfaces after the retreat of the ice about 12,000 years ago. Old soils with paleo-argillic horizons are therefore uncommon in much of Wales except in the extreme south which remained uncovered during the last glaciation. Youthful soils are extensive on marine and river alluvium and other soils with weak horizon differentiation occur on sand dunes around the coasts. the coasts.

Man has influenced the soil indirectly by modifying the native vegetation and, more directly, through agriculture. Study of preserved pollen in soil has shown that clearance of much upland deciduous forest by Neolithic man led ultimately to the development of heathland. Loss of woodland altered the microclimate and broke the sequence of nutrient recycling favourable to brown earth formation, thus leading to acidification and podzolization with the invasion of heather. Once heather became established, acid humus and, in the wetter sites, thin peat formed, encouraging further podzolization. As deforestation was relatively late in Wales (p.44), the widespread presence of brown podzolic soils and brown earths below 600 m O.D. may be partly attributable to the long period of uninterrupted soil development under woodland.

In the lowlands, and to a lesser extent in the uplands, ploughing, draining, marling and the use of lime and fertilizers have partly altered some soil profile features but many natural relationships between the different soils and their environment remain clear.

§ 6. SOIL CLASSIFICATION

The soil classification for England and Wales used on the map legend and throughout this book is fully described by Avery (1980) and Clayden and Hollis (1984). It provides a consistent and systematic basis for differentiating the properties and relationships of soils in the two countries. It is also a general purpose classification which groups soils that behave in a similar and therefore predictable way. Each group of soils, or *soil class*, has a limited and defined range of *diagnostic properties* that distinguish it from all other classes and ensure its consistent recognition in different parts of the country. The diagnostic properties used to differentiate classes reflect practical characteristics of soil behaviour. For classification purposes the soil profile is conceived as a three dimensional body with lateral dimensions large enough to determine the diagnostic properties of soil horizons. Relatively permanent properties, not easily changed by farming practices, are used to characterize the soils. These properties can either be observed directly in the field, or inferred from field examination.

The classification has a hierarchical structure, resembling a family tree, that assists the identification of classes. Soils are defined at four categorical levels, the highest level having a few broadly defined classes which are progressively subdivided to give more restrictively defined classes at lower levels. The four categories are, in descending order,

Table 6
Soil Classification: Higher Categories

Major soil group	Soil group	Soil subgroup
1. **Terrestrial raw soils** Mineral soils in very recently formed material, with no soil horizons other than a superficial organic or organo-mineral layer less than 5 cm thick, unless buried beneath a recent deposit more than 30 cm thick	1.1 *Raw sands* Non-alluvial sandy soils found mainly on dune sands	Not subdivided below group level
	1.2 *Raw alluvial soils* In recent alluvium	
	1.3 *Raw skeletal soils* With bedrock or very stony rock rubble at 30 cm or less	
	1.4 *Raw earths* In unconsolidated, non-alluvial loamy or clayey deposits	
	1.5 *Man-made raw soils* In artificially disturbed material such as mining spoil	
2. **Raw gley soils** Mineral soils in material that has remained waterlogged since deposition. Prominently mottled or greyish above 40 cm depth. Mainly confined to intertidal flats and saltings that represent stages in the development of mature salt marshes	2.1 *Raw sandy gley soils* In sandy material with no distinct topsoil	Not subdivided below group level
	2.2 *Unripened gley soils* With permanently waterlogged loamy or clayey alluvium (soft mud) at 20 cm or less	

Table 6 (contd.)
Soil Classification: Higher Categories

Major soil group	Soil group	Soil subgroup
3. **Lithomorphic (A/C) soils** Shallow, with a distinct, humose or peaty topsoil, but no subsurface horizons more than 5 cm thick (other than a bleached horizon). Normally over bedrock, very stony rock rubble or little altered soft unconsolidated deposits within 30 cm depth	3.1 *Rankers* Loamy or clayey, with non-calcareous topsoil over bedrock (including massive limestone), non-calcareous rock rubble or soft non-calcareous deposits	3.11 *Humic rankers* (with humose or peaty topsoil) 3.12 *Gleyic rankers* (faintly mottled and permeable) 3.13 *Brown rankers* (distinct topsoil and unmottled subsoil–if present) 3.14 *Podzolic rankers* (with bleached subsurface horizon) 3.15 *Stagnogleyic rankers* (faintly mottled over slowly permeable material)
	3.2 *Sand-rankers* In non-calcareous, unconsolidated soft non-alluvial sandy deposits	3.21 *Typical sand-rankers* (unmottled with no bleached subsurface horizon) 3.22 *Podzolic sand-rankers* (with bleached subsurface horizon) 3.23 *Gleyic sand-rankers* (faintly mottled)
	3.3 *Ranker-like alluvial soils* In non-calcareous recent alluvium	3.31 *Typical ranker-like alluvial soils* (unmottled) 3.32 *Gleyic ranker-like alluvial soils* (faintly mottled)
	3.4 *Rendzinas* Calcareous, over chalk, or extremely calcareous rock rubble or soft unconsolidated deposits	3.41 *Humic rendzinas* (with humose or peaty topsoil) 3.42 *Grey rendzinas* (with distinct topsoil that is extremely calcareous or greyish) 3.43 *Brown rendzinas* (with brownish distinct topsoil that is not extremely calcareous) 3.44 *Colluvial rendzinas* (in colluvium more than 40 cm thick)

3.45 *Gleyic rendzinas* (mottled with distinct topsoil over permeable material)
3.46 *Humic gleyic rendzinas* (mottled with humose or peaty topsoil over permeable material)
3.47 *Stagnogleyic rendzinas* (mottled over slowly permeable material)

3.5 *Pararendzinas*
Loamy or clayey over moderately calcareous (1 to 40% CaCO₃) non-alluvial material

3.51 *Typical pararendzinas* (unmottled with distinct topsoil)
3.52 *Humic pararendzinas* (unmottled with humose or peaty topsoil)
3.53 *Colluvial pararendzinas* (in colluvium more than 40 cm thick)
3.54 *Stagnogleyic pararendzinas* (faintly mottled over slowly permeable material)
3.55 *Gleyic pararendzinas* (faintly mottled over permeable material)

3.6 *Sand-pararendzinas*
In calcareous, unconsolidated soft sandy deposits other than alluvium (normally dune sands)

3.61 *Typical sand-pararendzinas* (unmottled)
3.62 *Gleyic sand-pararendzinas* (faintly mottled)

3.7 *Rendzina-like alluvial soils*
In little altered calcareous alluvium, lake marl or tufa, at least 30 cm thick

3.71 *Typical rendzina-like alluvial soils* (unmottled)
3.72 *Gleyic rendzina-like alluvial soils* (mottled with distinct topsoil)
3.73 *Humic gleyic rendzina-like alluvial soils* (mottled with humose or peaty topsoil)

4. **Pelosols**
Non-alluvial clayey soils that crack deeply in dry seasons, but are slowly permeable when wet. They have a coarse blocky or prismatic structure and no prominently mottled non-calcareous subsurface horizons within 40 cm depth

4.1 *Calcareous pelosols*
With calcareous subsoil and no clay-enriched subsurface horizon

4.11 *Typical calcareous pelosols*

Table 6 (contd.)
Soil Classification: Higher Categories

Major soil group	Soil group	Soil subgroup
	4.2 *Non-calcareous pelosols* Non-calcareous to at least 80 cm depth and no clay-enriched subsurface horizon	4.21 *Typical non-calcareous pelosols*
	4.3 *Argillic pelosols* With a clay-enriched subsurface horizon	4.31 *Typical argillic pelosols*
5. **Brown soils** With dominantly brownish or reddish subsoils and no prominent mottling or greyish colours (gleying) above 40 cm depth. They are developed mainly on permeable materials at elevations below about 300 m O.D. Most are in agricultural use	5.1 *Brown calcareous earths* Non-alluvial, with calcareous loamy or clayey subsoils without significant clay enrichment	5.11 *Typical brown calcareous earths* (unmottled) 5.12 *Gleyic brown calcareous earths* (faintly mottled with permeable subsoil) 5.13 *Stagnogleyic brown calcareous earths* (faintly mottled with slowly permeable subsoil) 5.14 *Colluvial brown calcareous earths* (in colluvium more than 40 cm thick)
	5.2 *Brown calcareous sands* Non-alluvial, with calcareous sandy subsoils without significant clay enrichment	5.21 *Typical brown calcareous sands* (unmottled) 5.22 *Gleyic brown calcareous sands* (faintly mottled)
	5.3 *Brown calcareous alluvial soils* In calcareous recent alluvium more than 30 cm thick	5.31 *Typical brown calcareous alluvial soils* (unmottled) 5.32 *Gleyic brown calcareous alluvial soils* (faintly mottled with permeable subsoil) 5.33 *Pelogleyic brown calcareous alluvial soils* (faintly mottled and clayey with slowly permeable subsoil)

5.4 Brown earths
Non-alluvial, with non-calcareous loamy or clayey subsoils without significant clay enrichment

5.41 *Typical brown earths* (unmottled)
5.42 *Stagnogleyic brown earths* (faintly mottled with slowly permeable subsoil)
5.43 *Gleyic brown earths* (faintly mottled with permeable subsoil)
5.44 *Ferritic brown earths* (unmottled with bright ochreous iron-rich subsoil)
5.45 *Stagnogleyic ferritic brown earths* (faintly mottled with bright ochreous iron-rich slowly permeable subsoil)
5.46 *Gleyic ferritic brown earths* (faintly mottled with bright ochreous iron-rich permeable subsoil)
5.47 *Colluvial brown earths* (in colluvium more than 40 cm thick)

5.5 Brown sands
Non-calcareous sandy or sandy gravelly

5.51 *Typical brown sands* (unmottled with no clay-enriched subsoil)
5.52 *Gleyic brown sands* (faintly mottled with permeable subsoil without significant clay enrichment)
5.53 *Stagnogleyic brown sands* (faintly mottled with slowly permeable subsoil)
5.54 *Argillic brown sands* (unmottled with clay-enriched subsoil)
5.55 *Gleyic argillic brown sands* (faintly mottled with permeable clay-enriched subsoil)

5.6 Brown alluvial soils
In non-calcareous loamy or clayey alluvium more than 30 cm thick

5.61 *Typical brown alluvial soils* (unmottled)
5.62 *Gleyic brown alluvial soils* (faintly mottled with permeable subsoil)
5.63 *Pelogleyic brown alluvial soils* (faintly mottled and clayey with slowly permeable subsoil)

5.7 Argillic brown earths
Loamy or clayey with an ordinary clay-enriched subsoil

5.71 *Typical argillic brown earths* (unmottled)
5.72 *Stagnogleyic argillic brown earths* (faintly mottled with slowly permeable subsoil)

Table 6 (contd.)
Soil Classification: Higher Categories

Major soil group	Soil group	Soil subgroup
		5.73 *Gleyic argillic brown earths* (faintly mottled with permeable subsoil)
	5.8 *Paleo-argillic brown earths* Loamy or clayey, with an ancient reddish or reddish mottled, clay-enriched subsoil formed, at least in part, before the last (Devensian) glacial period	5.81 *Typical paleo-argillic brown earths* (unmottled)
		5.82 *Stagnogleyic paleo-argillic brown earths* (faintly mottled with slowly permeable subsoil)
		5.83 *Gleyic paleo-argillic brown earths* (faintly mottled with permeable subsoil)
6. **Podzolic soils** With black, dark brown or ochreous humus and iron-enriched subsoils formed as a result of acid weathering conditions. Under natural or semi-natural vegetation, they have an unincorporated acid organic layer at the surface	6.1 *Brown podzolic soils* With a dark brown or ochreous, iron-enriched subsoil and no overlying bleached horizon	6.11 *Typical brown podzolic soils* (unmottled with distinct topsoil)
		6.12 *Humic brown podzolic soils* (unmottled with humose or peaty topsoil)
		6.13 *Paleo-argillic brown podzolic soils* (unmottled with an ancient reddish clay-enriched subsoil)
		6.14 *Stagnogleyic brown podzolic soils* (faintly mottled with slowly permeable subsoil)
		6.15 *Gleyic brown podzolic soils* (faintly mottled with permeable subsoil)
	6.2 *Humic cryptopodzols* With humose or peaty topsoil over a thick humose, humus-enriched subsoil and no bleached subsurface horizon	6.21 *Typical humic cryptopodzols* (with no iron-enriched subsurface horizon)
		6.22 *Ferri-humic cryptopodzols* (with an iron-enriched subsurface horizon)

6.3 Podzols
Well drained, with a bleached subsurface horizon and no thin ironpan

6.31 *Humo-ferric podzols* (with black or dark brown humus and iron-enriched layer beneath the bleached horizon)

6.32 *Humus podzols* (with black humus-enriched layer containing little iron, below the bleached horizon)

6.33 *Ferric podzols* (with a dark brown or ochreous, iron-enriched layer containing little humus, below the bleached horizon)

6.34 *Paleo-argillic podzols* (with ancient, reddish clay-enriched subsoil)

6.4 Gley-podzols
With a bleached subsurface horizon over a dark coloured humus or iron-enriched subsoil directly over a periodically wet, prominently mottled or greyish horizon

6.41 *Typical gley-podzols* (with humus-enriched subsoil containing little iron)

6.42 *Humo-ferric gley-podzols* (with black or dark brown humus and iron-enriched subsoil)

6.43 *Stagnogley-podzols* (with slowly permeable subsoil)

6.5 Stagnopodzols
With a peaty topsoil and periodically wet bleached subsurface horizon over an iron-enriched subsoil. Mainly found in uplands

6.51 *Ironpan stagnopodzols* (with a thin ironpan)

6.52 *Humus-ironpan stagnopodzols* (with humus-enriched subsurface horizon over a thin ironpan)

6.53 *Hardpan stagnopodzols* (with a bleached hardpan)

6.54 *Ferric stagnopodzols* (with dark brown or ochreous iron-enriched subsoil containing little humus)

7. Surface-water gley soils
Non-alluvial, seasonally waterlogged slowly permeable soils, formed above 3 m O.D. and prominently mottled above 40 cm depth. They have no relatively permeable material starting within and extending below 1 m of the surface

7.1 Stagnogley soils
With a distinct topsoil. They are found mainly in lowland Britain

7.11 *Typical stagnogley soils* (with ordinary clay-enriched subsoil)

7.12 *Pelo-stagnogley soils* (clayey)

7.13 *Cambic stagnogley soils* (with no clay-enriched subsoil)

7.14 *Paleo-argillic stagnogley soils* (with ancient reddish or reddish mottled clay-enriched subsoil)

7.15 *Sandy stagnogley soils* (with sandy topsoil)

Table 6 (contd.)
Soil Classification: Higher Categories

Major soil group	Soil group	Soil subgroup
	7.2 *Stagnohumic gley soils* With a humose or peaty topsoil. They are mainly upland soils, intermediate between stagnogley soils and peat soils	7.21 *Cambic stagnohumic gley soils* (with no clay-enriched subsoil) 7.22 *Argillic stagnohumic gley soils* (with ordinary clay-enriched subsoil) 7.23 *Paleo-argillic stagnohumic gley soils* (with ancient, reddish or reddish mottled, clay-enriched subsoil) 7.24 *Sandy stagnohumic gley soils* (with sandy subsurface horizon)
8. Ground-water gley soils Seasonally waterlogged soils affected by a shallow fluctuating groundwater-table. They are developed mainly within or over permeable material and have prominently mottled or greyish coloured horizons within 40 cm depth. Most occupy low-lying or depressional sites	8.1 *Alluvial gley soils* With distinct topsoil, in loamy or clayey recent alluvium more than 30 cm thick	8.11 *Typical alluvial gley soils* (loamy with non-calcareous subsoil) 8.12 *Calcareous alluvial gley soils* (loamy with calcareous subsoil) 8.13 *Pelo-alluvial gley soils* (clayey with non-calcareous subsoil) 8.14 *Pelo-calcareous alluvial gley soils* (clayey with calcareous subsoil) 8.15 *Sulphuric alluvial gley soils* (extremely acid subsoil within 80 cm)
	8.2 *Sandy gley soils* Sandy, with distinct topsoil and no clay-enriched subsoil	8.21 *Typical sandy gley soils* (with non-calcareous subsoil) 8.22 *Calcareous sandy gley soils* (with calcareous subsoil)
	8.3 *Cambic gley soils* Non-alluvial, loamy or clayey with a distinct topsoil and no clay-enriched subsoil	8.31 *Typical cambic gley soils* (loamy with non-calcareous subsoil) 8.32 *Calcaro-cambic gley soils* (loamy with calcareous subsoil) 8.33 *Pelo-cambic gley soils* (clayey)

8.4 *Argillic gley soils*
With a distinct topsoil and a clay-enriched subsoil

    8.41 *Typical argillic gley soils* (with loamy topsoil)

    8.42 *Sandy argillic gley soils* (with sandy topsoil)

8.5 *Humic-alluvial gley soils*
With a humose or peaty topsoil in loamy or clayey recent alluvium more than 30 cm thick

    8.51 *Typical humic-alluvial gley soils* (with non-calcareous subsoil)

    8.52 *Calcareous humic-alluvial gley soils* (with calcareous subsoil)

    8.53 *Sulphuric humic-alluvial gley soils* (with extremely acid subsoil within 80 cm)

8.6 *Humic sandy gley soils*
Sandy, with humose or peaty topsoil and no clay-enriched subsoil. Intermediate between sandy gley soils and lowland peat soils

    8.61 *Typical humic-sandy gley soils* (with non-calcareous subsoil)

    8.62 *Calcareous humic-sandy gley soils* (with calcareous subsoil)

8.7 *Humic gley soils*
Non-alluvial loamy or clayey with humose or peaty topsoil. Intermediate between cambic and argillic gley soils and lowland peat soils

    8.71 *Typical humic gley soils* (with non-calcareous subsoil lacking significant clay enrichment)

    8.72 *Calcareous humic gley soils* (with calcareous subsoil lacking significant clay enrichment)

    8.73 *Argillic humic gley soils* (with clay-enriched subsoil)

9. Man-made soils
With a thick man-made topsoil or a disturbed subsurface layer (containing disturbed fragments of soil horizons) to at least 40 cm depth. They result from the addition of earth containing manures, or the restoration of soil material after mining or quarrying

9.1 *Man-made humus soils*
With a thick man-made topsoil

    9.11 *Sandy man-made humus soils* (sandy or sandy skeletal)

    9.12 *Earthy man-made humus soils* (loamy or clayey)

9.2 *Disturbed soils*
With a distinct topsoil and and a disturbed subsurface layer to at least 40 cm depth

Not subdivided below group level

Table 6 (contd.)
Soil Classification: Higher Categories

Major soil group	Soil group	Soil subgroup
10. **Peat soils** With more than 40 cm of organic material in the upper 80 cm or with more than 30 cm of organic material over bedrock or very stony rock rubble	10.1 *Raw peat soils* In undrained organic material that has remained wet to within 20 cm of the surface	10.11 *Raw oligo-fibrous peat soils* (mainly fibrous or semi-fibrous with pH less than 4.0 throughout) 10.12 *Raw eu-fibrous peat soils* (mainly fibrous or semi-fibrous with pH 4.0 or more in some part) 10.13 *Raw oligo-amorphous peat soils* (mainly humified with pH less than 4.0 throughout) 10.14 *Raw eutro-amorphous peat soils* (mainly humified with pH 4.0 or greater in some part)
	10.2 *Earthy peat soils* Normally drained, with a well aerated and structured earthy topsoil or ripened mineral surface layer	10.21 *Earthy oligo-fibrous peat soils* (mainly fibrous or semi-fibrous with pH less than 4.0 throughout) 10.22 *Earthy eu-fibrous peat soils* (mainly fibrous or semi-fibrous with pH 4.0 or 10.23 *Earthy oligo-amorphous peat soils* (mainly humified with pH less than 4.0 throughout) 10.24 *Earthy eutro-amorphous peat soils* (mainly humified with pH 4.0 or more in some part) 10.25 *Earthy sulphuric peat soils* (with an extremely acid subsurface horizon containing pale yellow crystals or streaks, above 80 cm)

major soil group, soil group, soil subgroup and soil series. At major soil group level, divisions are based on the presence or absence of major diagnostic horizons that have an important agronomic, hydrological, ecological or engineering significance. Soil groups and subgroups are subdivisions based on general properties inherited from the soil parent material, or on the presence of subsidiary diagnostic horizons. They reflect broad differences of soil behaviour within major soil groups. A soil series, the lowest category in the system, is a subdivision of a subgroup based on narrowly defined diagnostic properties inherited from the soil parent material. In most cases soil series represent small, but specific differences of soil characteristics within subgroups.

Major groups, groups and subgroups are named using words which describe the features used to distinguish them. At soil series level such a nomenclature would be unwieldy, so for ease of communication soil series are labelled by the names of places where they were first described. The diagnostic properties of major groups, groups and subgroups are described by Avery (1980) and the classification is outlined in Table 6. Criteria for the recognition of soil series (Clayden and Hollis 1984) are given below.

Definition of soil series
In the past, soil series were differentiated partly by the stratigraphic age of their presumed parent material. This produced many soil series nationally because of the large number of geological formations in England and Wales. Some of the distinctions have since proved to be of little practical value, particularly those based largely on stratigraphic age. These have been abandoned and soil series are now defined using intrinsic properties of practical significance. To maintain continuity with previous work, soils developed in soft pre-Quaternary formations are still distinguished from those in unconsolidated Quaternary and recent drifts.

Soil series are defined (Clayden and Hollis 1984) using a combination of three main properties, the broad type of parent material present (substrate type), the texture of the soil material (textural grouping) and the presence or absence of material with a distinctive mineralogy.

Substrate type
Five broad types of parent material are recognized which are subdivided according to their lithological properties, or, in the case of organic soils, their botanical composition (Table 7).

Textural grouping
The texture of the soil profile is defined from the single predominant or two main contrasting textural groupings present within standard reference depths depending on substrate type (Table 8). Textural groupings depend upon organic matter content, $CaCO_3$ content and mineral particle-size distribution and are defined in detail by Clayden and

Table 7
Definition and Lithology of Substrate Types

Substrate group	Substrate type	Lithology
Lithoskeletal substrates With bedrock or very stony rock rubble at or within 80 cm depth	*Lithoskeletal* With bedrock or very stony rock rubble within 40 cm depth and no more than 30 cm of superficial material with less than 15% stones	Acid crystalline (igneous, gneiss and pyroclastic) rocks Basic crystalline rocks Ultra-basic crystalline rocks Acid schist Basic schist Chert or quartzite (including quartzitic sandstone)
	Over lithoskeletal With bedrock or very stony rock rubble at or below 40 cm depth, or with at least 30 cm of superficial material with less than 15% stones	Sandstone, siltstone, mudstone, shale or slate (including phyllite):- choice of lithology depends on texture of overlying weathered material Ironstone (ferruginous sedimentary rock) Limestone Chalk
Gravelly substrates With very stony gravel (rounded stones) starting at or above and extending below 80 cm depth	*Gravelly* With gravel within 40 cm depth and no more than 30 cm of superficial material with less than 15% stones	Acid crystalline stones Basic crystalline stones Ultra-basic crystalline stones Very hard siliceous stones Sandstones, siltstones, mudstones or slates Limestones Chalk stones
	over gravelly With gravel at or below 40 cm depth	Non-calcareous (with non-calcareous stones and matrix) Calcareous (with calcareous stones or matrix)

Soft pre-Quaternary substrates With weathered or little-altered *soft* pre-Quaternary material at or within 80 cm depth	*drift..passing to* With an upper drift layer of contrasting texture (in clayey soils this is recognized by the presence of a lithological discontinuity below 30 cm depth marked by an increase in clay of a least 10%) *passing to* With no upper drift layer of contrasting texture	Acid crystalline rock Basic crystalline rock Ultra-basic crystalline rock Acid schist Basic schist Clay, silt, sand, mudstone, siltstone, sandstone or shale:- choice of lithology depends on texture of overlying weathered material Ironstone Soft chalk
Thick drift substrates With no very stony or pre-Quaternary material at or within 80 cm, unless covered with at least 30 cm of recent alluvium or 40 cm of colluvium	Not subdivided	River alluvium, lake marl or tufa Marine alluvium Colluvium Stoneless drift Chalky drift (chalk stones present in the parent material) Drift with limestones (limestones are the predominant stone in the material) Drift with siliceous stones (with non-calcareous stones predominant and no chalk in the material)
Peat substrates In peat soils only	Not subdivided	Loamy peat Sedimentary peat Humified peat (in amorphous subgroups) *Sphagnum* peat *Sphagnum-Eriophorum* peat *Eriophorum-Sphagnum* peat *Molinia* peat Grass-sedge peat

Table 8
Textural Grouping for Different Substrate Types

Substrate type	Reference section	Textural definition of the soil class
Lithoskeletal and Gravelly	Upper 30 cm of the profile	All soil classes defined by the predominant textural group (sandy, loamy or clayey)
over gravelly	To the top of the gravelly layer	Lithomorphic soil classes defined on the predominant textural group All other soil classes defined on the predominant textural subgroup (sandy, coarse loamy, coarse silty, fine loamy, fine silty or clayey)
over lithoskeletal	To bedrock or the top of the very stony layer	1. Lithomorphic soil classes defined on the predominant textural group
Soft pre-Quaternary substrates and thick drift	Upper 80 cm of the profile	2. Stagnopodzols and stagnohumic gley soil classes defined on the predominant textural group or the two main contrasting textural groups 3. All other soil classes defined on the predominant textural subgroup or the two main contrasting textural subgroups
Peat	30 to 90 cm or the 60 cm directly above a mineral substrate or, if less than 60 cm thick, from the surface to the top of the mineral substrate	All soil classes defined on the predominant organic texture (peat or loamy peat) or the main contrasting organic and mineral textures.

Hollis (1984). The main combinations of textural groupings considered to be contrasting are:

Sandy over fine loamy,
Sandy over fine silty,
Sandy over clayey,
Sandy over peaty,

Coarse loamy over clayey,[1]
Coarse silty over clayey,[1]
Fine loamy over clayey,[1]
Fine silty over clayey,[1]

Coarse silty over sandy,
Fine loamy over sandy,
Fine silty over sandy,

Clayey over sandy or coarse loamy,
Clayey over coarse silty,
Clayey over peaty,

Coarse loamy over peaty,
Coarse silty over peaty,
Fine loamy over peaty,
Fine silty over peaty.

Peaty over lithoskeletal material,
Peat with interstratified mineral material,

[1]These are grouped as Loamy over clayey for some soil classes.

Mineralogical properties

Soil series containing material with a distinctive mineralogy, likely to have a significant effect on soil behaviour, are separated from otherwise similar soils with the same textural and lithological definitions. Twelve diagnostic mineralogical distinctions are made, but some apply only to series with specific textural properties:

Distinctions made in soil series of any texture

Carbonatic (the material is extremely calcareous in the less than 2 mm fraction),
Ferruginous (the material contains a large amount of iron oxides relative to the clay content),
Glauconitic (at least 5 per cent glauconite is present as greenish grains),
Serpentinitic (the sand fraction is dominated by magnesium-rich serpentine minerals or their weathering products),
Saline (marine alluvium with a relatively large content of soluble salts),
Sulphidic (waterlogged mineral material that becomes extremely acid when drained),
Sulphuric (extremely acid material with significant amounts of yellow jarosite crystals and mottles).

Distinctions made only in loamy and clayey soils

Grey siliceous (non-carbonatic greyish material containing relatively little iron),
Reddish (reddish material rich in haematite).

Distinctions made only in clayey soils

Swelling clayey (material with a relatively large potential to shrink or swell at high or low moisture contents).
Series defined as swelling clayey are further differentiated as:
Brownish (if they are in dominantly brownish parent material resulting from the oxidation of pyrite-rich deposits),
Kaolinitic (material rich in weakly swelling kaolinitic clay minerals).

The actual form of the soil series definition varies according to substrate type. Different combinations of textural and lithological terms are used for different substrates, with the diagnostic mineralogical properties usually denoted by a term prefixed to the series definition. Examples of different kinds of series definitions are given below, with their soil subgroup numbers and mineralogical prefixes in italics:

Lithoskeletal soils
  3.11 Bangor series. *Loamy or peaty; lithoskeletal acid crystalline rocks*
  3.43 Andover series. *Silty; lithoskeletal chalk*

Over lithoskeletal soils with non-contrasting texture
  5.44 Banbury series. *Ferruginous* fine loamy over lithoskeletal ironstone

Over lithoskeletal soils with contrasting texture
  5.81 Nordrach series. Fine silty over clayey over lithoskeletal limestone

Gravelly soils
  6.34 Southampton series. Sandy gravelly; very hard siliceous stones
  6.11 Baschurch series. Loamy gravelly; sandstones, siltstones, mudstones or slates

Over gravelly soils
  5.71 Sutton series. Fine loamy over calcareous gravelly
  8.41 Hurst series. Coarse loamy over non-calcareous gravelly

Drift passing to soils
  6.43 Holidays Hill series. Sandy drift over fine loamy passing to clay and sand
  7.12 Lawford series. Clayey drift passing to clay or mudstone

Passing to soils with non-contrasting texture
   7.12 Windsor series. *Swelling clayey* passing to *brownish* clay or mudstone

Passing to soils with contrasting textures
   5.71 Harwell series. *Grey siliceous* fine loamy or fine silty over clayey passing to interbedded clay and sandstone

Drift soils with non-contrasting texture
   8.12 Wisbech series. Coarse silty; marine alluvium
   7.12 Crewe series. *Reddish* clayey; stoneless drift
   5.51 Newport series. Sandy; drift with siliceous stones

Drift soils with contrasting textures
   3.72 Willingham series. Loamy *carbonatic* over peaty; lake marl or tufa
   7.11 Beccles series. Fine loamy over clayey; chalky drift
   5.71 Tickenham series. *Reddish* fine loamy over clayey; drift with limestones.

Definitions of all the soil series currently recognized by the Soil Survey of England and Wales are given in the above form by Clayden and Hollis (1984) and similar definitions introduce the selected representative profile descriptions given in Appendix 2 (p.309)

§ 7. TERMINOLOGY

The terminology used in this book is of three main kinds: that used in soil profile descriptions and more general description of soils; that applied to the interpretation of the results of soil survey in Chapters 3, 4 and 5; and that employed in Chapter 3 in the keys to soil series within the associations. Much of the terminology in the first two categories is technical and specific but that used in the keys has been deliberately simplified and generalized to encourage non-specialist readers to identify soils for themselves.

Terms used in soil profile description
The terms used in profile description are outlined in the introduction to Appendix 1 (p.301). They are more fully described by Hodgson (1976), Thomasson (1975a) and Avery (1980). They are used also in the abbreviated and generalized descriptions given in Chapter 3 for the main soil series within each soil association and are used elsewhere in the general text. The horizon notation employed both in Appendix 2 (p.309) and in the summary profiles in the main text is that fully described by Avery (1980).

Terminology used in interpretations of the soil map
In Chapter 3, the main soils of each association are classified in terms of soil water regime, workability and droughtiness. Although much of the terminology is self-explanatory, the underlying concepts and methods used in making these interpretations require explanation, as they provide a framework for drainage and other agricultural assessments in Chapters 4 and 5.

Soil water regime

Soil water regime is the cyclical seasonal variation of wet, moist or dry soil states. The duration and degree of waterlogging are described by the system of *wetness classes* (Table 9), grading from Wetness Class I, well drained, to Wetness Class VI, almost permanently waterlogged within 40 cm depth (Hodgson 1976). The incidence of waterlogging depends on soil and site properties, underdrainage and climate.

The main property affecting the soil's natural water regime and its response to drainage measures is its permeability. Field assessments of permeability are made from estimates of soil texture, structure and packing density, which are refined using laboratory measurements of macroporosity (Hall *et al.* 1977) and the limited data available on hydraulic conductivity. Dense clayey, fine loamy or fine silty subsoil horizons with a horizontal saturated hydraulic conductivity (Thomasson 1975a) of less than 0.1 m per day are classified as *slowly permeable*. These are effectively impermeable in terms of their contribution to the movement of water to field drains (Luthin 1957). In soils with a thick slowly permeable substratum, movement of excess water in the upper horizons is mainly lateral. The term *permeable* is used for sandy, loamy or well structured clayey and peaty soils that are assumed from available data to have a horizontal saturated hydraulic conductivity greater than about 0.6 m per day. Such soils are either naturally well drained or respond well to drainage measures. The term *moderately permeable* is restricted to soils which lack clear evidence to place them in either the permeable or slowly permeable categories or in which hydraulic conductivity is variable. Fine-textured alluvial soils are often placed in this category, as are some clayey soils which, because they have a permeable unsaturated substratum at depth such as chalk or shattered rock, have dominantly vertical, but somewhat slow, disposal of excess water.

Table 9
Soil Wetness Classes

Wetness class	Duration of waterlogging
I	The soil profile is not waterlogged within 70 cm depth for more than 30 days[1] in most years[2].
II	The soil profile is waterlogged within 70 cm depth for 30–90 days in most years.
III	The soil profile is waterlogged within 70 cm depth for 90–180 days in most years.
IV	The soil profile is waterlogged within 70 cm depth for more than 180 days, but not waterlogged within 40 cm depth for more than 180 days in most years.
V	The soil profile is waterlogged within 40 cm depth for 180–335 days, and is usually waterlogged within 70 cm for more than 335 days in most years.
VI	The soil profile is waterlogged within 40 cm depth for more than 335 days in most years.

[1] The number of days specified is not necessarily a continuous period.
[2] *In most years* is defined as more than 10 out of 20 years.

Gley morphology described on page 49, is an essential part of soil classification and usually indicates some degree of waterlogging. Soils lacking either gley features or slowly permeable horizons within 70 cm depth are usually considered to be well drained

(Wetness Class I). Soils with gley features *and* a slowly permeable horizon are almost certainly waterlogged for some time in most years, though the duration and severity of waterlogging depend on the efficiency of any drainage measures and on the local climate. In soils without a slowly permeable horizon within 1 metre depth, gley features usually indicate waterlogging caused by groundwater maintained by seepage water from higher ground or nearby rivers. In districts where the main drainage channels provide sufficient depth of outfall, the groundwater-tables in these relatively permeable soils are easily controlled by conventional ditch and pipe drainage and here the gley morphology often bears little relationship to the actual soil water regime.

Climate exerts a profound influence on the incidence of waterlogging. A useful measure of its influence on soil wetness is the median value of *field capacity* (§ 3). As described in Chapter 1, this is used as a purely meteorological concept calculated from potential transpiration and rainfall (Smith 1967). The meteorological field capacity period, mainly confined to the winter half of the year, is a condition when there is zero soil moisture deficit (Smith and Trafford 1976). During this period, however, actual field conditions vary markedly depending on soil properties and the hydrology of the site. In winter, the water content of well drained soils fluctuates following each rainfall event, but waterlogging occurs for only a few hours during or immediately after heavy rain. As drainage ceases a condition of *soil profile field capacity* is reached, representing the maximum water content that can be retained against gravity (Veihmeyer and Hendrickson 1931). Water content does not decline further until loss occurs by evaporation or by transpiration through plants. Soils with a slowly permeable substratum or with groundwater at shallow depth are commonly waterlogged throughout the meteorological field capacity period. Horizontal disposal of surplus water during rain-free periods is too slow to drain the soil profile adequately. Under such conditions soil profile field capacity is a poorly defined state in terms of water content, or hydraulic potential. Waterlogging in subsoil horizons of these soils is eventually alleviated mainly by evaporation and transpiration of water to the atmosphere, rather than by downward or lateral movement.

The meteorological field capacity period (measured in F.C. days) is a valuable indicator of the climatic influences affecting a piece of land, but to predict the duration and severity of soil waterlogging (Wetness Class) it must be considered with soil and site properties, including the presence or absence of field drainage measures and the effectiveness of arterial drainage. A large body of data on the actual duration of waterlogging in specific soils in wet and dry years (Robson and Thomasson 1977) is available to improve judgements from observations accumulated during field surveys under a range of weather conditions. Table 10 illustrates the features used to assess wetness classes of soils described in Chapter 3. In the lowlands, it is assumed that appropriate field drainage has been installed and that slowly permeable horizons are waterlogged for longer than the meteorological field capacity period. Areas where arterial drainage does not provide sufficient outfall are usually noted in Chapter 3 or 4.

Table 10
Relationships of Wetness Class to Field Capacity Days and Depth to Slowly Permeable Horizon

Average Field Capacity Days	Gleyed within 70 cm depth				Ungleyed within 70 cm depth
	Depth to slowly permeable horizon				
	<40 cm	40–80 cm	>80 cm		>80 cm
			Drainage outfalls limiting	Drainage outfalls not limiting	
<100	(x)	II	II–VI	I	I
100–125	(x)	II–III[1]	III–VI	I	I
125–150	(x)	II–III[1]	III–VI	I	I
150–175	(x)	III–IV[1]	III–VI	I	I
175–200	IV	III–IV[1]	IV–VI	I	I
200–225	V	III–IV[1]	V–VI	I–II	I
225–250	V	IV–V[1]	V–VI	II	I
250–300	V–VI	V	V–VI	III	I
>300	VI	VI	VI	IV	I

[1] The drier of the two wetness classes indicated is likely to occur either on slopes or in soils where the slowly permeable horizon is between 60 and 80 cm depth. Soils in these circumstances are normally not gleyed within 40 cm depth.

(x) In climates with less than 175 F.C. days subsoiling or other soil loosening techniques are usually effective to 40 cm depth. In this Table it is assumed that permeability has been improved to at least that depth.

Workability and trafficability

Workability (Thomasson 1982) of the soil depends on interactions between climate and soil physical properties. For example, good working conditions on clayey soils are commonly restricted to brief periods when the soil is neither too wet nor too dry for a good tilth to be obtained. Similarly, *trafficability* of land by machinery and stock is restricted by soil wetness. The influence of topsoil properties, permeability and wetness is integrated as shown in Table 11 to give assessments of workability and trafficability which are applied to the main arable soils in Chapter 3. For example, assessment *aa* is used for very coarse well drained soils and assessment *f* is used for extremely wet, heavy or peaty soils. The restriction of workability and trafficability increases from *aa* to *f*. The wetness class, in the first column of Table 11, is given on the assumption that, where appropriate, feasible drainage improvements have been undertaken.

On much land, the field capacity period sets broad limits to good ground conditions for tillage and trafficability with conventional machinery. In the diagrams in Chapter 3 illustrating the effects of soil and climate on landwork (for example Fig.22, § 22) the median return dates of the field capacity period are used to estimate the limits of the landwork periods in a normal year. Related quartile dates are used similarly for wet years which occur with a frequency of one in four. From these dates, and the soil assessments for workability in Table 11 and their weightings in Table 12, potential *machinery work*

Table 11
Soil Assessment for Workability

Wetness Class	Depth to slowly permeable horizon (cm)	Mineral topsoils Retained Water Capacity			Humose or Peaty topsoils
		Low	Medium	High	
I	>80 (sandy)	aa	—	—	—
	>80	a	a	a	a
II	>80	a	ab	b	a
	40–80	b	b	bc	b
III	>80	b	c	c	b
	40–80	c	c	cd	c
	<40	c	cd	d	d
IV	>80	c	d	d	d
	40–80	c	d	de	e
	<40	d	de	e	f
V	All depths	e	f	f	f
VI	All depths	f	f	f	f

days (Smith 1977) are calculated to provide a measure of the number of days when the land can be worked with acceptable risk of damage to soil structure during the main autumn and spring activities of harvesting, tillage and drilling.

Table 12
Soil Weightings applied to Field Capacity Data for estimating Machinery Work Days

Soil assessment	Soil weightings (days)		
	Autumn (1 Sept–31 Dec)	Spring (1 March–30 April)	Total
aa	+30	+20	+50
a	+20	+10	+30
ab	+10	+5	+15
b	0	0	0
bc	–10	–3	–13
c	–20	–5	–25
cd	–25	–8	–33
d	–30	–10	–40
de	–35	–13	–48
e	–40	–15	–55
f	–50	–20	–70

The following example shows how to calculate the average number of *machinery work days* in autumn for clay land with a *d* soil assessment:

Median date of return to field capacity   5 November

1 September to 5 November   66 days
Soil weighting (Table 12)   –30 days

Average number of machinery work days   36 days

This assessment does not mean that on average the period of good machinery work days ends 36 days after 1 September (on 6 October). In practice, during the progressive rewetting of the soil profile in autumn, some days in September are likely to be unsuitable and a few days after 6 October may be acceptable. The concept is one of increasing frequency of wet soil conditions as the autumn progresses. The diagrammatic presentations of machinery work days in Chapter 3 are intended to show broad differences in workability in autumn and spring for selected soils in representative localities. It is assumed that soils classified as *a* and *aa* are able to dispose of excess rainfall sufficiently rapidly to allow some landwork within the meteorological field capacity period, particularly near the beginning or towards the end when, during dry days, evaporation from surface soil is appreciable. It is assumed also that any slowly permeable horizon within 40 cm depth has been loosened by subsoiling in those districts where the duration of field capacity is less than 175 days. Soil series with *e* and *f* assessments are rarely in continuous arable use and the autumn and spring weightings for machinery work days on such soils are necessarily tentative.

   It is anticipated that improvements in agricultural machinery or techniques in the foreseeable future are unlikely to alter the *relative* differences in workability identified in this Bulletin. In practice, land is often worked at unsuitable times with consequent damage to soil structure. Such deterioration of soil structure is not readily quantified in terms of reduced crop yields, whereas the penalties for not being able to establish or harvest a crop are beyond doubt. The estimates of machinery work days for each soil series are a basis for the evaluation of existing or new cropping systems, the need for advanced techniques, or measures to relieve soil compaction.

Soil droughtiness
The requirements and use of water by crops involve complex interactions between climate, soil and the rooting habit of the crop. In Chapter 3 estimates are given, for most soil series in lowland districts, of the soil water reserves available for common crops. These calculated reserves, called *crop-adjusted profile available water* (A.P. mm, crop-adjusted) vary from crop to crop and are given for individual crops, for example A.P. mm, winter wheat. The laboratory and field procedures are described in Thomasson (1979) and Hall *et al.* (1977). The estimates are made on the assumption that crop rooting depths are shallowest for potatoes, intermediate for cereals and grass, and deepest for sugar beet, so in soils without rock or a soil horizon impenetrable to roots at shallow depth, soil water reserves increase accordingly. To estimate crop water requirements, the

A.P. is compared with the relevant meteorological parameter, the potential soil moisture deficit (P.S.M.D.). This is defined as the accumulated sum of potential transpiration minus rainfall during the growing season and represents the water requirements from the soil by a short green crop which completely covers the ground surface and is transpiring freely. Grass crops need most water because of their long growth period with full ground cover, so the average maximum P.S.M.D. for grass, usually abbreviated to M.D. (grass), is normally larger than that for other crops with shorter growing periods. The adjusted P.S.M.D. values for cereals – M.D. (Winter wheat), M.D. (Spring barley), etc. – are normally appreciably smaller than maximum P.S.M.D. because these crops do not cover the ground completely in the early growing season and, in most parts of the country, ripen a few weeks before the annual maximum value of P.S.M.D. is reached. Potatoes and sugar beet, although continuing to grow well into the autumn, do not achieve full ground cover until about the end of June. Consequently, adjusted P.S.M.D. values for these crops – M.D. (Potatoes), M.D. (Sugar beet) – are smaller than for grass to compensate for lack of transpiration during spring and early summer (Thomasson 1979). This effect is evident when examining the soil profile during a dry June in neighbouring fields under cereals and root crops. The average crop-adjusted M.D. values used in Chapter 3 were calculated from the same data-set as maximum P.S.M.D.

By comparing appropriate A.P. and M.D. data, average droughtiness for common crops can be estimated for most soils and in various climatic environments. This is expressed as:

Average droughtiness = A.P. (crop-adjusted) minus M.D. (crop-adjusted) (mm)

Table 13
Droughtiness Classes

A.P.-M.D. (mm)	Class
	Non-droughty
+50	
	Slightly droughty
0	
	Moderately droughty
-50	
	Very droughty

It is assumed that the A.P. for a particular crop is more or less constant from year to year. The standard deviation of M.D. (about 1 year in 6 frequency) is approximately 50 mm for arable crops but somewhat greater for grass in dry eastern areas. Droughtiness classes are given in Table 13.

Terms used in the keys to soil series
The terms used in the keys to soil series are intended for the non-specialist reader. Many

technical terms used elsewhere in this book have been replaced in the keys by everyday words. So simplified, the keys should help the layman to identify most soil series with some degree of confidence. It has not always been possible to find adequate substitute terms, so some technical terms survive which are explained below in a special glossary.

Each key gives successive choices numbered on the left. Opposite each choice there is a series name (or names) or a further numbered choice. In the few cases where more than one series is named, further differentiation usually needs more specialized interpretation which sometimes requires laboratory analyses. The soil series named in the map legend are printed in capital letters in the keys and, where possible, these are identified early.

Where a rock type is indicated in the keys it can be either undisturbed solid rock or a very stony rubble derived from it. Geological maps can aid the identification of rocks or other deposits.

Glossary of terms used in the keys

Alluvium. Unconsolidated material deposited by rivers on *floodplains* or by the sea on coastal and estuarine levels. The land may flood unless embanked.

Bleached subsurface horizon. Light grey, mineral horizon which may be *mottled.* Paleness is due to loss of organic matter and iron. This is the E horizon of Avery (1980) and is common in many podzolic soils.

Brightly coloured. Comprises strong brown (7.5YR 5/6 or 5/8), reddish yellow (7.5YR 6/6, 6/8, 7/6 or 7/8) and yellowish brown (10YR 5/6 and 5/8) colours of the Munsell Color Charts. Horizons of this kind are commonly found in podzolic soils, ferritic soil subgroups, some argillic soil subgroups of sandy soils (Avery 1980) and ferruginous soil material (Clayden and Hollis 1984).

Calcareous. The soil material, excluding stones, contains lime which reacts audibly when 10 per cent hydrochloric acid is added (Hodgson 1976, p.56-7). See also *extremely calcareous.*

Chalk. Fine grained, very pure *limestone,* often containing flints.

Cherty. Very stony, the stones being of chert, a very hard siliceous rock.

Clay-enriched. An horizon below about 50 cm depth which is more clayey than those above it. This is the argillic Bt horizon of Avery (1980) and results from gradual downward movement of clay particles. Field identification is often uncertain except in *sandy* or *coarse loamy* soils.

Clayey. Of clay, silty clay or sandy clay texture (p.303). When moistened the soil is sticky and shines when smeared.

Distinct topsoil. Surface layer, usually brown or grey, with insufficient organic matter (humus) to qualify as *humose* or *peaty.*

Disturbed. Artificially mixed soil horizons and rock layers, lacking discernible natural order.

Earthy topsoil. Peaty surface layer with few visible plant remains and having granular or blocky structure (Hodgson 1976), normally resulting from cultivation or drying.

Extremely acid with pale yellow streaks. Horizon containing the mineral jarosite, usually associated with very low pH – the sulphuric horizon of Avery (1980).

Extremely calcareous. Contains at least 40 per cent lime. Extremely calcareous alluvium, lake marl or tufa are very pale grey, of low density and feel smooth and silty. *Chalk* and *limestone* are extremely calcareous rocks.

Floodplain. Flat riverside land (see *alluvium*).

Flush. Wet ground downslope of a spring.

Gravel. Unconsolidated drift deposit with abundant, usually rounded stones.

Greenish speckled. Materials with green colours and dark grains, indicating the presence of the mineral glauconite.

Gritstone. Hard sandstone with coarse, angular grains.

Humose topsoil. Dark topsoil with smooth, silky or soapy feel, the organic matter content being intermediate between *peat* and mineral material (Avery 1980). Common on uncultivated land, but can form during cultivation if peaty topsoil or surface litter is incorporated with underlying mineral subsoil.

Humus-enriched subsoil. Dark subsoil horizon containing redeposited organic matter which is sometimes cemented. Equivalent to the Bh horizon of Avery (1980). Typical of many podzolic soils, most often it underlies a *bleached subsurface horizon.*

Igneous rocks. Hard usually crystalline rocks often originating as magma which can be *acid*, with much free quartz – includes granite and rhyolite; *basic*, largely quartz-free – includes basalt, dolerite and gabbro; *ultrabasic*, dominated by ferro-magnesian minerals – includes serpentine (peridotite). For help in identifying these and other rock types reference can be made to Tables 14 and 17 of Hodgson (1976) and to geological maps.

Ironpan (thin). A reddish brown to black band a few millimetres thick, enriched with iron and carbon and forming a barrier to roots and water. Found in some soils with *podzolic features*, it generally occurs beneath a *bleached subsurface horizon.*

Ironstone. Iron-enriched consolidated rock. Extensive ironstones include the Middle Lias Marlstone and Northampton Sand.

Limestone. Extremely calcareous consolidated rock.

Loamy. A textural group which can be subdivided as follows:

> *Coarse loamy* comprises the texture classes sandy loam and sandy silt loam. When moist these usually feel slightly or moderately abrasive, will mould into a ball and are cohesive, but with little stickiness.
>
> *Fine loamy* comprises texture classes sandy clay loam and clay loam. When moistened these usually feel both moderately sticky and slightly or moderately abrasive; they bind together and will exhibit a shine.
>
> *Coarse silty* corresponds to the silt loam texture class, which feels smooth, silky or soapy, without appreciable stickiness when moistened in the hand.
>
> *Fine silty* corresponds to the silty clay loam texture class in which moderate stickiness is combined with a smooth, silky or soapy feel if worked in the hand after moistening. The soil is cohesive and can exhibit a shine. For very stony and some other soils, broader *loamy* or *silty* textural groups are used. See also p.65.

Man-made topsoil. Dark surface horizon at least 40 cm thick resulting from addition of waste, earthy manure or compost or from deep cultivation of humus-rich soil. It often contains pottery and other artefacts. Equivalent to the thick man-made A horizon of Avery (1980).

Mineral soils. Soils with less than 40 cm peat in the upper 80 cm of the profile.

Mottled. With spots or blotches of colour of varying intensity, commonly the result of periodic waterlogging, either currently or before drainage.

> *Subsoil faintly mottled above 60 cm or distinctly mottled between 40 and 80 cm:* equivalent to the gleyic features of Avery (1980).
>
> *Prominently mottled or greyish above 40 cm:* equates with the gleyed subsurface horizon of Avery (1980). The absence of mottles in brownish or reddish soil generally indicates that the horizon or profile is rarely waterlogged.

Mudstone. Consolidated rock of mostly clay-sized particles in which there are few partings. Often interbedded with *shale* or *siltstone.*

Non-calcareous. No observable reaction when dilute hydrochloric acid is added; lacking free lime.

Peat. Dark, organic material (plant residues) formed or deposited under wet conditions, fully defined by Avery (1980).

Peat soils. Soils with more than 40 cm organic material in the upper 80 cm, or more than 30 cm if directly over *rock*. They can be *amorphous*, usually black and decomposed (humified) without identifiable plant remains, or *fibrous* with less decomposed material and somewhat lighter in colour or browner. Fibrous peat soils include the semi-fibrous peats of Hodgson (1976) and are further divided (Avery 1980, p.60–2) by identified plant remains.

Peaty topsoil. Topsoil of peat less than 40 cm thick over mineral material or less than 30 cm thick over *rock.*

Podzolic features. The combination of a *bleached subsurface horizon* with an underlying *humus-enriched* or *brightly coloured* subsoil or *thin ironpan.*

Raw topsoil. A peaty surface layer without granular or blocky structure, in which there may be visible plant remains.

Reddish. Of Munsell hue 5YR or redder, distinguishing red parent material in all but *shallow* and *sandy* soils. Over non-red parent material such colours can indicate paleo-argillic horizons.

Rock. This term is used for bedrock *or* very stony rubble.

Sandstone. Consolidated rock formed of sand-sized particles. *Hard sandstone* often gives stony soils and the bedrock cannot be dug readily. *Soft sandstone* forms few stones and is easily penetrated with an auger or spade.

Sandy. A textural group comprising the sand and loamy sand classes – sometimes divided into fine, medium or coarse. Moist soil feels abrasive but lacks stickiness.

Serpentine. See *igneous rocks.*

Shale. Consolidated rock of predominantly clay or silt particles, with thin laminations.

Shallow soils. Soils in which bedrock or very stony rubble usually occurs within 30 cm depth. These include many of the rankers and rendzinas of Avery (1980).

Siltstone. Consolidated deposit, mainly of silt-sized particles, with few partings.

Silty. See *loamy* above.

Slate. Fine grained, hard rock with many parallel partings.

Stoneless. Containing less than 1 per cent stones in more than half the upper 80 cm.

Surface leaf mould. Thin layer of fresh or decomposed litter resting on top of mineral soil. Equated with the L, F and H horizons (Avery 1980); under semi-natural vegetation it occurs in place of a distinct topsoil.

Terrace. A platform of land on a valley side or plain.

Unmottled. Absence of mottles, see *mottled.*

3

DESCRIPTION OF SOIL ASSOCIATIONS

The ninety-one soil associations in Wales are described below in alphabetical order. Their map symbols, areas and proportionate extents are given in Table 14. The descriptions are mainly standard but twenty-eight soil associations which are inextensive are described only briefly.

Each of the sixty-three fuller descriptions has an introductory paragraph outlining the kinds of soil included and describing geology, landscape features and national distribution. This is generally followed by brief typical profile descriptions of the soil series named on the map legend, but in some cases the reader is referred elsewhere in the book to save repetition. References to full representative soil profile descriptions are given

Table 14
Map symbol, Area and Proportionate Extent of Soil Associations

Soil association	Map symbol	Area km²	%	Soil association	Map symbol	Area km²	%
Adventurers' 1	1024a	17	0.08	Conway	811b	295	1.42
Agney	812c	5	0.02	Crewe	712f	7	0.03
Altcar 1	1022a	7	0.03	Crowdy 1	1013a	241	1.16
Alun	561c	40	0.19	Crowdy 2	1013b	441	2.12
Anglezarke	631a	156	0.75	Crwbin	313c	41	0.20
Arrow	543	29	0.14	Denbigh 1	541j	2,019	9.73
Bangor	311e	171	0.82	Denchworth	712b	56	0.27
Barton	541l	25	0.12	Disturbed soils 3	92c	89	0.43
Blackwood	821b	11	0.05	Eardiston 1	541c	80	0.39
Brickfield 1	713e	421	2.03	Eardiston 2	541d	308	1.48
Brickfield 2	713f	795	3.83	East Keswick 1	541x	517	2.49
Brickfield 3	713g	234	1.13	East Keswick 3	541z	322	1.55
Bridgnorth	551a	<2	<0.01	Ellerbeck	541u	<2	<0.01
Bromsgrove	541b	<2	<0.01	Enborne	811a	25	0.12
Bromyard	571b	180	0.87	Escrick 1	571p	15	0.07
Cegin	713d	1,494	7.19	Escrick 2	571q	44	0.21
Clifton	711n	68	0.33	Everingham	821a	29	0.14
Compton	813e	23	0.11	Fforest	713c	101	0.48

Table 14 (contd.)
Map symbol, Area and Proportionate Extend of Soil Associations

Soil association	Map symbol	Area km²	%	Soil association	Map symbol	Area km²	%
Fladbury 1	813b	5	0.02	Pinder	711q	5	0.03
Fladbury 3	813d	4	0.02	Revidge	311a	82	0.40
Flint	572l	15	0.07	Rheidol	541v	87	0.42
Foggathorpe 1	712h	75	0.36	Rivington 2	541g	31	0.15
Foggathorpe 2	712i	<2	<0.01	Rockcliffe	811d	31	0.15
Gelligaer	654c	281	1.36	Rowton	571A	13	0.06
Goldstone	631e	10	0.05	Salop	711m	259	1.24
Hafren	654a	1,320	6.36	Salwick	572m	43	0.21
Hallsworth 1	712d	37	0.18	Sandwich	361	122	0.59
Hexworthy	651b	70	0.33	Skiddaw	311b	28	0.14
Hollington	811c	4	0.02	Stanway	711a	13	0.06
Laployd	871a	29	0.14	Ston Easton	571a	139	0.67
Lugwardine	561d	82	0.40	Tanvats	811e	85	0.41
Lydcott	654b	174	0.84	Teme	561b	240	1.16
Malham 2	541p	89	0.43	Trusham	541n	10	0.05
Malvern	611a	121	0.58	Unripened gley soils	22	43	0.21
Manod	611c	3,744	18.04	Wallasea 1	813f	31	0.15
Middleton	572b	11	0.06	Waltham	541q	23	0.11
Midelney	813a	4	0.02	Wenallt	721e	229	1.10
Milford	541a	1,064	5.13	Wetton 2	311d	12	0.06
Moor Gate	612b	132	0.63	Wick 1	541r	355	1.71
Moretonhampstead	611b	55	0.27	Wigton Moor	831c	<2	<0.01
Neath	541h	263	1.27	Wilcocks 1	721c	899	4.33
Nercwys	542	23	0.11	Wilcocks 2	721d	424	2.04
Newchurch 2	814c	73	0.35	Wisbech	812b	53	0.26
Newnham	541w	13	0.06	Withnell 1	611d	502	2.42
Newport 1	551d	59	0.29	Worcester	431	21	0.10
Oglethorpe	541D	126	0.61	Urban		617	2.97
Parc	612a	139	0.67				
				Total		20,765	

for readers who would like more detail. These references are mainly to previous Soil Survey publications but some are published for the first time in Appendix 2. Information is then given on other soils likely to be encountered in the association and their distribution and inter-relationships. At this point a key to the component soils is provided. The soil water regime and the suitability of the land for agriculture and, where relevant, horticulture, forestry and recreation are then described. Where appropriate a table is included showing profile available soil water, crop-adjusted mean moisture deficit and droughtiness class for local crops. For selected associations a figure is given showing the periods when soils are at field capacity and when they are accessible for landwork without risk of damage to soil structure.

The shorter descriptions of the twenty-eight soil associations do not include soil profile descriptions, keys, figures or tables.

§ 8. ADVENTURERS' 1 ASSOCIATION
1024a

This association comprises amorphous and semi-fibrous peat soils formed mainly in reed and sedge peat, often with wood fragments from carr. The most widespread soil is the Adventurers' series, earthy eutro-amorphous peat soils, in which the subsoils are predominantly humified peat and contain few identifiable plant remains. The locally important Altcar series of earthy eu-fibrous peat soils is less humified and its subsoils are mainly semi-fibrous or fibrous with abundant recognizable plant material. Both series have well decomposed organic topsoils. The association covers 308 km² mainly in the Fenland of East Anglia, but it is scattered widely elsewhere in England and Wales. In Northern England it occurs mainly east of the Pennines. It is found also in glaciated morainic landscape near Wem in Shropshire and in Lancashire and Cheshire. In Wales it occurs on Anglesey and in valleys bordering Cardigan Bay. There is a small area in the Arun valley in West Sussex. The land is flat and permanently waterlogged in its natural state. It is now ditched, without hedges, and generally with few trees. In the Fens, at or slightly below sea level, and in other similar sites in England, water-tables are controlled by pumping and the land is arable. Elsewhere, drainage is mainly by gravity, its effectiveness depending on the gradient and clean maintenance of the outfalls. Where drainage is efficient and the climate is sufficiently dry the land is cultivated but much in Wales is under grass.

Adventurers' series
(Full description Seale 1975, p.84)

0–25 cm — Op
Black, stoneless humified or loamy peat.

25–50 cm — Oh1
Black, stoneless humified peat; weak angular blocky or moderate medium prismatic structure.

50–100 cm — Oh2
Dark reddish brown, stoneless humified peat; massive structure.

100–120 cm — Cg
Grey, stoneless clay loam or sandy loam; massive structure.

Altcar series
(Full description Jarvis, R.A. 1973, p.56)

0–25 cm — Op
Black, stoneless humified or loamy peat.

25–35 cm — Om1
Dark reddish brown, stoneless semi-fibrous peat; moderate coarse angular blocky.

35–85 cm — Om2
Black, stoneless semi-fibrous peat; weak coarse platy structure.

85–100 cm — Cg
Olive grey, stoneless humose clay loam or sandy loam; weak medium prismatic structure.

The association covers 17 km² in Wales. It occurs on the coastal lowland between Harlech and Barmouth, bordering the Mawddach, Dysynni (Plate 14) and Dovey estuaries and in Anglesey. In Wales Altcar series is almost as extensive as Adventurers' series. In Anglesey the peat, originally mapped as Cadarn series (Roberts 1958) is found in shallow depressions in the Carboniferous limestone. The sites were originally lakes and the peat,

R. Hartnup

Plate 14. View eastwards up the Dysynni valley to the volcanic crag of Craig-yr-Aderyn. Cader Idris is in the background. Lowland peat soils of the Adventurers' 1 association are in the foreground, with the Manod association on valley sides. The rocky ground on the crags is in the Bangor association.

60 to 120 cm or occasionally 180 cm thick, rests on white algal marl. The pH values of the peat are between 6 and 7. In the estuaries the soils are formed in thick fen or wood peat where, although the groundwater seeping from adjacent soils is relatively mineral-rich (Rudeforth 1970), pH values are below 4.5 in places.

Key to component soil series

Amorphous peat	ADVENTURERS'
Fibrous peat with grass and sedge remains	ALTCAR

Soil water regime

Drainage is effected by gravity through pipes and ditches and is of varying efficiency. In Anglesey drainage schemes have been largely unsuccessful because outlets are either dammed by rock formations or there is insufficient gradient (Roberts 1958). Thus soils are waterlogged for long periods in winter and there is risk of flooding.

Land use

The land is almost entirely under permanent grass used for summer grazing although the persistently high water-table makes them liable to poaching. Limited arable cropping has been attempted in the Dovey valley where outfall flaps prevent excessive rise of the water-table at times of high tides.

Some sites such as Cors Erddreiniog and Cors Goch in Anglesey are of particular interest to naturalists, being peatland with high pH in a mild wet western district where most peats are much more acid. They have a distinctive flora (Ratcliffe 1977).

§ 9. AGNEY ASSOCIATION
812c

Agney association consists mainly of calcareous alluvial gley soils including Agney and Wisbech series developed in marine alluvium on flat reclaimed land at 2–8 m O.D. near the coast of parts of Lincolnshire and Essex, Humberside and north-east Wales. The soils are stoneless and silty with a brownish plough layer over greyish brown mottled horizons with blocky or relic laminar structure.

Typically, Agney association has about half Agney and one-third Wisbech soils; but in the Dee Estuary, the non-calcareous Tanvats series (§ 85) is the main associate, and the association contains a narrow strip of unreclaimed mature saltings. The land is drained by ditches but is waterlogged in winter, in some parts for long periods (Wetness Class III and IV). A few arable crops, mostly winter cereals, are grown but grassland is the main use. A full description of this association is given by Hodge *et al.* (1984).

§ 10. ALTCAR 1 ASSOCIATION
1022a

The Altcar association is extensive on the Somerset Moors, in the Fenland of East Anglia and in Lancashire. There are also small areas elsewhere in Northern England and along the Welsh border. It covers about 223 km² at heights usually less than 10 m above sea level. The association is composed of fen peat, most of which has been reclaimed using pumped drainage schemes. The Altcar series, earty eu-fibrous peat soils in grass sedge peat, is the main soil, with the humified Adventurers' series, earthy eutro-amorphous peat soils, in association. The association is found on Fenn's and Whixhall Mosses, extending to about 600 ha. This is a Site of Special Scientific Interest and is under semi-natural vegetation, although some peat is extracted for horticultural use and some is afforested.

When undrained these soils are permanently waterlogged at shallow depth (Wetness Class VI) but in most areas groundwater levels are lowered to facilitate cultivation. Depending on the efficiency of arterial and field drainage, waterlogging may vary from slight to severe (Wetness Class II to V). These soils are not droughty and high yields, particularly of grass, are possible. The soils poach easily when under grass and have a limited period when they can be cultivated, especially in spring. They are nevertheless capable of high yields and pumped drainage would make intensive cropping possible. A full description of the association is given by Ragg *et al.* (1984).

§ 11. ALUN ASSOCIATION
561c

The Alun association consists of coarse and fine loamy brown alluvial soils. It is widespread in Northern England and occurs in the Midlands, Wales and Devon. The Alun series of coarse loamy typical brown alluvial soils occupies two-thirds of the association, commonly next to the rivers. The Enborne series, fine loamy typical alluvial gley soils and the Trent series, fine loamy gleyic brown alluvial soils are found in slight hollows, including abandoned meander channels. Brief profile description of the main series and a key to the component series are given below.

Alun series
(Full description Wright 1980, p.97)

0–25 cm — Ap
Dark brown, stoneless sandy loam or sandy silt loam.

25–60 cm — Bw1
Brown, stoneless sandy loam; moderate medium subangular blocky structure.

60–100 cm — Bw2
Brown, stoneless sandy loam; weak medium subangular blocky structure.

Enborne series
(Full description Palmer 1982, p.117)

0–20 cm — Apg
Dark greyish brown, slightly mottled,stoneless clay loam.

20–50 cm — Bg1
Brown, mottled, stoneless clay loam; moderate medium angular blocky structure.

50–80 cm — Bg2
Light brownish grey, mottled, stoneless clay loam; moderate coarse prismatic structure; common black ferri-manganiferous concretions.

80–100 cm — Cg
Light brownish grey, mottled, stoneless or slightly stony clay loam; massive structure.

Trent series
(Full description Kilgour 1979, p.53)

0–25 cm — Ap
Dark brown, stoneless clay loam.

25–50 cm — Bw
Brown, stoneless clay loam; moderate coarse subangular blocky structure.

50–100 cm — Bg
Brown, mottled, stoneless clay loam or sandy loam; weak medium angular blocky structure.

Typical Alun profiles have a brown to dark brown sandy loam topsoil which merges into a yellowish brown or brown sandy loam subsurface horizon. The deeper subsoil ranges in texture from loamy sand to sandy loam and in some parts is faintly mottled at depth. Most profiles are stoneless or slightly stony with very small to small rounded sandstone fragments. In Wales it is mapped over 40 km^2 mainly in narrow valleys draining the Carboniferous sandstone uplands. There is also a small area on the Cothi at Llansawel. In many places there are coarse loamy Tavy (Hogan 1977) and sandy Aled (Wright 1980) soils with gravel at less than 80 cm.

Key to component soil series

	Unmottled soils	1
	Mottled soils	3
1.	Gravel within 80 cm; coarse loamy	Tavy
	Deeper soils	2
2.	Coarse loamy	ALUN or Wick
	Fine loamy	Wharfe
3.	Subsoil faintly mottled above 60 cm or distinctly mottled between 40 and 80 cm	4
	Prominently mottled or greyish above 40 cm	5
4.	Fine loamy	TRENT
	Coarse loamy	Ty Gwyn
5.	Fine loamy	ENBORNE
	Clayey	Fladbury

Soil water regime

Alun soils are freely draining (Wetness Class I) and readily absorb winter rainwater. Flooding occurs in places and this can affect long-term management. The Enborne and Trent soils (Wetness Class IV and III) are affected by groundwater and, because they are in depressions, suffer surface ponding in winter. The soils are not droughty for most crops.

Cultivation and cropping

The soils are easily worked but all the land is under grass because of climate, the small field size or the risk of flooding. There is little poaching risk and the land is accessible throughout the year except immediately after heavy rain. In some places the high proportion of silt and fine sand can lead to structural instability, but if cultivation is avoided during wet periods or following heavy rain the damage is minimal.

§ 12. ANGLEZARKE ASSOCIATION
631a

The Anglezarke association is found throughout the Pennines from Staffordshire to Northumberland, on the North York Moors, in Wales and in the Forest of Dean. It is composed mainly of the Anglezarke series, humo-ferric podzols, and the Revidge series, humic rankers, over sandstones, grits and related Head. The ground is mostly gently to strongly sloping but includes some steeper valley-sides and craggy, irregular escarpments. Brief profile descriptions of the main soils and a key to the component series are given below.

Anglezarke series
(Full description Carroll and Bendelow 1981, p.69)

0–5 cm — H or Oh
Black, humified peat; weak medium angular blocky structure.

5–15 cm — Ah
Black, slightly stony humose sandy loam; weak subangular blocky structure.

15–30 cm — Ea
Brown, slightly stony loamy sand; single grain or weak fine angular blocky structure.

30–40 cm — Bh
Dark reddish brown, slightly stony humose sandy loam; weak fine angular blocky structure.

40–60 cm — Bs
Dark brown, moderately stony sandy loam; weak fine subangular blocky structure.

At 60 cm — R
Hard sandstone.

Revidge series
(Full description Carroll and Bendelow 1981, p.31)

0–20 cm — Oh or Ah
Dark reddish brown to black, humified peat or dark reddish brown, stoneless humose sandy loam; moderate fine blocky structure.

At 20 cm — R or Cu
Hard sandstone or grit or extremely stony sandy loam.

In Wales quartzitic sandstones and grits of the basal beds of the Millstone Grit outcrop form the soil parent material. The association is mapped in south Wales from Mynydd y Garreg near Kidwelly to the Blorenge north of Blaenavon and also in north Wales on Ruabon Mountain. The hard rocks form prominent, rugged relief rising to over 600 m O.D. on the Black Mountain. Bare rock and boulders dominate the landscape and in places the surface is deeply cratered by solution hollows and collapsed caverns in the underlying limestone.

Between the bare rock and boulders the ground is peaty and the Revidge series is recognized where the peat is shallower than 30 cm over the rock. Beneath the surface boulders however there is usually the coarse textured Anglezarke series. A thin ironpan occurs in places below the humus-enriched horizon, as in the Maw series (Carroll and Bendelow 1981, p.91). Wilcocks series (§ 95) and Fordham series (§ 95) occur in patches of drift, usually in depressions, and there is some Crowdy series (§ 28) in blanket and basin peat.

Key to component soil series

Shallow peaty soils, rock within 30 cm	REVIDGE
Deeper soils	1

1. With bleached subsurface horizon and dark, humus-enriched subsoil or ironpan; coarse loamy   2
   Other soils   4

2. With humus-enriched subsoil; no ironpan   ANGLEZARKE
    With thin ironpan   3

3. With humus-enriched subsoil   Maw
    No humus-enriched subsoil   Belmont

4. Mottled throughout; loamy with dark, humose or   Wilcocks or
    peaty topsoil   Fordham
    Unmottled; coarse loamy   5

5. Brown subsoil   Rivington
    Brightly coloured subsoil   Withnell

Soil water regime
The climate is wet and cold, annual precipitation exceeding 1,300 mm everywhere and ranging to more than 2,000 mm. The field capacity period is long and excess of winter rainwater is not readily absorbed by the peaty topsoils, so the amount of run-off is large.

Land use
The land is almost entirely open moorland with a little forestry. The rocky nature of the land and the wet climate preclude reclamation and its main use is rough grazing. The vegetation is mainly heather moor, with *Empetrum nigrum*, but there is less *Molinia* than on the adjacent Wilcocks association. Where burning and grazing have removed heath species there is *Nardus* grassland, while wet hollows contain cotton-grass bog. These communities provide poor grazing, and frequent rock exposures reduce available grazing further so that this is among the worst land in Wales.

§ 13. ARROW ASSOCIATION
543

The Arrow association, which covers 29 km², is mapped in south Glamorgan, north-east of Cowbridge on glaciofluvial deposits with kettle-hole topography.

The irregular, hummocky terrain with enclosed hollows has a complicated soil pattern described by Crampton (1972) as the Hensol complex. Most of the soils are coarse loamy, locally with gravelly subsoils. Arrow series (gleyic brown earths) is most extensive but Quorndon series, cambic gley soils, and the peaty-topped Laployd series (§ 55) are found commonly, but not exclusively, on low-lying ground, while the Wick (§ 93) and Hall series (p.315) usually occupy hilltops.

The component soils have different water regimes since the groundwater-table fluctuates to various depths in this undulating land. Arrow series (Wetness Class II to III) has a regime intermediate between the seasonally waterlogged Quorndon series (Wetness Class IV) and the well drained Wick and Hall series (Wetness Class I).

The complexity of the soils is reflected in the pattern of land use. Much of the land is under conifers in Hensol Forest, while on farmland there are often marked contrasts between improved grassland on the better drained soils and rough grazing on wet ground. Some of the land has been drained but improving the deep, enclosed hollows would be uneconomic.

§ 14. BANGOR ASSOCIATION
311e

The Bangor association is extensive in the Lake District and is also found in the Cheviot Hills and in north and central Wales. It covers 533 km², forming moderate to steep craggy ground, from about 300 m to 1,000 m O.D. in the Lake District and north Wales (Fig.21)(Plate 3). The climate is cold, windy and wet, the most exposed parts having the harshest climate in England and Wales. Humic rankers consisting of shallow acid peat

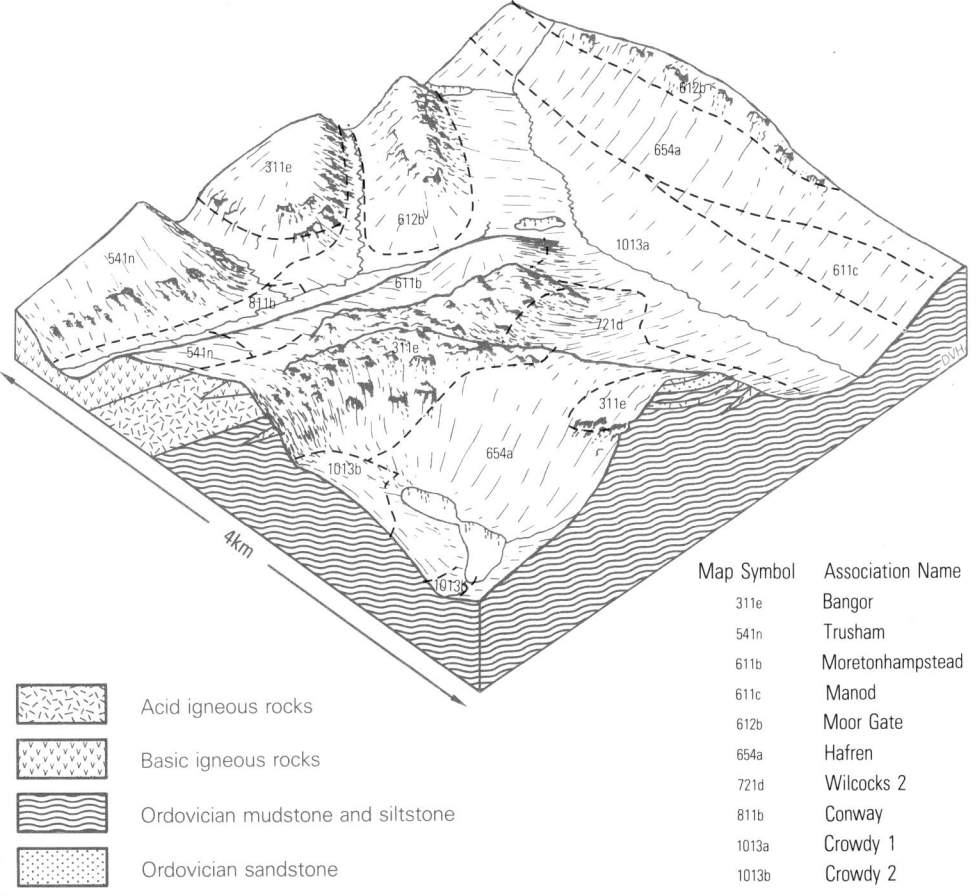

Map Symbol	Association Name
311e	Bangor
541n	Trusham
611b	Moretonhampstead
611c	Manod
612b	Moor Gate
654a	Hafren
721d	Wilcocks 2
811b	Conway
1013a	Crowdy 1
1013b	Crowdy 2

Acid igneous rocks

Basic igneous rocks

Ordovician mudstone and siltstone

Ordovician sandstone

Figure 21. Soil associations east of Snowdon

over acid crystalline rock, Bangor series, are the main soils. They are associated with raw oligo-amorphous peat soils belonging to Crowdy and raw oligo-fibrous peat soils, Winter Hill series, on gentle slopes. About a fifth of the land surface is bare rock, which is particularly extensive on hill tops. Brief profile descriptions of the main soils are given below.

Bangor series
(Full description p.309)

0–25 cm — Om or Ah
Black, slightly stony semi-fibrous peat or humose sandy loam; moderate medium subangular blocky structure.

25–35 cm — Cu
Black, extremely stony, humose coarse sandy loam.

At 35 cm — R
Rhyolite.

Crowdy series
(Full description p.311)

0–10 cm — Om
Black, semi-fibrous peat, moderate granular structure.

10–20 cm — Oh1
Dark brown, stoneless humified peat; massive structure.

20–100 cm — Oh2
Black, stoneless humified peat; massive structure.

Winter Hill series
(Full description Carroll *et al.* 1979, p.88)

0–10 cm — Om1
Black, semi-fibrous peat.

10–40 cm — Om2
Dark reddish brown, semi-fibrous or amorphous *Eriophorum-Sphagnum* peat; moderate coarse platy structure.

40–70 cm — Om3
Dark reddish brown, semi-fibrous *Eriophorum-Sphagnum* peat; weak coarse platy structure.

70–120 cm — Om4
Dark reddish grey, semi-fibrous *Eriophorum-Sphagnum* peat; massive structure.

There are minor inclusions of stagnopodzols where there is thick drift, humic cryptopodzols on level summits and, around the medieval town of Cefnllys near Llandrindod Wells, deep man-made humus soils. Small pockets of Bangor soils occur at unusually low altitudes near the town of Bangor.

Key to component soil series

Peat thicker than 40 cm		1
Other soils		2

1. Fibrous peat with remains of moss and cotton-grass — WINTER HILL
   Amorphous peat — CROWDY

2. Shallow peaty soils, rock within 30 cm — 3
   Deeper soils; with dark, humose or peaty topsoil — 4

3. Over acid igneous rock — BANGOR
   Over basic igneous rock — Preseli

4. With bleached subsurface horizon over unmottled subsoil; loamy — 5
   Without podzolic features; prominently mottled or greyish above 40 cm; coarse loamy — Laployd

| 5. | Over acid igneous rock | 6 |
| | Over basic igneous rock | 7 |

| 6. | With thin ironpan | Hexworthy |
| | Without ironpan | Rough Tor |

| 7. | With thin ironpan | Earle |
| | Without ironpan | Harthope |

Soil water regime

The shallow organic soils are quickly saturated with water so with an annual rainfall of 1,000 to 3,600 mm, they rapidly shed rainwater, and river catchments containing a large proportion of these soils often have sudden heavy floods. Nevertheless the shallow soils also dry rapidly and are waterlogged for only about one-third of the time, although the thicker Crowdy and Winter Hill soils remain waterlogged for much of the year (Wetness Class V). A moisture deficit sufficient to cause plants to wilt for short periods in summer occurs in an average of four years out of ten.

Grassland

The land is among the poorest in the country and there are few possibilities for improvement. The soil is being eroded in many places, partly due to an inevitable decline in the vigour of plant communities after centuries of sheep rearing with little or no replacement of plant nutrients. Where much-used footpaths cross steep or boggy ground the erosion is exacerbated by human trampling. The land is greatly valued for its landscape and most is protected by National Park legislation.

Most of the land is in heather moor, with pockets of blanket bog, often with pools of water. Summits like those of the Carneddau are very rocky with humic cryptopodzols and stony peat soils carrying *Rhacomitrium* moss, alpine club-moss or viviparous fescue. Patches of *Nardus* grassland with sedges and heath rush often indicate where snow patches persist in late spring. In the drier Llandrindod Wells district at about 300 m there are bent-fescue grasslands with much bracken and some *Nardus*, with permanent pastures on lower slopes. The arctic-alpine species of the Cader Idris cliffs and the ash woods of Craig y Benglog are examples of local base-rich flora. Even on acid rocks there is local seepage of water from base-rich rocks above.

Forestry

Most of the land is exposed and too high to be considered suitable for afforestation. Because of the shallow rooting depth, trees are very susceptible to being blown over. There is some potential for forestry on moraine at the heads of valleys where soils are generally deeper and exposure is less but yields would be low and some species would suffer frost damage.

§ 15. BARTON ASSOCIATION
5411

The Barton association consists of brown well drained fine silty soils developed over Lower Palaeozoic siltstones with thin interbedded silty shales and fine sandstones. It is locally extensive in the Welsh Borderland with almost 25 km² in Powys. The olive and buff-coloured rock is mostly decalcified and weathers to give moderately deep, very porous soils, often with stony subsoils. The land is between 120 and 450 m O.D. and mainly steeper than 7 degrees. This is a uniform association, most of the soils being typical brown earths. Barton soils are developed over clayey siltstone which occurs within 80 cm of the surface. The main subsidiary soils belong to the Munslow series, coarse silty typical brown earths, over coarser siltstones and interbedded fine sandstones. Silty shales give the fine silty Yeld series, stagnogleyic argillic brown earths with slight waterlogging. Sandstones influence the soils locally to give fine loamy profiles of the Denbigh series (§ 31) whereas hilltops and upper slopes often feature bare rock fringed with shallow Powys soils (§ 60).

Barton, Munslow and Denbigh soils are well drained (Wetness Class I) but the Yeld soils suffer some waterlogging in winter (Wetness Class III). Surplus winter rainfall can pass downwards through the permeable well drained soils though some of the water is shed from the steepest slopes.

Much of the land is under permanent grass which has been improved for intensive agricultural use but there are large commons, for example Hergest Ridge. Many steep slopes are bracken infested and currently provide only rough grazing. About a fifth of the land is wooded. Livestock rearing, mainly sheep, is the most important enterprise. Poaching is negligible except on the Yeld series where there can be a significant risk of damage.

Some of the land, particularly that with deep soils or on gentle slopes, is cultivated for cereals, mainly barley. Seed potatoes have been introduced recently to exploit the disease-free environment. The soils are easy to cultivate where slopes are less than 8 degrees and good seedbeds can be produced, even in spring. The large content of silt and fine sand leads to capping during heavy rain and run-off then causes erosion on slopes. The risks are greatest in spring before crop cover is established and during summer storms which follow dry spells.

The soils are generally well suited to trees so plantations of Scot's pine, European larch, Western hemlock and Sitka spruce flourish. There is some oak, sycamore and beech grown on the fringes of the big woods or as remnants of larger stands of mixed deciduous trees. Surface horizons of these soils under woodland can become strongly acid, particularly under conifers.

A full description of this association is given by Ragg *et al.* (1984).

§ 16. BLACKWOOD ASSOCIATION
821b

The Blackwood association is dominated by deep permeable sandy and coarse loamy soils in glaciofluvial drift. The drift has variable stone content and frequently overlies glaciolacustrine clay or till at depth. Past groundwater levels have been high enough for pronounced gleying to have developed but in most instances these have since been lowered by regional land drainage. The association occurs widely throughout Northern England, the Midlands, East Anglia and north Wales on gently undulating or flat ground, mainly between 1 and 37 m O.D. The most common soils are slightly stony typical sandy gley soils, the Blackwood series. Although prominent gleying in the subsoil suggests poor natural drainage, most profiles are now adequately drained and colours due to gleying are a relic feature. The associated coarse loamy Quorndon and stoneless sandy Formby series, cambic gley soils and typical sandy gley soils respectively, also have relic gleying. The fourth main soil, the sandy Ollerton series, gleyic brown sands, lacks relic gley colours because, being on slightly higher ground, it was less affected by groundwater than the other main soils before regional drainage was carried out.

The association is mapped on the Lleyn Peninsula at Tudweiliog and north-west of Llanbedrog on undulating glaciofluvial deposits. Hummocks and hollows, the latter containing peat, make for variable drainage. The pattern is usually simple but Isleham series (Clayden and Hollis 1984) and peat soils are common in hollows. Stagnogley soils develop where the underlying clayey drift is within profile depth. There are some inclusions of Brickfield series (§ 19) north of Llanbedrog. Where undrained, Blackwood, Quorndon and Formby soils are waterlogged for long periods in winter (Wetness Class IV) and Ollerton soils for short periods in winter (Wetness Class II). All suffer from fluctuating groundwater, which depends on the amount of seasonal rainfall, the depth of the impermeable material below, and lateral inflow from surrounding land. Where the regional water-table has been lowered, all soils are well drained (Wetness Class I) or only occasionally waterlogged (Wetness Class II).

The association is mostly under permanent grass with beef, dairy or sheep enterprises. Barley or ley grassland is found on the better drained land and rushy pasture occupies poorly drained hollows. Willow carr and wetland communities are found in the wettest areas. Land use depends upon whether drainage is feasible, seepage from surrounding areas being often a limiting factor. With efficient drainage, grass is available all year round except where the topsoil is rich in organic matter and susceptible to poaching. The climate is suited to grass, there being a long growing season (more than 300 days) and low moisture deficit (100 mm). The small risk of frost (more than 225 frost-free days) provides possibilities for delicate crops on favourable sites, and for horticulture. The association has been described fully by Ragg et al. (1984).

§ 17. BRICKFIELD 1 ASSOCIATION
713e

This association consists mainly of fine textured wet soils of the Brickfield, Cegin, Greyland and Wilcocks series and covers 420 km^2 in Wales with 47 km^2 in Cumbria. It is most extensive in the western and central parts of mid-Wales on the flanks of valleys and in basins where local drift has accumulated. The drift is largely till with some Head, the stones consisting mainly of mudstone and siltstone and occasionally sandstone and quartz. Developed on gentle slopes, the soils are slowly permeable and waterlogged for long periods in winter, and, especially where the topsoil is humose or peaty, well into the growing season. Most of the land is between 250 and 300 m O.D., but the association ranges from near sea level in Gwynedd to over 400 m O.D. on the slopes of Aran Benllyn. The Brickfield, Cegin and Greyland series are cambic stagnogley soils, and the Wilcocks series consists of stagnohumic gley soils with peaty or, less commonly, humose topsoils. In flushes, valleys and basins where soils are relatively permeable and affected by groundwater, the Freni series is found. At higher altitudes or in basins with little drainage outlet there are peat soils of the Crowdy or Winter Hill series. Brief descriptions of the Brickfield, Cegin and Wilcocks series and a key to the soils are given below. The Greyland series is described briefly in § 23.

Brickfield series	**Cegin series**	**Wilcocks series**
(Full description Hollis 1975, p.66)	(Full description Thompson 1982, p.39)	(Full description Hollis 1975, p.75)
0–20 cm — Ap	0–20 cm — Apg	0–20 cm — Oh or Ah
Very dark greyish brown, slightly stony clay loam.	Dark greyish brown, slightly stony silty clay loam.	Black, stoneless humified peat or humose clay loam.
20–50 cm — Bg	20–50 cm — Bg	20–50 cm — Bg
Greyish brown with many ochreous mottles, slightly stony clay loam; moderate medium subangular blocky structure.	Light brownish grey, mottled, slightly stony silty clay loam; moderate coarse prismatic structure.	Light brownish grey, mottled, slightly stony clay loam or sandy clay loam; weak subangular blocky structure.
50–100 cm — BCg	50–100 cm — BCg	50–100 cm — BCg
Grey, mottled, moderately stony clay loam; weak coarse angular blocky or prismatic structure; high packing density.	Light grey with many ochreous mottles, moderately stony silty clay loam; strong coarse prismatic structure becoming massive with depth.	Grey with many ochreous mottles, moderately stony clay loam; weak medium blocky or prismatic structure; high packing density.

The proportions of fine loamy Brickfield and fine silty Cegin soils vary from place to place, but Brickfield is dominant in most localities. In some districts Wilcocks series is widespread. Manod, Rheidol and Nefern profiles are generally inextensive, occasionally occurring on higher drier ground in the generally wet landscape. There is usually a distinct break of slope to steeper ground where the association meets the drier Manod association and in places it is marked by spring-lines with wet peaty flushes. Alluvial soils in narrow valleys, too small to separate at the map scale, are mostly typical and humic alluvial gley soils.

T.R.E. Thompson

Plate 15. Drainage of Cegin and Wilcocks soils. Here in the Conwy valley above Ysbyty Ifan, peat soils on the brow of Ffridd-y-Fedw give way to the Manod association on steep bracken-clad ground above the wet footslopes.

In parts of mid- and south-west Wales as at Crychell and Fedw-ddu periglacial modification of the drift has resulted in pingo scars (Bradley 1980). These form circular or arcuate ridges – the ramparts of former ice-cored mounds – which usually enclose small areas of Freni and Crowdy soils. The ramparts consist of compact drift on which shallow stagnogleyic rankers of the Eriviat series (Clayden and Hollis 1984) are developed.

Key to component soil series

Mineral soils		1
Peat thicker than 40 cm		3
1.	With dark, humose or peaty topsoil; fine loamy	WILCOCKS or Freni
	With grey, distinct topsoil	2
2.	Fine loamy	BRICKFIELD
	Fine silty	CEGIN
	Fine loamy over clayey	GREYLAND
3.	Amorphous peat	Crowdy
	Fibrous peat with moss and cotton-grass remains	Winter Hill

Soil water regime and land use

Soil wetness and the cool moist climate make arable use of the land uneconomic, so most is under permanent grass. Some is periodically reseeded but many once-improved grasslands have reverted to rushy pastures. Much of the wettest land is under *Molinia* which has little value for grazing. An efficient drainage system is essential for grassland production. Liver fluke is a hazard to stock where the land is undrained. Drainage measures in Brickfield and Cegin series probably improve their Wetness Class only from IV or V to III or IV so the soils remain susceptible to poaching and compaction; especially soils of the Wilcocks series with their peaty or humose surface horizons. The land is best suited to seasonal grazing in conjunction with adjacent drier ground such as that in the Manod association (Plate 15).

There is little forestry, although the land is moderately suited for this use. Surface wetness is a minor limitation for forestry on Brickfield and Cegin soils, but is more serious on Wilcocks series, because the longer periods of waterlogging restrict rooting. This reduces yield and seedling establishment, and makes trees susceptible to windthrow in exposed areas.

§ 18. BRICKFIELD 2 ASSOCIATION
713f

This association consists of fine loamy soils in till or Head mainly derived from Carboniferous shale and sandstone. It extends from Lancashire to the Scottish border and is also found in Wales, Shropshire and Derbyshire, covering a range of country from gently undulating till plains to drumlins interspersed with peaty or alluvial hollows. Elevations are generally 60 to 260 m O.D. Cambic stagnogley soils, Brickfield series, are predominant and are associated with stagnogleyic and typical brown earths, Nercwys (§ 67) and East Keswick series (§ 36) respectively. Brief profile descriptions of the main soils and a key to the component series are given below.

Brickfield series
(Full description Hollis 1975, p.66)

0–20 cm — Ap
Very dark greyish brown, slightly stony clay loam.

20–50 cm — Bg
Greyish brown with many ochreous mottles, slightly stony clay loam; moderate medium subangular blocky structure.

50–100 cm — BCg
Grey, mottled, moderately stony clay loam; weak coarse angular blocky or prismatic structure; high packing density.

Nercwys series
(Full description Rudeforth 1974, p.74)

0–25 cm — Ap
Dark brown, slightly stony clay loam.

25–50 cm — Bw
Dark yellowish brown, slightly stony clay loam; moderate medium subangular blocky structure.

50–100 cm — Bg
Yellowish brown, mottled, slightly stony clay loam; weak coarse angular blocky structure; high packing density.

East Keswick series
(Full description p.313)

0–25 cm — Ap
Dark brown, slightly stony clay loam.

25–70 cm — Bw
Brown, slightly stony clay loam; moderate medium subangular blocky structure.

70–100 cm — BCu
Dark yellowish brown, moderately stony clay loam; moderate coarse angular blocky or massive structure.

The association is mapped on drift from Carboniferous rocks in both the south Wales and the north Wales coalfields, where the Brickfield series was formerly identified as the Talog series (Clayden and Evans 1974, Thompson 1978). It generally occurs at a lower elevation and in a more dissected landscape than the soils of the related Wilcocks 1 association (§ 95). The drier Nercwys and East Keswick soils are found on the steeper slopes and convex hill tops, while the Brickfield series is usually found on gentle slopes. Clayey Hallsworth soils occur locally.

Key to component soil series

	Mottled soils	1
	Unmottled soils	4
1.	Prominently mottled or greyish above 40 cm	2
	Subsoil faintly mottled above 60 cm or distinctly mottled between 40 and 80 cm; fine loamy	NERCWYS
2.	With grey, distinct topsoil	3
	With dark, humose or peaty topsoil; fine or coarse loamy	Wilcocks
3.	Fine loamy	BRICKFIELD
	Clayey	Hallsworth
4.	Fine loamy	EAST KESWICK
	Coarse loamy	Wick

Soil water regime

The Brickfield series has slowly permeable subsurface horizons which cause prolonged or seasonal waterlogging (Wetness Class V or IV). Where drained, most still remain seasonally waterlogged (Wetness Class IV or III). The East Keswick series is well drained (Wetness Class I), whereas the Nercwys series is usually in Wetness Class III. Brickfield and Nercwys soils are quickly saturated and absorb little winter rainwater. It drains away laterally, mainly within the top 50 cm. By contrast, surplus rainwater passes downwards through the relatively permeable East Keswick series.

Cultivation and cropping

Annual rainfall on this association in Wales ranges from more than 1,600 mm in parts of the south to less than 800 mm in the north-east. In all areas however the soils are mainly under permanent grass or rush-infested rough grazing, and there is risk of severe poaching and liver fluke on undrained soils. Drainage measures, using permeable backfill followed by subsoiling or moling every four years, aid pasture improvement. Continuing care is needed, however, as a moderate poaching risk exists even after drainage. Flooding is another hazard on undrained Brickfield soils. Arable crops are largely restricted to forage, there being a better chance of success in the drier north even though the field capacity period is still more than 200 days. There are only a few days between September and May in most years when soil conditions are suitable for cultivation. Reserves of available moisture are usually adequate to ensure that the soils are non-droughty, although droughtiness can occur where topsoils have become compacted and rooting is restricted.

Forestry

Small areas of the association are used for forestry, Sitka spruce being the preferred species; site preparation involves deep ploughing and open drains to remove surplus water and encourage root penetration, so increasing resistance to the wind.

§ 19. BRICKFIELD 3 ASSOCIATION
713g

This association consists predominantly of loamy and clayey surface-water gley soils belonging to the Brickfield, Dunkeswick and Hallsworth series. It is widespread throughout Northern England, the north Midlands and north Wales, covering over 3,600 km², usually below 250 m O.D. on gentle to moderate slopes. The parent material is a greyish till or Head derived from Carboniferous and other Palaeozoic sandstones and shales. The predominant loamy textures are due to the preponderance of sandstone in the drift. Greater proportions of shale give rise to more clayey drift.

The Brickfield series of cambic stagnogley soils is loamy throughout and commonly contains many sandstones or locally, shales. At lower altitudes hard igneous and other erratic stones are also present. Stone content frequently increases with depth. The Dunkeswick series of typical stagnogley soils is distinguished from the Brickfield series by a clay horizon beginning between 40 and 80 cm depth, whereas the Hallsworth series of pelo-stagnogley soils is clayey to either the surface or the base of the topsoil. Brief profile descriptions of the main soils and a key to the component series are given below.

The soils were mapped over some 120 km² on till in the Lleyn Peninsula as the Dinas series by Hughes and Roberts (1958); they also occur near Wrexham, west of Newtown and between Saundersfoot and Newgale in south-east Dyfed. The Greyland series (§ 23) replaces the Dunkeswick series in mid-Wales, and soils in reddish till, belonging to the Clifton (§ 24) and Salop series (§ 79) are included locally in Clywd.

Key to component soil series

Fine loamy soils	1
Soils with clayey horizons	4

1. Subsoil faintly mottled above 60 cm or distinctly
   mottled between 40 and 80 cm   Nercwys
   Prominently mottled or greyish above 40 cm   2

2.	With dark, humose or peaty topsoil	Wilcocks
	With grey, distinct topsoil	3
3.	With greyish subsoil	BRICKFIELD
	With reddish subsoil	Clifton
4.	Clayey	HALLSWORTH
	Fine loamy over clayey	5
5.	With greyish subsoil	DUNKESWICK or
		Greyland
	With reddish subsoil	Salop

Brickfield series
(Full description Hollis 1975, p.66)

0–20 cm — Ap
Very dark greyish brown, slightly stony clay loam.

20–50 cm — Bg
Greyish brown with many ochreous mottles, slightly stony clay loam; moderate medium subangular blocky structure.

50–100 cm — BCg
Grey, mottled, moderately stony clay loam; weak coarse angular blocky or prismatic structure; high packing density.

Dunkeswick series
(Full description Hollis 1975, p.54)

0–20 cm — Ap
Very dark greyish brown, slightly stony clay loam or sandy clay loam.

20–40 cm — Eg
Greyish brown, mottled, slightly stony clay loam or sandy clay loam; weak medium subangular blocky structure.

40–60 cm — Btg
Yellowish brown with many grey mottles, slightly stony clay; moderate coarse prismatic structure.

60–100 cm — BCg
Grey, mottled, slightly or moderately stony clay; weak coarse prismatic or massive structure.

Hallsworth series
(Full description Bradley 1980, p.44)

0–20 cm — Apg
Dark greyish brown, slightly mottled, slightly stony clay loam or clay.

20–50 cm — Bg
Yellowish brown with many ochreous mottles, slightly stony clay; moderate coarse prismatic structure.

50–100 cm — BCg
Greyish brown, mottled, slightly stony clay; moderate coarse prismatic structure; high packing density.

Soil water regime

The soils are seasonally or severely waterlogged (Wetness Class IV or V) and fields are often rush infested. With underdrainage, Brickfield soils can be Wetness Class III in drier districts. The soils have slowly permeable loamy or clayey subsoils, which cause surface waterlogging. Excess winter rainwater either runs off or moves away at shallow depth. Profile available water in the Brickfield series averages 130 mm and is similar in the Dunkeswick and Hallsworth series, making them, at the most, only slightly droughty under grass.

Cultivations and cropping

Most farms rear stock or have dairy herds, so grass, much of it long term, is the principal crop. Yields are potentially high especially on the Lleyn Peninsula with its long growing season and small moisture deficit. However, there is risk of severe poaching so actual yields may be restricted. Liver fluke can infect stock where the land is not adequately drained. The land is often too wet for early fertilizer dressings and the grazing period is shortened by the early return to field capacity. Winter access is very restricted and on intensive farms slurry storage is necessary. Whilst unsuited to continuous arable cropping, breaks of barley and forage crops are incorporated in rotations before the land is resown to grass. Landwork is normally possible in the autumn but structural damage quickly results in wet conditions. Undrained, the land supports wet herb-rich grassland communities which are diverse on the Lleyn Peninsula, where there are soils with base-rich horizons.

Forestry
Little of the land is forested, although there are small oak woods on farms and oak is the dominant hedge tree inland. On the Lleyn Peninsula, exposure to salt-laden winds discourages tree planting. Elsewhere with adequate drainage, Sitka and Norway spruce grow well, although shallow rooting can lead to appreciable risk of windthrow on more exposed sites.

§ 20. BRIDGNORTH ASSOCIATION
551a

This association consists of well drained reddish sandy and coarse loamy soils developed in sandstone. It is very extensive throughout the Midlands and South West England and also occurs in Cumbria mainly on Permo-Triassic sandstone but locally over reddish Carboniferous sandstone. Small patches occur close to the Welsh border in Cheshire and Shropshire. Slopes are normally gentle to moderate but locally are steep bordering rivers; altitudes range from 50 to 200 m O.D.

The predominant soils belong to Bridgnorth series, typical brown sands with permeable sandy profiles developed in reddish sandstone which is hard and consolidated within 80 cm of the surface. Locally where the sandstone is soft and unconsolidated to at least 80 cm depth, Bromsgrove (§ 21) and Cuckney series (Ragg *et al* 1984) are found. The former is coarse loamy and belongs to the typical brown earths, whilst the latter accommodates similar typical brown sands previously mapped as part of the Bridgnorth series (Hollis 1978). Both soils pass to soft sandstone within 1.2 m of the soil surface. Pockets of deep sandy drift give soils of the Newport series (§ 70). The association is fully described by Ragg *et al.* (1984).

The porous sandy and coarse loamy soils of this association are naturally well drained (Wetness Class I) and readily absorb winter rainfall even on the occasional steep slopes. The soils are droughty for most crops but easily worked on gentle to moderate slopes which favour arable cultivation. Permanent grass is restricted to moderately steep slopes. Crops include cereals, potatoes, sugar beet, vegetables and soft fruit.

§ 21. BROMSGROVE ASSOCIATION
541b

This association consists of well drained reddish coarse loamy soils developed in soft Permian, Triassic and Carboniferous sandstones. It is widespread in the Midlands and South West England. Small patches occur close to the Welsh border in Cheshire and Shropshire. The fine and medium grained sandstones contain thin intercalations of siltstone and mudstone and are usually deeply weathered.

Dominating the association are typical brown earths of the Bromsgrove series with deep, permeable sandy loam profiles passing to soft sandstone. In places where the sandstones are hard and occur within 80 cm depth, Eardiston soils, also typical brown earths, are included. Where siltstones and mudstones contribute significantly to the parent material, stagnogleyic argillic brown earths with slowly permeable subsoils occur. These include the coarse loamy Staunton series (Hollis, n.d.) and the fine loamy Hodnet series (Beard 1984). Well drained coarse loamy over clayey soils of the Sellack series (Colborne 1981) are included locally where mudstones are thick enough and persistent enough to form discrete soil horizons. On some footslopes, where the sandstone is particularly deeply weathered, deep coarse loamy local drift gives well drained reddish Oglethorpe soils (§ 71). On flat low-lying ground, often adjacent to alluvium, seasonal waterlogging produces strongly mottled coarse loamy Greinton soils (Hollis, n.d.).

The Bromsgrove series and most of its associates developed on permeable sandstone are naturally well drained (Wetness Class I). They readily accept winter rainfall even on steep slopes. The Hodnet series with moderately permeable subsoil is occasionally waterlogged (Wetness Class II) but can be drier or wetter depending on slope or long term land use.

The level or gently undulating ground and easily worked soils encourage arable cropping. Grass is often restricted to the steeper slopes but short-term leys are sometimes sown especially on small farms as a part of an arable rotation. Cereals, sugar beet, potatoes, and horticultural crops including vegetables and soft fruit, are widely grown. Although the soils are deep, reserves of available water for grass are limited and growth is checked sometime during the summer in most districts.

A full account of the Bromsgrove association is given by Ragg et al. (1984).

§ 22. BROMYARD ASSOCIATION
571b

This association consists of well drained reddish fine silty soils over silty shales and soft siltstones and occasionally coarse loamy soils over sandstones. It is widespread and covers 1,436 km² on Devonian rocks in the Welsh Borderland, Wales and South West England. The soil pattern closely reflects the underlying bedrock which is micaceous and often slightly calcareous where unweathered. Thin, impersistent, soft, fine-grained sandstones are fairly common, as are greenish grey blotches and streaks.

The fine silty Bromyard soils, which belong to the typical argillic brown earths, are dominant on the silty shales and siltstones with the wetter Middleton series, stagnogleyic argillic brown earths, common on flat and gently sloping sites. Where the sandstones are thick and persistent, the Eardiston series (§ 34), a coarse loamy typical brown earth is recognized. Smaller inclusions of calcareous soils occur on impersistent rubbly argillaceous limestones (cornstones) and calcareous shales (Palmer 1976).

Brief profile descriptions of the main series are given below:

Bromyard series
(Full description Palmer 1976, p.43)

0–20 cm — Ap
Dark reddish brown, stoneless silty clay loam.

20–35 cm — Eb
Reddish brown, stoneless or slightly stony silty clay loam; weak coarse angular blocky structure.

35–65 cm — Bt
Reddish brown, stoneless or slightly stony silty clay loam; moderate medium prismatic structure.

65–100 cm — Cr
Reddish brown, blotched greenish grey, silt loam; weathered bedded silty shale and siltstone *in situ*. slightly calcareous.

Middleton series
(Full description p.318)

0–20 cm — Ap
Reddish brown, stoneless silty clay loam.

20–45 cm — Eb(g)
Reddish brown, stoneless silty clay loam; moderate medium subangular blocky structure.

45–70 cm — Btg
Reddish brown, mottled, stoneless silty clay loam; moderate medium prismatic structure.

70–85 cm — BCt(g)
Reddish brown, mottled, moderately stony silty clay loam; weak prismatic structure.

85–100 cm — Cr
Reddish brown, silty shale or siltstone.

Eardiston series
(Full description Palmer 1976, p.40)

0–20 cm — Ap
Dark reddish brown, stoneless or slightly stony sandy silt loam.

20–40 cm — Bw
Reddish brown, slightly stony sandy silt loam; moderate medium angular blocky structure.

40–60 cm — BCu
Reddish brown, slightly or moderately stony sandy loam; weak coarse angular blocky structure.

At 60 cm — R
Dark reddish grey, hard bedded micaceous sandstone.

The association is mapped west of Monmouth on broad gentle interfluves formed in Downtonian (Raglan Marl Group) and Dittonian (St Maughan's Group) rocks. There are smaller areas between Cardiff and Cwmbran, and near Presteigne. Sandstone outcrops are few and Eardiston soils are less frequent than in England. The wetter Middleton and Netchwood soils are found near springs and seepage lines. Some of the well drained soils contain greenish grey mottles inherited from the parent rock, which makes their distinction from similar but gleyed soils difficult.

Key to component soil series

Soils over siltstone and shale; fine silty	1
Soils over sandstone	2

1. Unmottled   BROMYARD
   Subsoil faintly mottled above 60 cm or distinctly mottled between 40 and 80 cm   MIDDLETON
   Prominently mottled or greyish above 40 cm   Netchwood

2. Soft sandstone within 80 cm; coarse loamy   Bromsgrove
    Hard sandstone within 80 cm   3

3. Coarse loamy   EARDISTON
    Fine loamy   Milford

Soil water regime

In dry districts the Bromyard series, with moderate permeability, is waterlogged for short periods only in winter, the duration depending on slope or long-term land use (Wetness Class I and II). Land under continuous cultivation will be more susceptible to surface wetness and waterlogging than that under permanent grassland because of the inevitable compaction and reduction of permeability induced by this management system. Eardiston series is underlain by permeable sandstone and so is well drained (Wetness Class I). The Middleton series is seasonally waterlogged (Wetness Class III) but responds to artificial drainage (Wetness Class II). In the wet climate of the fringes of the Black Mountains, where the field capacity period is greater than 200 days, Bromyard series is seasonally waterlogged (Wetness Class III) while the Middleton series is waterlogged for long periods in the winter (Wetness Class IV).

The association rapidly absorbs winter rainfall although on steep slopes or on compacted arable land run-off is greater.

Cultivation and cropping

This association provides good mixed farming land most of which is under ley grassland, with permanent grass on steep slopes. Cattle and sheep rearing are important. Grass and cereals yield well in most seasons. Wheat and barley, the main arable crops, are most usual on the lower ground west of Monmouth. Here temperatures are sufficient to maintain grass growth for over 8 months (Smith 1976) although droughtiness (Table 15) in drier summers slows growth. In good years, livestock can graze for over 6 months without damaging the sward or soil structure. The adequate surface bearing strength of Bromyard and Eardiston soils enables permanent pastures to be used with little risk of poaching.

Nevertheless Bromyard and Middleton series are water retentive and have low bearing strength when at or near field capacity. Cultivations should be avoided at such times. Unless mismanaged or compacted, Bromyard soils can be worked in autumn or spring after a rain-free period of 5 days. In spring, such periods are few, so autumn cultivations are preferred. Early autumn ploughing is essential on the wetter Middleton soils as, in wet seasons, it is usually impossible to plough without causing structural damage.

Because of their large silt and fine sand contents, recently ploughed or sparsely vegetated soils of this association are particularly susceptible to capping and subsequent erosion, especially under long term cultivation where organic matter contents are small. The risks are greatest when heavy rain falls in autumn or winter on already saturated ground or when summer thunder storms follow a dry spell. Periodic sheet and gully erosion over the centuries have given the characteristic pattern of shallow soils on brows and on headlands at the tops of fields, and deep colluvial soils on the upper sides of hedge banks and on footslopes. The areas of shallow soils in fields give droughty patches with relatively small crop yields.

Figure 22 shows the effects of soil properties and climate on landwork of the main constituent soils.

Bromyard association soils become acid in the surface where unlimed, but pH increases with depth and is often neutral within 1 m, where the substratum is slightly calcareous. They

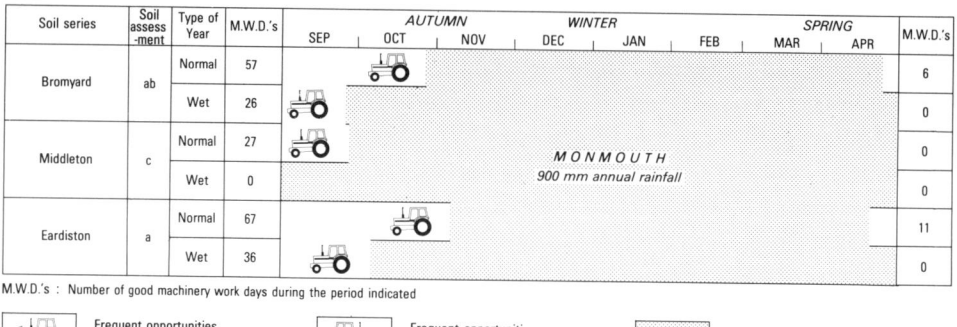

M.W.D.'s : Number of good machinery work days during the period indicated

Frequent opportunities for Autumn landwork

Frequent opportunities for Spring landwork

Little opportunity for landwork

Figure 22. The effects of soil and climate on landwork, Bromyard association

Table 15
Profile Available Water (A.P. mm), Crop-adjusted Mean Moisture Deficit (M.D. mm) and Droughtiness Class for extensive crops—Bromyard Association

Location Grid Ref.	Bromyard series Monmouth SO499130	Eardiston series Monmouth SO499130	Middleton series Monmouth SO499130
Grass			
A.P.	145	115	150
M.D.	113	113	113
Droughtiness	slightly droughty	slightly droughty	slightly droughty
Winter wheat			
A.P.	155	110	145
M.D.	74	74	74
Droughtiness	non-droughty	slightly droughty	non-droughty
Spring barley			
A.P.	155	110	145
M.D.	71	71	71
Droughtiness	non-droughty	slightly droughty	non-droughty

usually have low or very low available phosphorus and although available potassium can also be low, Bromyard soils have considerable slowly available reserves (Arnold and Close 1961). Eardiston soils tend to have moderate levels of potassium which is unusual for a coarse loamy soil. Adequate magnesium is available and deficiency symptoms are rare. Boron deficiency is sometimes noted in swedes.

§ 23. CEGIN ASSOCIATION
713d

Seasonally waterlogged loamy and clayey cambic stagnogley soils belonging to the Cegin, Brickfield and Greyland series with their slowly permeable subsoils are the chief components. They are intractable for much of the year unless artificially drained. Sannan and Denbigh series occur on steeper slopes or where the drift is thin over bedrock giving better drainage. The association which covers 1,622 km², is widespread over Silurian and Ordovician sedimentary rocks in Wales but also occurs in the Midlands and Northern England, commonly on undulating till-covered lowlands or on footslopes and valley floors. Hallsworth (§ 52), Nercwys (§ 67), Barton (§ 15) and East Keswick series (§ 36) are minor components occurring locally.

Brief descriptions of the main series are given below. The Sannan and Denbigh series are briefly described in section § 31.

Cegin series
(Full description Thompson 1982, p.39)

0–20 cm — Apg
Dark greyish brown, slightly stony silty clay loam.

20–50 cm — Bg
Light brownish grey, mottled, slightly stony silty clay loam; moderate coarse prismatic structure.

50–100 cm — BCg
Light grey with many ochreous mottles, moderately stony silty clay loam; strong coarse prismatic structure becoming massive with depth.

Greyland series
(Full description p.315)

0–25 cm — Ah
Dark greyish brown, slightly mottled, slightly stony clay loam.

25–50 cm — Bg
Pale olive, mottled, slightly stony clay loam; moderate coarse angular blocky structure.

50–80 cm — 2Bg
Light grey, mottled, moderately stony clay; weak coarse prismatic structure; high packing density.

80–100 cm — 2BCg
Light grey, mottled, moderately stony clay; massive structure; high packing density.

Brickfield series
(Full description Hollis 1975, p.66)

0–20 cm — Ap
Very dark greyish brown, slightly stony clay loam.

20–50 cm — Bg
Greyish brown with many ochreous mottles, slightly stony clay loam; moderate medium subangular blocky structure.

50–100 cm — BCg
Grey, mottled, moderately stony clay loam; weak coarse angular blocky or prismatic structure; high packing density.

This is the main association of stagnogley soils in Wales. It is most extensive around Newtown and Llandrindod Wells where it occupies hill and valley land. To the south it follows the Tywi valley (Wright 1980). In west Dyfed it is in valleys and depressions affected by seepage. South of Aberystwyth it occupies drift covered plateaux and wet footslopes. In northern Anglesey thick drift and low relief combine to give an extensive uninterrupted area of the association.

The nature of Cegin soils varies with topography and parent material (Thompson 1982). On convex hilltops compact slowly permeable material may be near the surface as a result of truncation. On footslopes, profiles have permeable finely structured upper horizons in colluvium overlying the more compact layers characteristic of stagnogley soils. In some Cegin soils there is a thin clayey horizon possibly formed by *in situ* weathering. Soils with thicker clayey horizons are classified as fine loamy over clayey Greyland series or similar unnamed fine silty over clayey stagnogley soils. These two soils are components of this association throughout Wales, being most prominent in eastern counties where bedrock is soft and more easily weathered, and over the mudstones of south Wales. Clayey Hallsworth soils (pelo-stagnogley soils) have a similar distribution. Brickfield soils are fine loamy throughout and are most common where more sandstone and gritstone are incorporated in the drift. Stagnogleyic brown earths of the Sannan series and well drained brown earths of the Denbigh (§ 31), Barton (§ 15) or East Keswick series (§ 36) are most common on steeper slopes where topography varies rapidly over short distances.

Key to component soil series

	Mottled soils	1
	Unmottled soils	4
1.	Prominently mottled or greyish above 40 cm	2
	Subsoil faintly mottled above 60 cm or distinctly mottled between 40 and 80 cm	3
2.	Fine silty	CEGIN
	Fine loamy	BRICKFIELD
	Fine loamy over clayey	GREYLAND
	Clayey	Hallsworth
3.	Fine silty	SANNAN
	Fine loamy	Nercwys
4.	Fine silty	Barton
	Fine loamy	5
5.	Rock within 80 cm	DENBIGH
	Deeper soils	East Keswick

Soil water regime

In wet districts Cegin, Brickfield and Greyland soils are waterlogged for long periods in the growing season (Wetness Class V) and even with artificial drainage they can remain wet throughout the winter (Wetness Class IV). Where the field capacity period is less than 200 days, the soils are naturally drier (Wetness Class IV) and their drainage regime can be improved to Wetness Class III. Denbigh soils are naturally well drained (Wetness Class I) and Sannan soils seasonally waterlogged (Wetness Class III). Estimated figures for profile available water for Cegin and Greyland soils in two places are given in Table 16.

Table 16
Profile Available Water (A.P. mm), Crop-adjusted Mean Moisture Deficit (M.D. mm)
and Droughtiness Class for extensive crops—Cegin Association

	Cegin series		Greyland series	
Location	Arddleen	Llanerchymedd	Arddleen	Llanerchymedd
Grid Ref.	SJ260160	SH420845	SJ260160	SH420845
Grass				
A.P.	140	140	125	125
M.D.	125	88	125	88
Droughtiness	slightly droughty	non-droughty	moderately droughty	slightly droughty
Winter wheat				
A.P.	140	140	130	130
M.D.	85	66	85	66
Droughtiness	non-droughty	non-droughty	slightly droughty	non-droughty
Spring barley				
A.P.	140	140	130	130
M.D.	85	63	85	63
Droughtiness	non-droughty	non-droughty	slightly droughty	non-droughty

Assessments of available water from Brickfield and Sannan profiles are similar. Locally, lateral seepage from nearby slopes supplements the amounts shown. The data are from drier districts which are most used for arable cropping. Where annual rainfall is more than 1,100 mm Cegin, Brickfield, Greyland and Sannan series are non-droughty. The Cegin association absorbs only a small proportion of winter rainwater.

Cultivation and cropping

Grass, much of it long term, is the main crop but some barley and roots are grown in east Wales. Potential grass yields are large because growth is rarely or only slightly restricted by droughtiness but surface wetness can delay early fertilizer applications and the land may remain too wet for grazing cattle many weeks after growth starts. Liver fluke is a hazard to stock where the soils are undrained. The autumn flush of growth potentially provides useful late grazing but it cannot always be used as the grazing season is several weeks shorter than the growing season. Grazing, silage harvesting and slurry spreading on wet land all lead to poaching or compaction of surface horizons with consequent deterioration of sward composition, soil drainage and yield. The risk of surface run-off following slurry spreading is severe particularly on slopes. To avoid river pollution, slurry should be stored and applied only during dry periods.

Figure 23 shows periods when the soils in east Wales and Anglesey are drier than field capacity and can be cultivated without risk of damage. Landwork is best done in autumn and, in normal years, opportunities for spring cultivation are very limited although some is carried out on adequately drained land. Poor weather often delays cereal harvesting so preventing timely autumn sowing of the next crop. Fungal diseases transmitted on stubble infect susceptible cereal crops and limit productivity.

Some of this land is afforested. For new coniferous plantations, the Forestry Commission recommends deep double mouldboard ploughing at 4 m spacing downslope with connecting cross drains before planting either Sitka or Norway spruce. Trees normally respond to phosphorus fertilizers given either at planting or as a subsequent top-dressing. Unless controlled, grass growth smothers young trees. These soils inhibit deep rooting so trees are liable to windthrow before reaching maturity. Where the climate is suitable most hardwoods will grow on this land and, in particular, there are notable stands of oak, which thrive in the poorly drained soils as at Crowther Coppice near Welshpool.

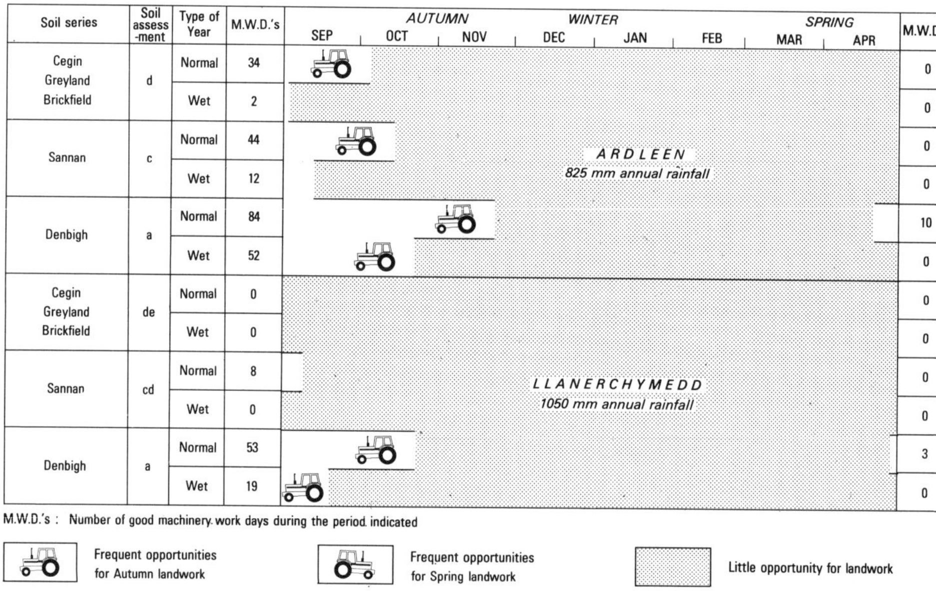

Soil series	Soil assess-ment	Type of Year	M.W.D.'s	AUTUMN			WINTER			SPRING		M.W.D.'s
				SEP	OCT	NOV	DEC	JAN	FEB	MAR	APR	
Cegin Greyland Brickfield	d	Normal	34									0
		Wet	2									0
Sannan	c	Normal	44				*ARDLEEN*					0
		Wet	12				*825 mm annual rainfall*					0
Denbigh	a	Normal	84									10
		Wet	52									0
Cegin Greyland Brickfield	de	Normal	0									0
		Wet	0									0
Sannan	cd	Normal	8				*LLANERCHYMEDD*					0
		Wet	0				*1050 mm annual rainfall*					0
Denbigh	a	Normal	53									3
		Wet	19									0

M.W.D.'s :   Number of good machinery work days during the period indicated

Frequent opportunities for Autumn landwork   Frequent opportunities for Spring landwork   Little opportunity for landwork

Figure 23. The effects of soil and climate on landwork, Cegin association

§ 24. CLIFTON ASSOCIATION
711n

The Clifton association consists of seasonally waterlogged soils developed in reddish fine loamy till and related glaciofluvial deposits. It usually forms gently undulating terrain, but the land is deeply incised by rivers and streams in places. The association is extensive south and west of the Pennines, from south Staffordshire and Clwyd to the Scottish border. East of the Pennines it is restricted to Teesside. The till, mainly of Devensian age, is derived from Permo-Triassic sandstones and mudstones and is non-calcareous or decalcified to at least 80 cm depth. It is generally dense and slowly permeable, but contains occasional pockets of sand and gravel and, in many areas is overlain by coarse loamy glaciofluvial deposits, usually less than 70 cm thick with an irregular lower boundary. Because of the slowly permeable nature of the till, the soils are mainly typical stagnogley soils. Clifton series is developed where the coarse superficial drift is absent or relatively thin (less than 40 cm thick) and the Claverley series, formerly classified as a deep sandy loam phase of the Clifton series (Hollis 1978), is found where the glaciofluvial drift is between 40 and 80 cm thick. Less mottled, fine loamy stagnogleyic argillic brown earths of the Salwick series occur on upper slopes where glacial outwash is absent and surface run-off of water is rapid. Coarse loamy ground-water gley soils of the Quorndon series are found mainly along streams and in hollows, where outwash deposits are thicker than 80 cm. Small areas of Arrow (§ 13), Wick (§ 93) or Newport (§ 70) soils are also included. usually on isolated glacial sand and gravel deposits in hummocky terrain. Brief descriptions of three of the main series are given below, and of Salwick series in (§ 80).

Clifton series
(Full description p.311)

0–25 cm — Ap
Dark greyish brown slightly stony clay loam or sandy clay loam.

25–35 cm — Eg
Greyish brown, mottled, slightly stony sandy loam or sandy clay loam; weak medium subangular blocky structure.

35–80 cm — Btg
Reddish brown, mottled, slightly stony clay loam or sandy clay loam; moderate coarse prismatic structure.

80–100 cm — BCtg
Reddish brown, mottled, slightly stony clay loam; weak coarse prismatic or massive structure; high packing density.

Claverley series
(Full description p.310)

0–25 cm — Ap
Dark brown, slightly stony sandy loam or sandy silt loam.

25–50 cm — Eg
Greyish brown, mottled, slightly stony sandy loam or sandy silt loam; weak medium subangular blocky structure.

50–80 cm — Btg
Reddish brown, mottled, slightly stony sandy clay loam or clay loam; moderate coarse prismatic structure.

80–100 cm — BCtg
Reddish brown, mottled, slightly stony clay loam or sandy clay loam; weak coarse prismatic or massive structure; high packing density.

Quorndon series
(Full description p.321)

0–20 cm — Ap
Very dark greyish brown, stone-less or slightly stony sandy loam.

20–40 cm — Bg1
Light olive brown with many ochreous mottles, slightly stony sandy loam; weak coarse sub-angular blocky structure.

40–80 cm — Bg2
Greyish brown with many ochreous mottles, slightly stony sandy loam; weak coarse angular blocky structure.

80–100 cm — Cg
Grey, mottled, slightly or moderately stony sandy loam or loamy sand; very weak medium angular blocky or single grain structure.

This association covers 52 km² in Clwyd with a further 16 km² in the Vale of Glamorgan, and occurs from sea level to 130 m O.D. It is extensive south-west of the Dee estuary from Hawarden to Holywell passing over a ridge westwards into the Alun valley near Mold. Small areas occur in the Vale of Clwyd (Lea and Thompson 1978), and in shallow depressions in morainic drift around Cardiff. Clifton and Salop soils (§ 79) are dominant but greyish surface-water gley soils of the Brickfield (§ 19), Cegin (§ 23) and Greyland series (§ 23) are also present. Locally, Clifton soils have coarse loamy surface horizons but deep coarse loamy profiles of the Claverley and Quorndon series are rare. Salwick soils are usually found on steeper slopes, but also fringe small isolated deposits of glaciofluvial drift which carry freely drained permeable brown earths of the Wick (§ 93) and East Keswick series (§ 36). Pinder (Colborne 1981, p.64) and Rufford series (Whitfield and Beard 1980, p.89) are also rare.

Key to component soil series

	Soils without clayey horizons	1
	Soils with clayey horizons; with reddish subsoil	7
1.	Unmottled; coarse loamy	Wick
	Mottled	2
2.	Prominently mottled or greyish above 40 cm	3
	Subsoil faintly mottled above 60 cm or distinctly mottled between 40 and 80 cm	6
3.	With reddish subsoil	4
	With greyish subsoil	5
4.	Fine loamy	CLIFTON
	Coarse loamy	CLAVERLEY
5.	Coarse loamy	QUORNDON
	Fine loamy	Brickfield or Pinder
	Fine silty	Cegin
6.	Reddish; fine loamy	SALWICK
	Brownish; coarse loamy	Arrow
7.	Coarse loamy over clayey	Rufford
	Fine loamy over clayey	Salop

Soil water regime

The parent till is slowly permeable, causing intermittent wetness in all the soils. Clifton and Claverley series have slowly permeable subsoils and their upper horizons are seasonally waterlogged (Wetness Class IV). Drainage measures significantly reduce the duration of waterlogging in Clifton profiles (Wetness Class III), but have an even greater effect on Claverley soils which remain wet for short periods only (Wetness Class II). Like the Clifton series, Salwick soils are slowly permeable but, being on slopes, shed more water by surface run-off and their upper horizons do not stay wet for quite as long (Wetness Class III). Quorndon soils suffer from seasonal waterlogging by fluctuating groundwater, usually held up over till at depths of 1.2 to 3 m, and their lower horizons stay wet for most of the winter (Wetness Class IV). Underdrainage is normally very effective, however, ensuring that the soils are well drained in all but the wettest seasons (Wetness Class I).

In most of the soils, water moves laterally through the topsoil or immediate subsoil (above 40 cm depth) and, in general, they tend to shed excess winter rain.

Cultivations and cropping

Much of this association is under cereals and short term grass for dairying, although some maincrop potatoes are grown for local markets. The small acreage of sugar beet on soils of the association in the Vale of Clwyd is amongst the most westerly in Britain.

Although rooting is restricted in the compact, coarsely structured subsoil of Clifton, Claverley and Salwick soils, moisture deficits are not large and crops suffer only slightly from drought. Quorndon soils, which have relatively large water reserves augmented by groundwater are usually non-droughty (Table 17).

Table 17
Profile Available Water (A.P. mm), Crop-adjusted Mean Moisture Deficit (M.D. mm) and Droughtiness Class for extensive crops—Clifton Association

	Clifton and Salwick series	Claverley series	Quorndon series
Location	Flint	Flint	Flint
Grid Ref.	SJ240720	SJ240720	SJ240720
Grass			
A.P.	125	130	155
M.D.	120	120	120
Droughtiness	slightly droughty	slightly droughty	slightly droughty
Winter wheat			
A.P.	125	135	170
M.D.	85	85	85
Droughtiness	slightly droughty	slightly droughty	non-droughty
Spring barley			
A.P.	125	135	170
M.D.	80	80	80
Droughtiness	slightly droughty	non-droughty	non-droughty
Potatoes			
A.P.	110	110	120
M.D.	70	70	70
Droughtiness	slightly droughty	slightly droughty	slightly droughty

The interactions of soils and climate and the effect on landwork are shown in Figure 24. For the main soils, cultivations need to be carefully timed to minimize structural damage and plough pan formation. It is easy to obtain a good tilth without special machinery or winter frost, but subsoils are often waterlogged in autumn and spring. Cultivations carried out under these conditions produce compaction below the plough sole which further impedes natural drainage. Due to the very slow drying of subsoils, spring working

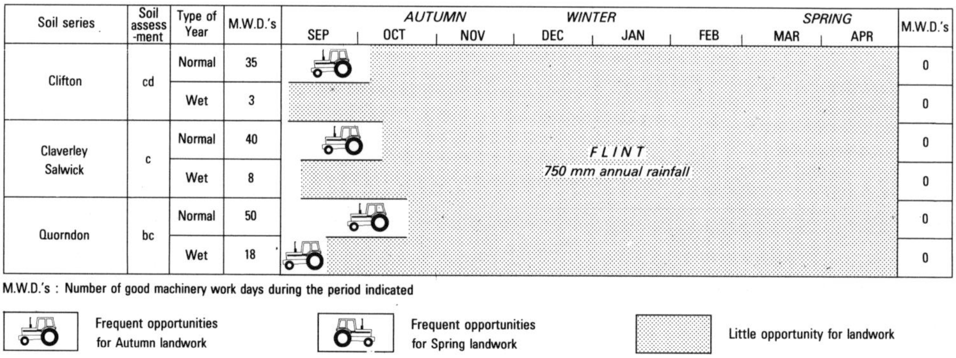

Soil series	Soil assess -ment	Type of Year	M.W.D.'s	AUTUMN			WINTER			SPRING		M.W.D.'s
				SEP	OCT	NOV	DEC	JAN	FEB	MAR	APR	
Clifton	cd	Normal	35									0
		Wet	3									0
Claverley Salwick	c	Normal	40									0
		Wet	8				_FLINT_					0
							750 mm annual rainfall					
Quorndon	bc	Normal	50									0
		Wet	18									0

M.W.D.'s : Number of good machinery work days during the period indicated

Frequent opportunities for Autumn landwork — Frequent opportunities for Spring landwork — Little opportunity for landwork

Figure 24. The effects of soil and climate on landwork, Clifton association

time cannot be guaranteed and cultivations are best undertaken in autumn. Compaction can be destroyed by regular subsoiling and this is especially recommended if root crops have been harvested late. Timing is important, for subsoils should not be wet.

Under grassland, most topsoils, especially those with more than 3 per cent organic carbon, retain large amounts of water. This and the possibility of waterlogging render the soils liable to poaching if stock are grazed too early or late in the year. There are 160–200 safe grazing days in a normal year, the season being May to September. Before or after these months, access to the land is not guaranteed and on intensive livestock farms, slurry storage is necessary.

The soils generally have a moderate cation exchange capacity and are inherently fertile. Although the parent material is often calcareous, with high percentage base saturation, soils are usually decalcified to below one metre depth and topsoils need occasional dressings of lime. Amounts of naturally available phosphorus and potassium are small, especially in Quorndon and Claverley soils, but with regular fertilizer applications deficiencies are rare. There is normally a moderate and well balanced supply of trace elements, but manganese deficiency sometimes occurs where liming has raised the pH above neutral.

§ 25. COMPTON ASSOCIATION
813e

The Compton association comprises clayey, severely waterlogged soils developed in reddish and greyish river alluvium subject to seasonal flooding. It occupies low-lying flat ground along the floodplains of rivers and streams draining red mudstone outcrops and is most common in the Midlands and the South West. A small area is also mapped in south Wales and east of Wrexham.

The association is dominated by reddish clayey alluvial gley soils of the Compton series, which have prominently mottled reddish and greyish subsoils which usually pass downwards into permanently waterlogged, greyish clay within 1 m depth. The reddish colours come from nearby reddish rocks of Triassic or Devonian age. Along narrow valleys cut in red mudstone, the association consists almost entirely of Compton soils. On broader river floodplains however, greyish clayey alluvial gley soils of the Fladbury series are found in backland areas and hollows. Here, prolonged waterlogging at shallow depth has produced a predominantly grey subsoil. Fladbury soils also occur where local streams drain outcrops of non-reddish rocks such as Lias clay or Carboniferous shale. Both Compton and Fladbury soils often become coarser at depth, with fine loamy or fine silty horizons below 50 cm depth.

The Dosthill series (Whitfield and Beard 1980, p.120), a reddish clayey typical humic alluvial gley soil, forms a minor but consistent component of the association. It has a similar, reddish and greyish mottled profile to that of the Compton series, but is distinguished by its dark coloured, humose topsoil and usually occupies wet hollows in narrow valleys, or fringes Fladbury soils on the broader river floodplains.

Brief descriptions of the Compton and Fladbury series are given below.

Compton series
(Full description Palmer 1982, p.121)

0–15 cm — Ah
Dark brown, stoneless silty clay or clay.

15–35 cm — Bg1
Reddish brown, mottled, stoneless clay or silty clay; moderate medium angular blocky structure.

35–60 cm — Bg2
Reddish brown, mottled, stoneless silty clay or clay; moderate coarse prismatic structure.

60–100 cm — Cg
Grey, mottled, stoneless silty clay, clay, silty clay loam or clay loam; weak coarse prismatic or massive structure.

Fladbury series
(Full description p.314)

0–15 cm — Apg
Dark greyish brown, mottled, stoneless clay.

15–60 cm — Bg
Greyish brown with many ochreous mottles, stoneless clay; strong coarse prismatic structure.

60–100 cm — Cg
Grey, mottled, stoneless clay; moderate angular blocky or massive structure.

The association covers some 23 km² in Clwyd, Gwent and South Glamorgan. It is found along the River Dee east of Wrexham, near Penarth, around the mouth of the River Rhymney and north-east of Usk. Occasional profiles of the Midelney series (§ 62) with clayey alluvium overlying peat within 80 cm depth are included along the Dee. Clyst (Clayden 1971), Tewkesbury (Cope 1973) and Stixwould series (Reeve and Thomasson 1981) are rare.

Key to component soil series

Soils prominently mottled or greyish above 40 cm	1
Subsoils faintly mottled above 60 cm or	
distinctly mottled between 40 and 80 cm; clayey	Clyst

1. Clayey throughout — 2
   With lighter horizons — 4

2. With dark, humose topsoil — Dosthill
   With grey or brown, distinct topsoil — 3

3. Reddish — COMPTON
   Greyish — FLADBURY

4. Fine silty — Hollington
   Fine loamy — Tewkesbury
   Clayey over coarse loamy or sandy — Stixwould
   Clayey over peat — Midelney

Soil water regime

The association occupies flat low-lying floodplains and the soils are subject to prolonged waterlogging by groundwater at shallow depths. They also suffer occasional damaging floods which can be extensive and protracted during the winter. When drained, both Compton and Fladbury soils can be improved to Wetness Class III or IV depending on their height above the river level. Underdrainage is often impractical, however, because of the lack of adequate fall and in such cases the soils are severely waterlogged (Wetness Class V). Subsoil waterlogging caused by rising groundwater is accentuated by seasonal surface wetness. Although the soils have moderate or strong structure under old pasture, there are few pores within peds and excess water percolates through profiles mainly along fissures and earthworm channels. For most of the summer, when they are slightly moist or dry the soils are well fissured and relatively permeable, but their clayey peds swell rapidly as they become wetter during the autumn. Fissures and cracks then close, making the soils slowly permeable and subject to severe waterlogging in the autumn, winter and spring.

Cultivations and cropping

The soils are almost all under permanent grass or rough grazing. Prolonged waterlogging at shallow depths and the risk of damaging seasonal floods preclude arable crops in all but the highest parts of the floodplain. Grassland is used mainly for stock rearing because waterlogging, winter floods and the large amounts of moisture retained by topsoils, make cattle grazing safe only in summer. All the soils have a serious poaching risk and grassland needs to be very carefully managed to achieve consistent growth.

§ 26. CONWAY ASSOCIATION
811b

Deep stoneless fine silty soils dominate this association which is found on the floodplains of rivers and streams. The soils are usually greyish brown or grey with yellowish brown mottles and affected by high groundwater levels. Browner soils occur where the water-table is lower. The association occurs in nearly all parts of England and Wales, as strips along major valley floors. Altitudes range from near sea level to about 260 m O.D. in central Wales.

Fine silty typical alluvial gley soils, Conway series, dominate the association. Similar fine loamy Enborne (Palmer 1982), coarse loamy Eversley (Clayden and Hollis 1984) and coarse loamy over gravelly Kettlebottom soils (Clayden and Hollis 1984) occur locally. Fladbury series, which belongs to the pelo-alluvial gley soils, and fine silty gleyic brown alluvial soils of the Clwyd series are common whilst the fine loamy Trent series (Kilgour 1979, p.53) and the well drained Teme series (§ 86) are occasional, especially on levées. Wetter soils of the Dovey series (Lea 1975, p.56) with peaty or, more often, humose topsoils are found in places in western Britain. Peat soils are rare, although important in places. Gravelly subsoils are common where there has been torrential alluvial deposition, usually in upper reaches of the rivers.

Brief descriptions of the main soils and a key are given below.

Conway series
(Full description Thompson 1982, p.58)

0-20 cm — Apg
Dark greyish brown, stoneless silty clay loam.

20-80 cm — Bg1
Light brownish grey, mottled, stoneless silty clay loam; moderate coarse prismatic structure.

80-120 cm — Bg2
Light grey with many ochreous mottles, stoneless silty clay loam; moderate coarse prismatic structure.

Clwyd series
(Full description Thompson 1982, p.51)

0-20 cm — Ap
Dark greyish brown, stoneless silty clay loam.

20-50 cm — Bw
Yellowish brown, stoneless silty clay loam; moderate fine sub-angular blocky structure.

50-75 cm — Bw(g)
Yellowish brown, slightly mottled, stoneless silty clay loam; moderate course angular blocky structure.

75-100 cm — Bg
Brownish grey, mottled, stoneless silty clay loam; moderate coarse prismatic structure.

Fladbury series
(Full description p.314)

0-15 cm — Apg
Dark greyish brown, mottled, stoneless clay.

15-60 cm — Bg
Greyish brown with many ochreous mottles, stoneless clay; strong coarse prismatic structure.

60-100 cm — Cg
Grey, mottled, stoneless clay; moderate angular blocky or massive structure.

The association covers nearly 300 km² in Wales, the largest area being in the Severn valley between Shrewsbury and Welshpool. Part of this is described in detail by Thompson (1982). It is also found along the valleys of the Conwy, Glaslyn, Mawddach, Dee, Dovey, Rheidol, Ystwyth, Aeron and Teifi, and many smaller rivers and streams. It is mostly absent from the Old Red Sandstone areas of Dyfed, Powys and Gwent where the river valleys have reddish alluvium. The better drained Clwyd series occurs in many valleys, especially on low levées near the main rivers. Clayey profiles of the Fladbury series are also frequent, especially in the Severn valley and in South Dyfed. The Dovey series is more common in Wales than elsewhere, as in parts of the Aran valley near Dolau. Gravelly alluvial soils although uncommon are sometimes included as in the Ystwyth and Rheidol valleys of north Dyfed (Rudeforth 1970). Peat soils are included in places with larger areas in the valleys of the Cefni (Anglesey), Conwy, Glaslyn and Dwyryd (Gwynedd), Gwaun, Dovey, Aeron and Teifi (Dyfed), and in Burton Meadows between Wrexham and Chester.

Key to component soil series

	Unmottled soils; fine silty	Teme
	Mottled soils	1
1.	Prominently mottled or greyish above 40 cm	2
	Subsoil faintly mottled above 60 cm or distinctly	
	mottled between 40 and 80 cm	5
2.	With dark, humose or peaty topsoil; fine silty	Dovey
	With grey, distinct topsoil	3
3.	Over gravel within 80 cm; coarse loamy	Kettlebottom
	Deeper soils	4
4.	Fine silty	CONWAY
	Clayey	FLADBURY
	Coarse loamy	Eversley
	Fine loamy	Enborne
5.	Fine silty	CLWYD
	Fine loamy	Trent

Excess winter rain is absorbed fairly slowly on the level ground, but it reaches the rivers relatively quickly because they are close by. Winter floods are very common and summer flooding also occurs in many areas after heavy rain. This, with the high groundwater-table usually precludes arable cropping and most of the association is under permanent grass, with some rough grassland and scrub. High groundwater and slow infiltration of water through Conway and Fladbury soils makes them seasonally or severely waterlogged (Wetness Class IV or V), and greatly increases the risk of poaching (Harrod 1979) in wet periods. Field drainage can reduce the risk to some extent, but outfall flaps have to be fitted to prevent water backing up the drains into the fields when the river is full. The associated Clwyd series (Wetness Class II or III) is waterlogged for a shorter time. The soft wet surface soil limits the value of this association for recreation (Carroll *et al.* 1979), but in some districts it occupies the only level land available for playing fields. Clwyd, Trent or Teme series, if locally extensive enough, are the most suitable component soils for this purpose.

Wetter parts sometimes provide wildlife habitats but are becoming increasingly rare with the extension of field drainage. In backswamps, farthest from the river, field drainage is not possible and such areas can usefully be managed for conservation of wildlife.

§ 27. CREWE ASSOCIATION
712f

The Crewe association consists of stagnogley soils in reddish, stoneless till and Head. It has been mapped extensively in North Yorkshire, Cleveland, Durham, and from Cheshire to Warwickshire. In Wales it occupies nearly 7 km², in valleys and basins adjacent to the Milford association between Monmouth and Usk. The Crewe series (Jones 1983, p.61), pelo-stagnogley soils, occupies two-thirds of the area, mainly on gentle slopes. Fine loamy over clayey, typical stagnogley soils, Salop series (§ 79) are included, together with other soils in reddish till.

In most years Crewe soils are waterlogged to the surface throughout the winter and into the growing season, because permeability is slow. Effective drainage measures can restrict the period of waterlogging to winter (Wetness Class IV) but the slow permeability of the topsoil and the proximity of the clayey subsoil are not easily remedied. When not at field capacity these clayey soils do not hold much water available to plants. They can be slightly droughty for most arable crops and moderately droughty for grass, even though the overall retained water capacity is large. Winter run-off is rapid from the saturated soils.

The association is mainly under permanent pasture, cereals generally being sown in autumn because the soils are usually too wet to work in spring. Cultivations must be carefully timed to avoid severe damage to the soil structure and these soils are difficult to work. There is great risk of poaching by stock when the topsoil is wet because of the large retained water capacity. The risk is greatest in winter although summer poaching is possible on intensively stocked farms during wet seasons. The association has been fully described by Ragg *et al.* (1984).

§ 28. CROWDY 1 ASSOCIATION
1013a

Amorphous raw peat soils of the Crowdy series (Staines 1976) dominate this association with stagnopodzols belonging to the Hafren series (§ 51) and stagnohumic gley soils of the Wilcocks series (§ 96). The association covers 240 km² on plateaux, gentle slopes and basins in the uplands of mid- and north Wales. Some fibrous peat soils of the Winter Hill series (§ 14) are included as well as thin peaty soils over rock belonging to the Hepste series (Clayden and Hollis 1984). Ironpan stagnopodzols (Hiraethog series, § 51), humic rankers (Skiddaw series, Appendix 2 p.323) and stagnohumic gley soils on rock (Mynydd series, Clayden and Hollis 1984) are also present. The main series are briefly described and there is a key to the soils below.

The association is distributed discontinuously across the main north to south watershed of Wales, generally above 350 m O.D., at the heads of streams and on saddles between the major hills. As well as the wide spreads of blanket peat, there are a few raised bogs (usually

Crowdy series
(Full description p.311)

0–10 cm — Om
Black, semi-fibrous peat; moderate granular structure.

10–20 cm — Oh1
Dark brown, stoneless humified peat; massive structure.

20–100 cm — Oh2
Black, stoneless humified peat; massive structure.

Hafren series
(Full description p.315)

0–10 cm — Om or Ah
Dark reddish brown, stoneless semi-fibrous peat or slightly stony humose sandy silt loam or clay loam.

10–30 cm — Eg
Greyish brown, mottled, slightly stony clay loam or sandy silt loam.

30–45 cm — Bs
Strong brown, slightly or moderately stony sandy silt loam or clay loam.

45–60 cm — BCu
Yellowish brown, moderately or very stony sandy silt loam or clay loam.

At 60 cm — Cr
Silty shale *in situ*.

Wilcocks series
(Full description Hollis 1975, p.75)

0–20 cm — Oh or Ah
Black, stoneless humified peat or humose clay loam.

20–50 cm — Bg
Light brownish grey, mottled, slightly stony clay loam or sandy clay loam; weak subangular blocky structure.

50–100 cm — BCg
Grey with many ochreous mottles, moderately stony clay loam; weak medium blocky or prismatic structure; high packing density.

with the Welsh name *Cors* or *Gors*) such as *Gors Lwyd* on the watershed between the rivers Ystwyth and Elan in west Powys. In these bogs the nutrient status is even lower than in blanket bog. The more continuous mantle of blanket bog is often pierced by bare rock, especially along ridges following the strike of the rocks (Rudeforth 1970).

Key to component soil series

Peat thicker than 40 cm	1
Mineral soils; with dark, humose or peaty topsoils	3
1. Fibrous peat with moss and cotton-grass remains	Winter Hill
Amorphous peat	2
2. Thicker than 80 cm	CROWDY
Over rock within 80 cm	Hepste
3. With bleached subsurface horizon and brightly coloured subsoil; fine loamy or fine silty	4
Without podzolic features	5
4. With thin ironpan	Hiraethog
Without ironpan	HAFREN
5. Topsoil over rock within 30 cm	Skiddaw
Deeper soils; fine loamy or fine silty, mottled throughout	6
6. Deeper than 80 cm	WILCOCKS
Rock within 80 cm	Mynydd

The climate is cold and wet with an average annual rainfall of about 2,000 mm. Most areas are exposed or very exposed, with prevailing winds from the south-west or west maintaining high relative humidity throughout the year, and moderating to some extent the severity of winters. Deep snow is common in the winter.

Climate and soil wetness limit agricultural use to rough grazing, and most of the land is under unenclosed semi-natural vegetation. Cotton-grasses *(Eriophorum* spp.*)* and purple moor-grass *(Molinia caerulea)* dominate the latter especially where water from surrounding rocks and mineral soils has carried in more nutrients. Heather moor is also extensive, particularly where the peat is hagged. The better drained stagnopodzols and rankers usually bear *Nardus* grassland or, where under-grazed, heather moor. Some of the thinner peats carry a wet form of *Nardus* grassland containing purple moor-grass.

Plate 16. Llanymawddwy, Gwynedd. Hagged peats of the Crowdy 2 association on the plateau in the foreground give way to soils of the Manod association in the valley, with the Hafren association on rolling hills in the background.

C.C. Rudeforth

Plate 17. Peat haggs on Plynlimon. Erosion by wind and water causes hagging in upland peats.

The wet regime of the soils, their low nutrient status and the risk of windthrow limit their use for forestry. Lodgepole pine and Sitka spruce are grown but yields are often only moderate or poor. The need for drainage and fertilizers, and for herbicides to suppress *Calluna*, increases the cost of production.

The semi-natural vegetation has a low or moderate grazing value but heather moor provides sheep with a valuable bite of young *Calluna* shoots in the winter when little else is available. Although the main peat soils are not easily improved, piecemeal improvement is possible on the drier Skiddaw, Hiraethog and Hafren soils (§ 51).

Many of the hills are water-gathering grounds for large reservoirs as at Nant-y-moch and Claerwen. The available water capacity of peat soils is about twice that of mineral soils (Rudeforth and Thomasson 1970), and the peat stores large amounts of water which is released slowly into the reservoirs, sustaining flow during dry periods. During long dry spells the peat contracts, so that heavy subsequent rains pass rapidly through cracks to the streams, producing extensive networks of erosion channels and peat haggs in places (Plates 16 and 17). In winter when the peat has swollen and the cracks have closed, excess rainwater flows rapidly from the saturated surface into the water channels.

The semi-natural vegetation communities of the rough grazings are valuable refuges for native plants and animals. In particular heather moorland, including bog heather moor, is now relatively rare internationally, so there is considerable need to conserve the best areas. Recreational use is limited because it is difficult to walk or ride over the soft wet land with its uneven and sometimes thick heathery vegetation, but the land provides a retreat for those seeking solitude in a relatively natural environment.

§ 29. CROWDY 2 ASSOCIATION
1013b

Raw acid peat soils dominate this association which occupies wide upland tracts of blanket bog and scattered peat-filled basins in Wales and South West England. A small area on the Black Mountains extends into Herefordshire. Well humified peats of the Crowdy series (Staines 1976) are most extensive while the chief associate is the more fibrous Winter Hill series (Carroll *et al.* 1979) with visible *Eriophorum-Sphagnum* remains.

The Crowdy and Winter Hill series are briefly described in § 14 and a key to the soils is given below.

Altitudes range from sea-level in west Wales to more than 600 m in the east. The famous lowland raised bogs at Borth and Tregaron (Plate 9) are included. These have floristic and pedological affinities with blanket bogs, the peat having both amorphous and semi-fibrous layers unlike the dominantly fibrous peats of raised bogs elsewhere. The association also includes some grass-sedge peats of the Floriston series in valley bottoms. Shallower soils, either peaty topped or mineral, are rare.

Key to component soil series

	Peat thicker than 40 cm		1
	Mineral soils; with dark, humose or peaty topsoils		3
1.	Amorphous peat		CROWDY
	Fibrous peat		2
2.	Blanket peat with moss and cotton-grass remains		WINTER HILL
	Basin peat with grass and sedge remains		Floriston
3.	With reddish, mottled subsoil; fine loamy or fine silty		Wenallt
	With greyish, mottled subsoil		4
4.	Very stony or rock within 80 cm; coarse loamy		Laployd or Princetown
	Deeper soils; fine loamy		Wilcocks or Freni

The climate of the blanket peat land is cold, wet and exposed. Average annual rainfall is typically around 2,000 mm. Relative humidity remains high throughout the year, with the prevailing moist winds from the south-west or west moderating the severity of the winter. Winter snowfalls are usual and often deep. By contrast the climate of the lowland bogs such as those at Borth and Tregaron is warmer, drier and more oceanic (Bendelow and Hartnup 1980).

The climate and soil wetness restrict land use to poor grazing, mainly by sheep, and most of the land remains unenclosed and carries a semi-natural vegetation. The nutrient status of the soils is very poor. The vegetation is blanket bog dominated by cotton-grasses or purple moor-grass, the latter being more common in basins where the nutrient status is higher. Bog heather moor occurs locally, particularly on the lowland bogs. Valley bogs usually contain more nutrients and this encourages tree growth to form acid carr vegetation with willows and alders.

The soils are more or less permanently wet (Wetness Class VI) and this usually prevents their economic improvement. The watersheds are important gathering grounds for reservoirs as at Lake Vyrnwy. The available water capacity of peat soils is about twice that of mineral soils (Rudeforth and Thomasson 1970) and the peats store large amounts of water which is released slowly into the reservoirs, sustaining flow in dry periods. During protracted dry spells the peat contracts so that heavy subsequent rains pass rapidly through cracks to the streams and rivers and locally form dendritic erosion channels in the peat itself (Plate 16). In winter when the peat has swollen and the cracks have closed, excess rainwater flows rapidly from the saturated surface.

The semi-natural vegetation communities of the rough grazings provide valuable refuges for wildlife. In particular heather moorland is now relatively rare and the ecosystems of the National Nature Reserves on Borth and Tregaron bogs are considered by conservationists to be of international importance for their assemblages of plants and animals, and for the evidence of past environments which is available in the form of pollen and other plant remains in the peat.

Recreational use is limited because it is difficult to walk or ride over the soft wet land with its uneven sometimes thick heathery vegetation (Plate 6), though the areas provide a retreat for those seeking solitude in a relatively natural environment, and the Offa's Dyke path crosses the association on the English border near Hay Bluff. Some moors are traditionally used for grouse shooting, but this is much less common than on similar ground in Northern England and Scotland.

§ 30. CRWBIN ASSOCIATION
313c

The Crwbin association consists mainly of loamy brown rankers of the Crwbin series (Clayden and Evans 1974) associated with typical brown earths over limestone belonging to Malham (Hollis 1975, p.35) and Waltham series (Courtney and Findlay 1978, p.44). It is found on limestone mainly of Carboniferous age in the South West, Wales, Northern England and the Peak District. The fine earth of almost all constituent profiles is decalcified although some calcareous profiles, Marian series (Reeve 1975) remain. Bare rock in the form of limestone pavements and small crags is common and many of the soils are stony.

Crwbin series
(Full description p.312)

0–25 cm — Ah
Dark brown, stoneless or slightly stony clay loam; weak fine subangular blocky structure; non-calcareous.

At 25 cm — R
Fractured limestone.

Malham series
(Full description Hollis 1975, p.35)

0–20 cm — Ap
Dark brown, stoneless silty clay loam.

20–70 cm — Bw
Brown, stoneless or slightly stony silty clay loam; moderate medium angular blocky structure.

At 70 cm — 2R
Limestone

Waltham series
(Full description Courtney and Findlay 1978, p.44)

0–20 cm — Ap
Dark brown, stoneless or slightly stony clay loam.

20–50 cm — Bw
Brown, stoneless or slightly stony clay loam; moderate medium angular blocky structure.

At 50 cm — R
Hard limestone

Plate 18. Limestone cliffs of Eglwyseg mountain. The steep screes are in the Crwbin association. The open moorland above includes Hafren and Anglezarke associations.

The pattern and proportions of the soils differ according to terrain. On level ground the Malham and Waltham series occur in drift-filled depressions in the underlying rock, whereas on sloping land they are commonly found in Head on footslopes and associated with screes (Plate 18). Since there is little weathered residue from the limestone, most profiles have a large drift component. The silty fine earth of Malham soils is of windblown origin, while the fine loamy Waltham series occurs in glacial drift containing sandstone. Where the drift source is Devonian or Triassic rock the soils are reddish (Wrington series, Clayden 1971, p.49). Clayey brown rankers (Torbryan series, Clayden and Hollis 1984) occur as minor inclusions. The Wetton series (Appendix 2, p.325), a humic ranker similar to the Marian series but non-calcareous, is also found. Most of the soils are permeable and well drained (Wetness Class I) so they readily absorb excess winter rain.

Key to component soil series

	Shallow soils, limestone within 30 cm	1
	Deeper soils	4
1.	With dark, humose topsoil; loamy	2
	With brown, distinct topsoil; non-calcareous	3
2.	Calcareous topsoil	Marian
	Non-calcareous topsoil	Wetton
3.	Loamy	CRWBIN
	Clayey	Torbryan
4.	Fine silty	MALHAM
	Loamy	5
5.	With bleached subsurface horizon and thin ironpan	Lonsdale
	Without podzolic features	6
6.	Brownish; fine loamy	WALTHAM
	Reddish; very stony, loamy	Wrington

Land use

Most of the land is used for rough grazing and the bent-fescue grassland gives valuable grazing, with little poaching risk where steep slopes and rock outcrops preclude cultivation or intensive grassland management. Scrub and woodland are fairly extensive, some such as Oxwich Wood on the Gower cliffs of south Wales being nature reserves because of their rich herb flora. Droughtiness is a serious limitation but where soil depth, stoniness and gradient permit there is some improved pasture and arable cultivation, the latter notably on the footslopes of the Gower cliffs where vegetables are grown. The cultivation and management of these deeper soils are summarized in § 58. Crwbin soils are too shallow for commercial afforestation and the limestone substratum produces chlorosis and inhibits growth in most conifers.

§ 31. DENBIGH 1 ASSOCIATION
541j

This association has brown stony well drained soils of moderate depth over Palaeozoic sedimentary rocks. It extends over more than 4,600 km² in the foothills of the Lake District, the Pennines, in South West England and Wales on hills and ridges from sea level to about 300 m O.D.

Fine loamy typical brown earths, Denbigh series, cover much of the land with loamy brown rankers, Powys series, fine loamy or fine silty typical brown podzolic soils, Manod series, fine silty stagnogleyic brown earths, Sannan series, and fine silty typical brown earths, Barton series, as the main subsidiary soils. Brief profile descriptions of three of the main series are given below, and those for Powys and Manod appear in § 60.

Denbigh series
(Full description p.312)

0–25 cm — Ap or Ah
Dark brown, slightly stony clay loam.

25–60 cm — Bw
Brown, slightly or moderately stony clay loam; moderate fine subangular blocky structure.

60–100 cm — BCu
Yellowish brown, very stony clay loam; moderate fine subangular blocky or massive structure or *in situ* slate or mudstone.

Sannan series
(Full description Lea 1975, p.79)

0–20 cm — Ap
Dark brown, slightly stony silty clay loam.

20–50 cm — Bw
Dark yellowish brown, slightly stony silty clay loam; moderate medium subangular blocky structure.

50–100 cm — Bg
Light brownish grey, mottled, slightly stony silty clay loam; weak coarse angular blocky structure; high packing density; passes to more stony BCg horizon with massive structure.

Barton series
(Full description Palmer 1976, p.80)

0–20 cm — Ap
Dark greyish brown, slightly stony silty clay loam.

20–70 cm — Bw
Greyish brown, slightly or moderately stony silty clay loam; moderate medium subangular blocky structure.

At 70 cm — Cr
Olive, thinly bedded siltstone or silty shale.

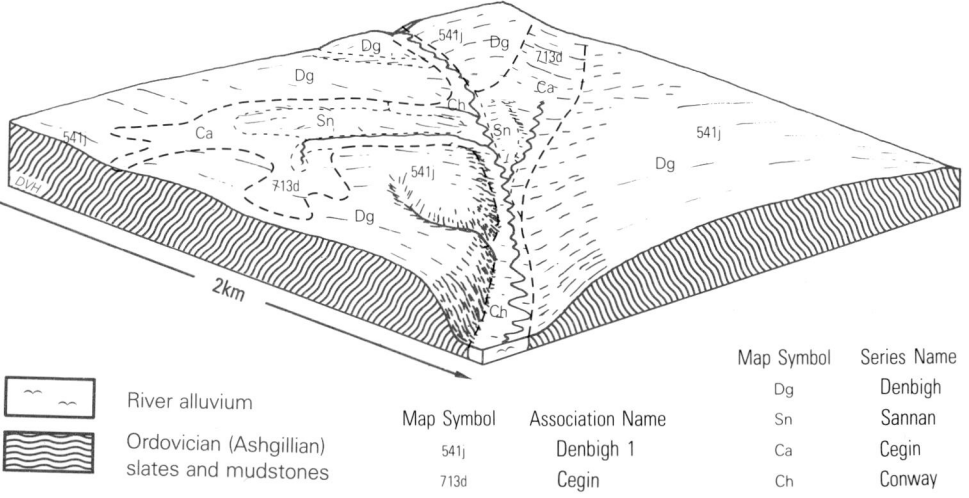

	Map Symbol	Series Name
	Dg	Denbigh
	Sn	Sannan
	Ca	Cegin
	Ch	Conway

River alluvium

Ordovician (Ashgillian) slates and mudstones

Map Symbol	Association Name
541j	Denbigh 1
713d	Cegin

Figure 25. Soil associations on Ordovician rocks near Haverfordwest, Dyfed

Denbigh soils are permeable clay loams on solid or shattered rock within 80 cm depth. A dark topsoil with fine subangular blocky structure overlies dull brown subsoil horizons with medium subangular blocky structure. Manod soils differ in having brightly coloured subsoil horizons with finer, often granular structure. Some Denbigh profiles also have bright subsoils similar to those of the Manod series but have less pyrophosphate extractable iron and aluminium (Thompson 1982) and usually a blockier structure. Powys soils are loamy and also free draining but shallow over massive or broken rock. The stones and rock consist mainly of Palaeozoic siltstone, mudstone, shale or slate with sandstone. While most of the soils are over rock, the silty Sannan series is in deep drift. This series is similar to the Denbigh series except below 40 cm depth where the subsoil has grey or rusty mottles indicating seasonal waterlogging and slow permeability. The East Keswick series (§ 36) in thick drift is also present locally and Cegin (§ 23) and similar soils, differing slightly in texture, are of minor extent.

The Denbigh and Sannan soils are usually on cultivable slopes up to about 7 degrees (Fig. 25). The silty Barton series is most common in the east. Manod soils often mantle steeper land while Powys series is on knolls or ridges often around rock outcrops, or on slopes eroded by cultivation or natural solifluction. Eriviat series (Clayden and Hollis 1984), on compact drift, occupies sites like those of the Powys soils. Rock outcrops and narrow valleys with alluvium and small river terraces occur locally.

Key to component soil series

	Unmottled soils	1
	Mottled soils; fine silty	5
1.	Shallow soils, rock within 30 cm; fine loamy	POWYS
	Deeper soils	2
2.	Rock within 80 cm	3
	Deeper soils; fine loamy	East Keswick
3.	Fine silty	BARTON
	Fine loamy	4
4.	With brown subsoil	DENBIGH
	With brightly coloured subsoil	MANOD
5.	Subsoil faintly mottled above 60 cm or distinctly mottled between 40 and 80 cm	SANNAN
	Prominently mottled or greyish above 40 cm	Cegin

Soil water regime

Most of the soils are permeable and naturally well drained (Wetness Class I). Sannan soils experience occasional waterlogging (Wetness Classes II to III) depending on situation and subsoil permeability. The soils accept most winter rain but temporary water storage capacity is limited by rock or, locally, by compact drift at less than 80 cm depth, which cause some run-off. For most of the land drought restrictions to crop production in average years are slight. Exceptions occur in drier districts where grassland and more rarely cereals suffer moisture stress on Denbigh and similar soils. Crops on the shallow Powys soils are susceptible to drought even in moist areas (Table 18, Plate 1).

Table 18
Profile Available Water (A.P. mm), Crop-adjusted Mean Moisture Deficit (M.D. mm)
and Droughtiness Class for extensive crops—Denbigh 1 Association

	Denbigh series		Powys series	
Location	Denbigh	Haverfordwest	Denbigh	Haverfordwest
Grid Ref.	SJ050660	SM955155	SJ050660	SM955155
Grass				
A.P.	120	120	50	50
M.D.	136	90	136	90
Droughtiness	moderately droughty	slightly droughty	very droughty	moderately droughty
Winter wheat				
A.P.	115	115	50	50
M.D.	95	63	95	63
Droughtiness	slightly droughty	non-droughty	moderately droughty	moderately droughty
Spring barley				
A.P.	115	115	50	50
M.D.	90	60	90	60
Droughtiness	slightly droughty	non-droughty	moderately droughty	moderately droughty

Cultivation and cropping

Long-term grass is traditionally the most common crop (Plate 19) although the soils are more frequently cultivated in the drier districts. The land is firm enough to carry farm vehicles and resist poaching. The growing season is usually seven to nine months (Smith 1976) but the livestock grazing period is often a month or two less to avoid damage to sward and soil structure. Rainfall is the main limitation to arable crops, particularly in the west where August is usually one of the wettest months. Apart from steeper land, the Denbigh soils are suitable for direct drilling. Figure 26 summarizes the effects of climate on land work for two representative districts. Patchy distribution of the wetter soils can reduce the time available for working some fields. Bare soil, even on gentle slopes, is eroded during heavy rain and capping occurs from raindrop impact where organic matter in the plough layer has been depleted by long continued arable use.

Conditions for disposal of slurry on Denbigh and Manod soils are generally favourable since there is little risk of pollution (Lea 1979), and soils carry traffic well, particularly when spreading is done during drier periods.

C.C. Rudeforth

Plate 19. Grassland on the Denbigh 1 association near Painscastle, Powys. The steeper slopes carry Manod association under woodland and permanent grass.

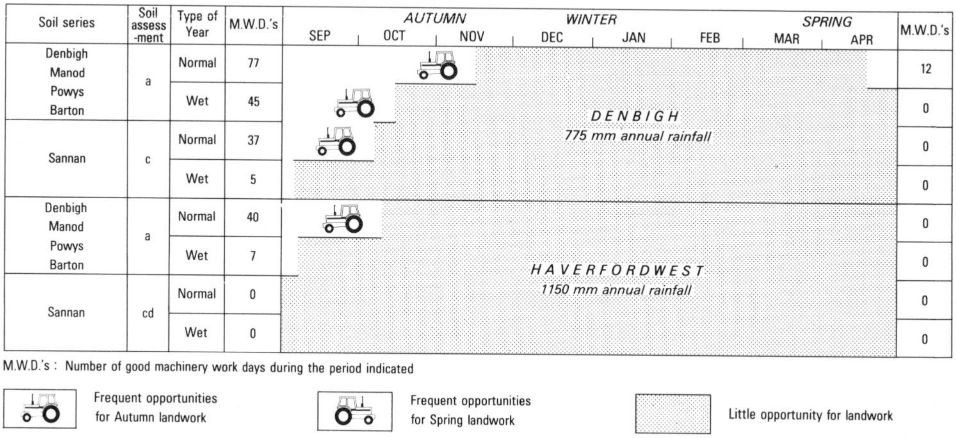

Soil series	Soil assess-ment	Type of Year	M.W.D.'s	AUTUMN		WINTER			SPRING		M.W.D.'s
				SEP OCT NOV		DEC JAN		FEB	MAR APR		
Denbigh Manod Powys Barton	a	Normal	77								12
		Wet	45			*DENBIGH* *775 mm annual rainfall*					0
Sannan	c	Normal	37								0
		Wet	5								0
Denbigh Manod Powys Barton	a	Normal	40								0
		Wet	7			*HAVERFORDWEST* *1150 mm annual rainfall*					0
Sannan	cd	Normal	0								0
		Wet	0								0

M.W.D.'s : Number of good machinery work days during the period indicated

Frequent opportunities for Autumn landwork

Frequent opportunities for Spring landwork

Little opportunity for landwork

Figure 26. The effects of soil and climate on landwork, Denbigh 1 association

Other land use

Forest clearance over the centuries has left a scatter of woods mainly on steep or otherwise inaccessible ground. Such land is valuable for wildlife and game cover and adds scenic interest. Although the land is well suited for trees, large new plantations are expensive to establish, needing stockproof fencing and control of vigorous weeds which can smother young trees. Sitka spruce and Douglas fir yield well and many other soft and hard wood species are also grown for scenic as well as timber value. On Powys soils where rooting is shallow, trees are liable to be blown over.

These mainly well drained loamy soils are well suited to a range of recreational and amenity uses, particularly in summer as camp, caravan and picnic sites (George and Jarvis 1979, Palmer and Jarvis 1979).

§ 32. DENCHWORTH ASSOCIATION
712b

The Denchworth association (Ragg *et al.* 1984) is extensive on Jurassic and Cretaceous clays and consists mainly of wet clayey soils belonging to Denchworth and Lawford series (both pelo-stagnogley soils). There are also calcareous pelosols of the Evesham series (Palmer 1982) fine loamy over clayey typical stagnogley soils of the Wickham series (Appendix 2, p.326) and stagnogleyic argillic brown earths of the Oxpasture series (Palmer 1982).

Denchworth soils are stoneless, clayey, strongly mottled and waterlogged for long periods in winter. Lawford soils, though also clayey throughout, have stones and small inclusions of sand in their topsoils and Wickham soils have loamy upper horizons. The other associates also have clayey subsoils but are waterlogged in winter for short periods only. These include Oxpasture soils with clay loam upper horizons similar to those of Wickham soils, and Evesham soils with stoneless calcareous clayey topsoils. There are also a few Hornton (Reeve 1978, p.86), Fladbury (§ 44) and Haselor soils (Whitfield, n.d.).

The association covers about 56 km² in south Wales on Jurassic rocks with small patches on thin strips of Rhaetic clays. It occurs mainly in a thin discontinuous band between Port Talbot and Barry in the Vale of Glamorgan and the component soils were originally mapped as Charlton Bank and Dyffryn series (Crampton 1972). A small area has also been mapped near Newport.

The landscape is mainly low-lying, gently sloping or level but Evesham and Hornton soils occur on moderately sloping ground. Some Evesham soils are slowly permeable and mottled above 40 cm but they are not extensive. Strips of alluvium along valleys carry Fladbury soils and adjacent thin drift gives Lawford, Wickham and Oxpasture soils. East of Port Talbot drift, of local origin, is an important component of the soil profiles and Lawford soils are common.

The association contains a good proportion of soils correlated with the Haselor series. They are most common on the map separates named Dyffryn by Crampton (1972).

Denchworth and Lawford soils are respectively slowly and moderately permeable in the topsoil and slowly permeable at depth. They are waterlogged for prolonged periods in the growing season (Wetness Class IV or V). Wickham soils have a moderately permeable topsoil but slowly permeable subsoil and are also waterlogged for long periods (Wetness Class IV). Denchworth, Lawford and Wickham soils are responsive to artificial drainage, and water regimes can be improved. Oxpasture and Evesham series have a slowly permeable subsoil that causes waterlogging in winter (Wetness Class III or IV) but after suitable drainage they have a drier regime. When waterlogged, disposal of excess rain is mainly by run-off and the land does not readily absorb excess winter rainfall.

With efficient drainage and careful management the soils yield moderately good crops of grass. Much of the grassland however is permanent pasture and in low-lying areas is difficult to drain effectively, so it is often poor and rush infested. On the better drained associate soils, some arable crops are grown but even with drainage improvements there is little opportunity to work the land in spring, and autumn sown crops are favoured. Because the topsoil takes a long time to dry out, timing of cultivations is critical and measures to reduce ground pressure, such as cage wheels, are desirable to protect against structural damage.

On grassland, surface wetness and weak soil bearing strength limit stocking density and grazing period, although good yields of grass are possible. The soils poach easily and where grazing is ill-timed yields are reduced. Stock is best put on to Oxpasture and Evesham soils early in the season and transferred to the Denchworth, Lawford and Wickham soils later.

The Denchworth soils are acid in the surface where unlimed, but pH increases gradually with depth and the soil is often neutral or alkaline within 1 m depth. The potassium status is usually good but phosphorus is held in forms not readily available to plants.

§ 33. DISTURBED SOILS 3
92c

In parts of the Midlands, Wales and Northern England, opencast mining of shallow coal seams has affected about 500 km² since the 1940s. Early sites were small, though numerous in some localities but recently, with advances in mechanization, sites have become larger (up to 1,000 ha) and some land has been worked again for deeper coal seams.

During mining, the topsoil and subsoil are stripped and stockpiled separately at the perimeter of the site. After mining is completed, recent practice has been to cover the overburden with soil 1.2 m thick. The nature of a restored soil reflects the characteristics of the soils stripped from the site – commonly those of the Brickfield (§ 17) and Wilcocks

(§ 95) associations – and the care and expertise in handling them. Thus a typical restored opencast soil has a loamy topsoil of variable thickness, depending on the amount of topsoil preserved, over a slowly permeable compact clay loam, silty clay loam or clay subsoil. This passes at depth to dark grey rock waste, chiefly of Coal Measures mudstone. Recently restored profiles show little structural development in the subsoil. Stoniness varies; angular sandstones, fragments of coal, and ironstones are common and often there are bricks, slag, cinders and even occasional items of discarded mining equipment.

Restored soils are extensive on the Middle and Lower Coal Measures from Pontypool and Brynmawr to Kidwelly, near Bridgend and there are small areas in north-east Wales near Wrexham. Quality of restoration is very variable, but generally recent techniques have led to improved restoration. At some sites no soil has been replaced and trees or grass grow directly in shale. Deep mines and their spoil heaps have been included within some delineations.

Key to component soil series

		RESTORED OPENCAST COAL WORKINGS
	Soils disturbed to below 40 cm	
		1
	Undisturbed soils; fine loamy	
1.	With dark, humose topsoil	Wilcocks
	With grey, distinct topsoil	Brickfield

Soil water regime

The compact, slowly permeable subsoil on most restored opencast sites severely impedes water movement and leads to lengthy periods of winter waterlogging (Wetness Class IV or V). Comprehensive drainage schemes are usually installed after restoration but subsoil compaction persists and its effects are compounded by structural collapse in the absence of adequate organic activity. The soils are therefore moled and subsoiled regularly to encourage deeper rooting. The compact subsoils also cause restored sites to shed winter rainfall and erode easily, particularly in early stages of restoration or subsequently under cultivation. The compaction also limits the amount of water available for plant growth in dry periods.

Land use

On restoration, the land is left under grass for five years and given regular dressings of fertilizer and, if available, farmyard manure. The grass is used for hay or silage or for carefully regulated grazing. The land is then reseeded and a comprehensive drainage system installed, or an earlier skeleton system completed, before it is returned to the farmer.

Grass is the best crop in the early years of rehabilitation as its rooting characteristics and moisture demand help to improve soil structure while the sward inhibits erosion.

Annual grass production from well-restored soils is often similar to that from adjacent undisturbed soils. Wetness and weak structure easily lead to poaching by livestock, even in summer during rainy periods. Hence low stocking density or light grazing after mowing is essential in the years immediately after restoration. Restrictions to rooting caused by compaction substantially reduce accessible water, resulting in droughtiness in summer. Despite heavy dressings of lime and fertilizer during restoration and rehabilitation, soils require periodic liming to correct acidity, regular applications of nitrogen and adequate inputs of phosphorus. Potassium deficiency is uncommon, but nutrient requirements are generally higher than on adjacent undisturbed land.

§ 34. EARDISTON 1 ASSOCIATION
541c

This association consists of reddish well drained coarse loamy and fine silty soils over hard sandstones interbedded with thin siltstones and silty shales. The sandstones are fine to medium grained, micaceous, and often slightly calcareous where unweathered. The association occurs on the Old Red Sandstone outcrop of the Welsh Borderland, Wales and South West England and on Permo-Triassic sandstones in west Cumbria.

The two main soil series are coarse loamy typical brown earths and differ in their depth to bedrock. The more extensive series, Eardiston, has hard sandstone between 40 and 80 cm depth whereas Bromsgrove soils are deeper. Reddish fine silty Bromyard soils (§ 22) are common on silty shales and siltstones, while in thick drift, often on footslopes, the coarse loamy Oglethorpe series (§ 71) is present. Shallow brown rankers (Newtondale series, Carroll and Bendelow 1981) occur on some ridges, convex brows and broader hilltops. In places thinly bedded sandstones and silty shales have weathered to give fine loamy profiles of the Milford series (§ 63). Most of the soils are slightly stony near the surface but the small and medium sandstones and siltstones become more abundant with depth.

Brief profile descriptions of the main series are given below:

In Wales, the soils occur along the Wye from Monmouth to the Severn over sandstones of the Brownstone and Tintern sandstone groups. Fine loamy Milford and Newbiggin (Appendix 2, p.320), fine silty Bromyard and coarse loamy Bromsgrove and Oglethorpe soils are associated with the Eardiston series. Locally, brown sandstones give rise to Neath and Rivington profiles. Small areas of Quartz Conglomerate bear either podzols of the Goldstone series or reddish sandy Bridgnorth soils on coarse sandstone bands. The steeper slopes of the Wye Gorge carry brown podzolic soils of the Whitcott and Withnell series but no Bromyard soils. On the Old Red Sandstone in north-east Anglesey, Newbiggin, Milford and Fforest soils (§ 43) are minor associates.

Eardiston series
(Full description Palmer 1976, p.40)

0–20 cm — Ap
Dark reddish brown, stoneless or slightly stony sandy silt loam.

20–40 cm — Bw
Reddish brown, slightly stony sandy silt loam; moderate medium angular blocky structure.

40–60 cm — BCu
Reddish brown, slightly or moderately stony sandy loam; weak coarse angular blocky structure.

At 60 cm — R
Dark reddish grey, hard bedded micaceous sandstone.

Bromyard series
(Full description Palmer 1976, p.43)

0–20 cm — Ap
Dark reddish brown, stoneless silty clay loam.

20–35 cm — Eb
Reddish brown, stoneless or slightly stony silty clay loam; weak coarse angular blocky structure.

35–65 cm — Bt
Reddish brown, stoneless or slightly stony silty clay loam; moderate medium prismatic structure.

65–100 cm — Cr
Reddish brown, blotched greenish grey, silt loam; weathered bedded silty shale and siltstone *in situ*. slightly calcareous.

Bromsgrove series
(Full description Hollis and Hodgson 1974, p.38)

0–30 cm — Ap
Dark reddish brown, stoneless sandy loam.

30–65 cm — Bw
Reddish brown, stoneless sandy loam; weak coarse subangular blocky structure.

65–90 cm — BCu
Reddish brown, slightly stony sandy loam; single grain structure.

At 90 cm — Cu
Soft weathered reddish brown sandstone.

Key to component soil series

Unmottled soils		1
Subsoils faintly mottled above 60 cm or distinctly mottled between 40 and 80 cm		4
1.	Shallow soils, sandstone within 30 cm; coarse loamy	Newtondale
	Deeper soils	2
2.	Rock within 80 cm	3
	Deeper soils; coarse loamy	Oglethorpe or Newent
3.	Over hard sandstone; coarse loamy	EARDISTON or Lilleshall
	Over soft sandstone; coarse loamy	BROMSGROVE
	Over silty shale or siltstone; fine silty	BROMYARD
	Over sandstone and shale; fine loamy	Milford
4.	Fine silty	Middleton
	Coarse loamy over clayey	Sellack

On Anglesey the Eardiston association seldom occurs above 80 m O.D. whereas it reaches 300 m in Tintern Forest and in Powys. On the higher land permanent grass predominates with occasional cereal crops although on steeper slopes broad-leaf and coniferous forests are extensive On lower ground, where slope allows, cereal crops are more extensive and there is a small acreage of soft fruit near Monmouth.

Table 19
Profile Available Water (A.P. mm), Crop-adjusted Mean Moisture Deficit (M.D. mm) and Droughtiness Class for extensive crops—Eardiston 1 Association

	Eardiston series	Bromsgrove series
Location	Llanvaches	Llanvaches
Grid Ref.	ST433917	ST433917
Grass		
A.P.	115	120
M.D.	84	84
Droughtiness	slightly droughty	slightly droughty
Winter wheat		
A.P.	110	120
M.D.	61	61
Droughtiness	slightly droughty	non-droughty
Spring barley		
A.P.	110	120
M.D.	58	58
Droughtiness	non-droughty	non-droughty

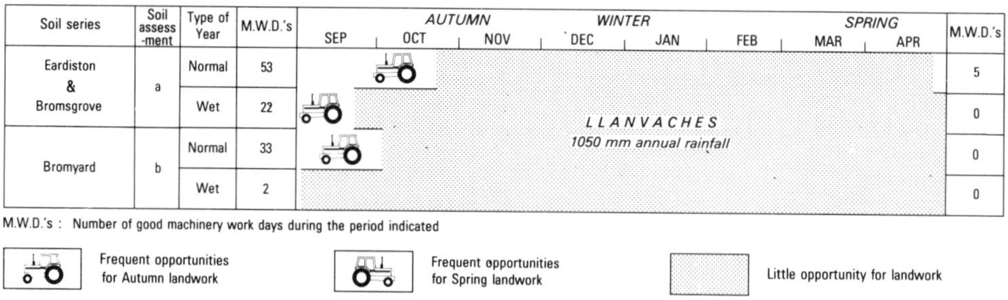

Soil series	Soil assess-ment	Type of Year	M.W.D.'s	SEP	AUTUMN OCT	NOV	WINTER DEC	JAN	FEB	SPRING MAR	APR	M.W.D.'s
Eardiston & Bromsgrove	a	Normal	53									5
		Wet	22									0
Bromyard	b	Normal	33			*LLANVACHES 1050 mm annual rainfall*						0
		Wet	2									0

M.W.D.'s : Number of good machinery work days during the period indicated

Frequent opportunities for Autumn landwork
Frequent opportunities for Spring landwork
Little opportunity for landwork

Figure 27. The effects of soil and climate on landwork, Eardiston 1 association

Profile available water varies according to depth to rock but, in most years only grass will suffer drought on soils deeper than 60 cm (Table 19). Excess rainwater percolates through these soils quickly and there is little risk of damage by poaching except after heavy rain. The grazing season is long with a usable autumn flush. Some days are available for slurry spreading during most months and early dressings of fertilizer are possible without damage to the sward.

The soils are easily cultivated either for reseeding or cereal crops. Autumn cultivations are most suitable (Fig.27), for in wet springs there is unlikely to be sufficient time for safe operations. Usually only one rain-free day is required for the topsoils to dry sufficiently for cultivation, although a slightly longer period is preferable to avoid subsoil compaction. High yields of timber are achieved on these soils which are in the 'Lowland brown earth' class of Pyatt (1977). Weediness is a problem at establishment but, even without fertilizer, Sitka spruce and Douglas fir yield well. The soils are also suitable for a wide range of broad-leaf species.

§ 35. EARDISTON 2 ASSOCIATION
541d

Well drained coarse loamy soils dominate this association which occupies 345 km² on reddish and brownish sandstones in south-east Wales, adjacent English counties and small parts of north Staffordshire. Most of the soils have solid or broken rock within 80 cm depth, but deeper soils occur, especially on concave lower slopes. Typical brown earths over sandstone occupy over half the land. They include Eardiston (Palmer 1976) and Rivington series (Whitfield and Beard 1980), the former is the more extensive and distinguished by its reddish colouring. Reddish fine loamy profiles of the Milford series (§ 63) are found locally. Shallow rankers of the Newtondale (Clayden and Hollis 1984) and Revidge series (§ 74), and brown podzolic soils of the Whitcott series occur especially on higher ground, whilst on footslopes there are Oglethorpe (§ 71) and Wick soils (§ 93). Melbourne series (Clayden and Hollis 1984) is found in a few small flushes and wetter hollows. Abbreviated descriptions of the main soils are given elsewhere (Eardiston § 34; Rivington § 76) and a key to the series is given below.

The association is extensive on undulating land formed by the Brownstone group of the

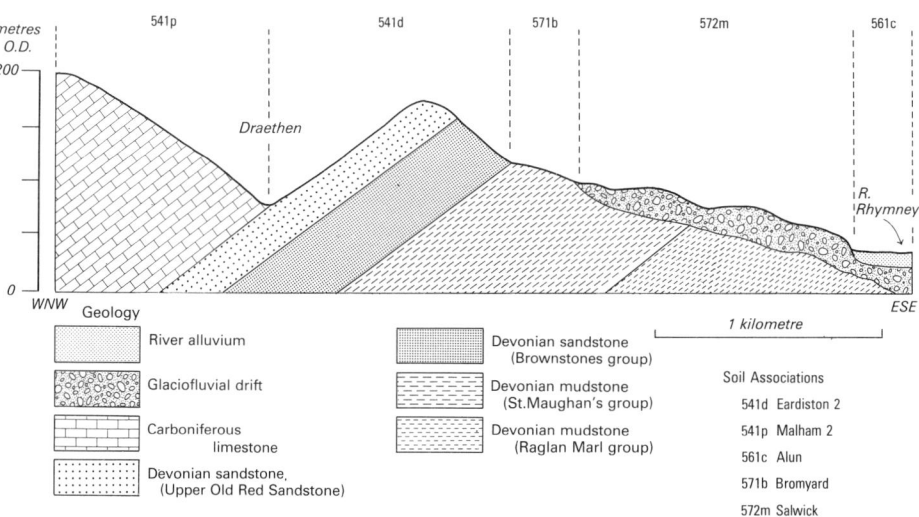

Figure 28. Soil associations on Devonian rocks and Carboniferous limestone near Cardiff

Lower Old Red Sandstone rocks in Gwent. It also forms a band around the eastern end of the south Wales coalfield from Cardiff (Fig.28) almost to Brecon. North of Abergavenny and Crickhowell the rocks rise into the Black Mountains (Plate 20). Here the association often occupies steep valley sides and moderate slopes up to 500 m O.D. (Plate 21). Brownish Rivington profiles rather than Eardiston series dominate the southern slopes of Sugar Loaf near Abergavenny. The summit of this hill has some ferric podzols. There is a larger proportion than usual of brown podzolic soils and rankers on the highest ground and the upper slopes of the deeply incised valleys also carry rankers and bare rock. There are minor inclusions of wet peaty-topped Wenallt soils (see § 91) in the heads of some of the tributary valleys which join the Grwyne Fechan from the north-east. The narrow valley floor of Cwm Coedcerrig, 6 km north of Abergavenny, includes about 10 ha of peat soils. Isolated calcareous profiles 3 km north of Crickhowell are related to springs from the limestones on Pen Cerrig-calch.

Plate 20. The Black Mountains near Crickhowell, Powys. The Grwyne Fechan valley has well drained soils of the Eardiston 2 association, with Oglethorpe association in the foreground. The high ridges carry the Lydcott association and patches of peat belonging to Crowdy 2 association.

R. Hartnup

Plate 21. The Rhian-goll valley at Tretower, Powys. Lugwardine association is on the flat land and the surrounding slopes on Devonian Sandstone have well drained soils of the Eardiston 2 association.

Key to component soil series

	Reddish soils	1
	Brownish soils	3
1.	Fine loamy; sandstone within 80 cm	Milford or Whitcott
	Coarse loamy	2
2.	Sandstone within 80 cm	EARDISTON or Lilleshall
	Deeper soils	Oglethorpe
3.	Shallow soils, rock within 30 cm	6
	Deeper soils; coarse loamy	4
4.	No rock within 80 cm	Wick
	Sandstone within 80 cm	5
5.	Unmottled	RIVINGTON
	Subsoil faintly mottled above 60 cm or distinctly mottled between 40 and 80 cm	Melbourne
6.	With dark, humose or peaty topsoil	Revidge
	With brown, distinct loamy topsoil	Newtondale

The soils are moderately acid and the land is used mostly for grazing sheep. The semi-natural vegetation is mainly bent-fescue grassland, though dry heather moor has developed on the higher parts of Mynydd Llangorse where the sward is undergrazed. Many of the steep valley sides are covered by dense bracken which reduces the value of the otherwise good grazing. Bilberry and bare rock on the steep upper slopes also provide poorer grazing. Gently sloping land below about 360 m O.D. is usually fenced and improved and so bears good permanent pasture. Winter rain is readily absorbed except on the steepest slopes, and there is little risk of poaching. The soils are slightly or moderately droughty depending mainly on their thickness.

There is much forestry around Wentwood between Usk and Chepstow. A wide range of species is grown including deciduous trees. Trees generally grow well, Sitka spruce and Douglas fir being most productive. Forestry is also important in the Black Mountains. The soils do not limit tree growth except where shallow but steep slopes restrict the use of machinery in places.

Recreation is an important secondary use of land in both Wentwood and the Black Mountains. The dry soils are convenient for nature trails and footpaths. Orienteering is currently a popular sport on the hills and there are large pony-trekking centres which use bridle-ways on the flanks of the mountains for most of the year. The Offa's Dyke long-distance path follows the north-eastern ridge of the mountains and crosses the association for some 6 km between Llanfihangel Crucorney and Hatterrall Hill.

§ 36. EAST KESWICK 1 ASSOCIATION
541x

This association comprises deep fine loamy brown earths with some wetter soils in drift. It has been mapped in North Yorkshire and Wales on gently undulating land. The well drained East Keswick series, typical brown earths in drift with siliceous stones, occupies approximately two-thirds of the association. The seasonally waterlogged Nercwys series, fine loamy stagnogleyic brown earths, and Arrow series, coarse loamy gleyic brown earths, occupy most of the remainder. Brief profile descriptions of the main soils and a key to the component series are given below.

The association occurs throughout Anglesey and Lleyn where the soils were previously mapped as Gaerwen and Arfon series (Roberts 1958), Hughes and Roberts 1958). The drift thickness varies considerably over short distances and there are isolated exposures of schist and gneissose granite. The soils are often stony, with hard metamorphic and igneous stones. The association is found in the Dwyfach valley and north of Pwllheli on low, hummocky, stony outwash deposits, often with fragipan. West of Llanbedrog, East Keswick soils occur in drift over igneous rocks which locally appear at the surface. Throughout north-west Wales the association occurs mostly below 50 m O.D. but reaches 200 m O.D. near Llanllyfni, where there are some Meline soils (Clayden and

East Keswick series
(Full description p.313)

0–25 cm — Ap
Dark brown, slightly stony clay loam.

25–70 cm — Bw
Brown, slightly stony clay loam; moderate medium subangular blocky structure.

70–100 cm — BCu
Dark yellowish brown, moderately stony clay loam; moderate coarse angular blocky or massive structure.

Nercwys series
(Full description Rudeforth 1974, p.74)

0–25 cm — Ap
Dark brown, slightly stony clay loam.

25–50 cm — Bw
Dark yellowish brown, slightly stony clay loam; moderate medium subangular blocky structure.

50–100 cm — Bg
Yellowish brown, mottled, slightly stony clay loam; weak coarse angular blocky structure; high packing density.

Arrow series
(Full description Palmer 1982, p.72)

0–25 cm — Ap
Dark brown, stoneless or slightly stony sandy loam.

25–50 cm — Bw
Dark yellowish brown, slightly stony sandy loam; weak medium angular blocky structure.

50–100 cm — Bg
Brown, mottled, slightly stony sandy loam or loamy sand; weak coarse angular blocky structure.

Hollis 1984). Along the English border on the Severn and Vyrnwy at 60 to 80 m O.D. small areas of glaciofluvial drift underlain by gravel are included. In south Wales the association is most extensive on Gower (Plate 7), where there is only a slight predominance of the East Keswick series. Nercwys and occasionally Brickfield soils (§ 18) occur where the drift is more compact and less permeable. Some topsoils are silty, reflecting a probable loessial contribution. Small areas of the association are mapped elsewhere, in drift over limestone near Merthyr Tydfil and in morainic material north of Swansea. Milton soils (Seale 1975) are rarely associated with this map unit in Wales.

Key to component soil series

Unmottled soils	1
Mottled soils	2
1. Fine loamy	EAST KESWICK
Coarse loamy	Wick
2. Prominently mottled or greyish above 40 cm; fine loamy	Brickfield
Subsoil faintly mottled above 60 cm or distinctly mottled between 40 and 80 cm	3
3. Fine loamy	NERCWYS
Coarse loamy	ARROW

Soil water regime

East Keswick soils are well drained (Wetness Class I), whereas seasonal waterlogging is a feature of lower horizons in Nercwys and Arrow soils. Natural drainage in the Nercwys is hampered by the slowly permeable subsoil. The land readily absorbs winter rainwater. Available water is adequate for arable crops in normal years in most places, but in Powys drought restricts grass growth in summer (Table 20).

Table 20
Profile Available Water (A.P. mm), Crop-adjusted Mean Moisture Deficit (M.D. mm)
and Droughtiness Class for extensive crops—East Keswick 1 Association

	East Keswick series	
Location	Newborough	Four Crosses
Grid Ref.	SH420660	SJ265175
Grass		
A.P.	130	130
M.D.	92	147
Droughtiness	slightly droughty	moderately droughty
Winter wheat		
A.P.	130	130
M.D.	69	100
Droughtiness	non-droughty	slightly droughty
Spring barley		
A.P.	130	130
M.D.	66	96
Droughtiness	non-droughty	slightly droughty
Potatoes		
A.P.	105	105
M.D.	44	100
Droughtiness	non-droughty	slightly droughty

Cultivation and cropping

The association is predominantly under medium to long-term grassland with about one-fifth under cereals. There is a small acreage of potatoes for local sale and, on the Gower and Anglesey, some early potatoes and field vegetables. The soils are well suited to grassland; the large retained water capacity of topsoils leads to a slight poaching risk but this can be minimized by control of grazing during wet weather. Potential yields are large in west and south Wales and there is a long autumn flush. In average years, the soils will be fit for slurry spreading for a few days in most winter months as well as throughout summer.

The field capacity period is mostly above 200 days and soil conditions for cultivation are best in autumn (Fig.29). The well developed structure, high organic matter content and free drainage of topsoils make cultivation possible for spells in winter and spring without damaging structure. Locally the hard stones cause machinery to wear rapidly and make harvesting root crops difficult. In normal years crops will not be affected by lack of moisture unless rooting depth is restricted by fragipan.

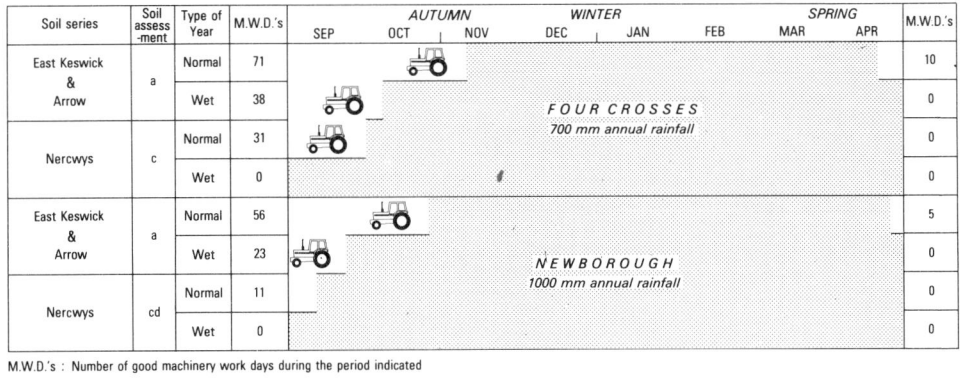

Soil series	Soil assess-ment	Type of Year	M.W.D.'s	AUTUMN			WINTER			SPRING		M.W.D.'s
				SEP	OCT	NOV	DEC	JAN	FEB	MAR	APR	
East Keswick & Arrow	a	Normal	71									10
		Wet	38									0
Nercwys	c	Normal	31				FOUR CROSSES 700 mm annual rainfall					0
		Wet	0									0
East Keswick & Arrow	a	Normal	56									5
		Wet	23				NEWBOROUGH 1000 mm annual rainfall					0
Nercwys	cd	Normal	11									0
		Wet	0									0

M.W.D.'s : Number of good machinery work days during the period indicated

Frequent opportunities for Autumn landwork

Frequent opportunities for Spring landwork

Little opportunity for landwork

Figure 29. The effects of soil and climate on landwork, East Keswick 1 association

§ 37. EAST KESWICK 3 ASSOCIATION
541z

East Keswick soils are typical brown earths in deep, well drained slightly stony fine loamy drift. They are mapped in this association with similar deep brown earths and shallower loamy soils over limestone in Dyfed, Anglesey, Clwyd and Shropshire. The association occurs on flat and gently undulating land where limestone exposures form a small but significant part of the landscape. The subsidiary typical brown earths include the reddish fine loamy Newbiggin series (p.320) in drift with siliceous stones and Barkston series (Clayden and Hollis 1984) in drift with limestones. These fine loamy soils in deep drift cover almost half the land. Most of the remainder has soils with limestone at less than 80 cm depth. Of these, the brownish fine loamy Waltham (§ 30), argillic Wilderhope (p.326) and reddish Wrington series (Clayden 1971) are most common. Coarse loamy soils of the Dinorben (Clayden and Hollis 1984) and Wick series (§ 93) and shallow limestone rankers of the Crwbin (§ 30) and humic Wetton series (§ 92) are minor associates. The association covers some 150 km² in Wales with a further 12 km² near the Welsh border in Shropshire. Brief profile descriptions of the most important series are given below.

In the East Keswick series stoniness increases with depth and the lower horizons can overlie fine loamy or loose coarse loamy drift or in places a compact fragipan. Most soils in this association are near-neutral from included limestone fragments. Wilderhope series occurs over limestone where the higher pH generated by the solution of the limestone acts as a barrier to the downward movement of clay particles suspended in percolating soil water so that a characteristic argillic horizon is formed.

East Keswick series
(Full description p.313)

0–25 cm — Ap
Dark brown, slightly stony clay loam.

25–70 cm — Bw
Brown, slightly stony clay loam; moderate medium subangular blocky structure.

70–100 cm — BCu
Dark yellowish brown, moderately stony clay loam; moderate coarse angular blocky or massive structure.

Wilderhope series
(Full description p.326)

0–25 cm — Ap
Dark brown, slightly stony clay loam or sandy silt loam.

25–40 cm — Eb
Dark yellowish brown, slightly stony clay loam; weak medium subangular blocky structure.

40–70 cm — Bt
Brown, slightly stony clay loam; moderate coarse subangular blocky structure.

At 70 cm — R
Jointed hard limestone.

Crwbin series
(Full description p.312)

0–25 cm — Ah
Dark brown, stoneless or slightly stony clay loam; weak fine subangular blocky structure; non-calcareous.

At 25 cm — R
Fractured limestone.

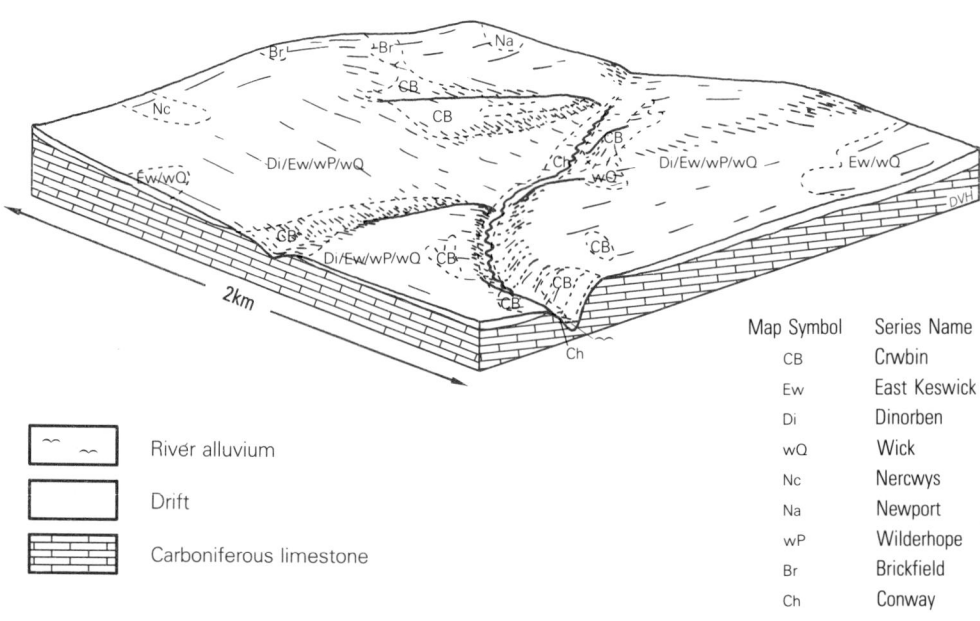

Map Symbol	Series Name
CB	Crwbin
Ew	East Keswick
Di	Dinorben
wQ	Wick
Nc	Nercwys
Na	Newport
wP	Wilderhope
Br	Brickfield
Ch	Conway

River alluvium

Drift

Carboniferous limestone

Figure 30. Soil series in the East Keswick 3 association north of Caerwys

The drift is thickest on Anglesey where soils now mapped with this association were previously mapped as the Pentraeth series (Roberts 1958). Here drift several metres thick covers the limestone. Land north of Brynteg has numerous small limestone outcrops and bedrock is at less than l m depth. Soils mapped here by Roberts (1958) as Gower series and his Rock Dominant areas are also included. Gower series is now classified as Crwbin series (Clayden and Hollis 1984). Southwards the drift thickens and the East Keswick

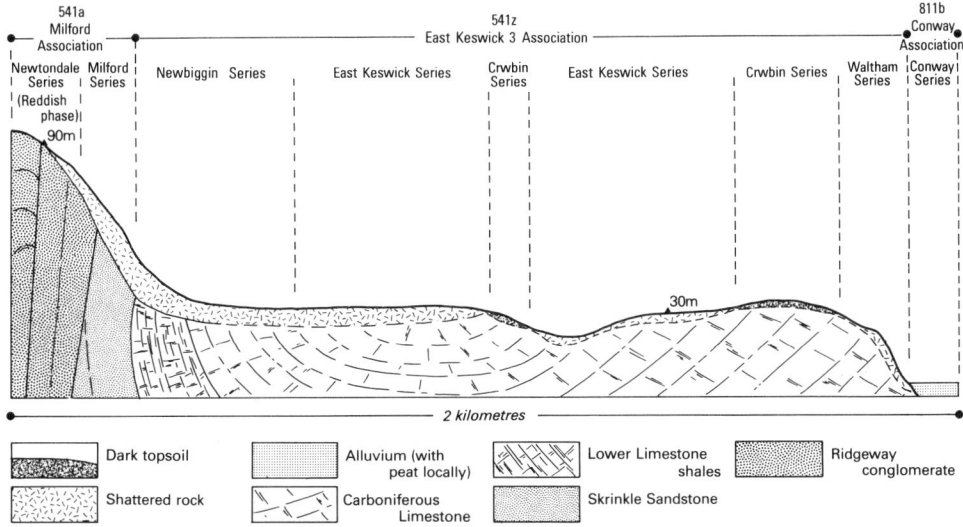

Figure 31. Distribution of soil series in the East Keswick 3 association and relationships with adjoining associations in south-west Dyfed

series predominates. There are small peat-filled basins with base-rich wetland communities of which Cors Goch is the largest. Isolated areas on Lower Limestone Shales formerly mapped as Dyfnan series and described as gley soils are now included in this association and reclassified as stagnogleyic brown earths of the Nercwys series (§ 67).

In Clwyd the main locality is the limestone platform north of the river Wheeler (Fig.30). The proportion of coarse loamy soils is higher here than elsewhere (Thompson 1978). South to Llandegla, the drift is fine loamy and the soils are similar to those on Anglesey but Manod (§ 60) and Brickfield series (§ 18) also occur. Relief is greatest where the limestone outcrop narrows and here Crwbin series and bare limestone pavements, marked by scrub of hawthorn *(Crataegus monogyna)* and blackthorn *(Prunus spinosa)*, cover more of the ground. Rich limestone plant communities survive on the shallow uncultivated soils.

In south-west Dyfed the soils also contain a variety of stone types, giving deep soils in which limestone is the dominant stone (Barkston series) and the reddish Newbiggin and Wrington series which contain material from the adjoining Devonian rocks (Fig.31). Associated argillic brown earths belong to Wilderhope, Ludford (Clayden and Hollis 1984) and the siltier Ston Easton series (§ 84). Cultivating the shallow soils brings limestone up into the topsoil so producing calcareous profiles classed as brown rendzinas (Elmton series, Hartnup 1977). As on Anglesey, Lower Limestone Shales are included and marked in places by stagnogleyic brown earths belonging mainly to Nercwys series (§ 67).

Key to component soil series

	Shallow soils, limestone within 30 cm; loamy	1
	Deeper soils; fine loamy	2
1.	With brown, distinct topsoil	CRWBIN
	With dark, humose topsoil	Wetton
2.	Limestone within 80 cm	3
	Deeper soils	4
3.	Brownish	WILDERHOPE or Waltham
	Reddish	Wrington
4.	Reddish	Newbiggin
	Brownish	5
5.	Stones mostly of limestone	Barkston
	Stones mostly of other kinds	EAST KESWICK or Ludford

Soil water regime

All the main soils are well drained (Wetness Class I) and artificial drainage is not necessary since excess winter rain passes readily through the soils and into widely jointed limestone. Figure 32 illustrates the effect of bedrock on the soil water available to crops

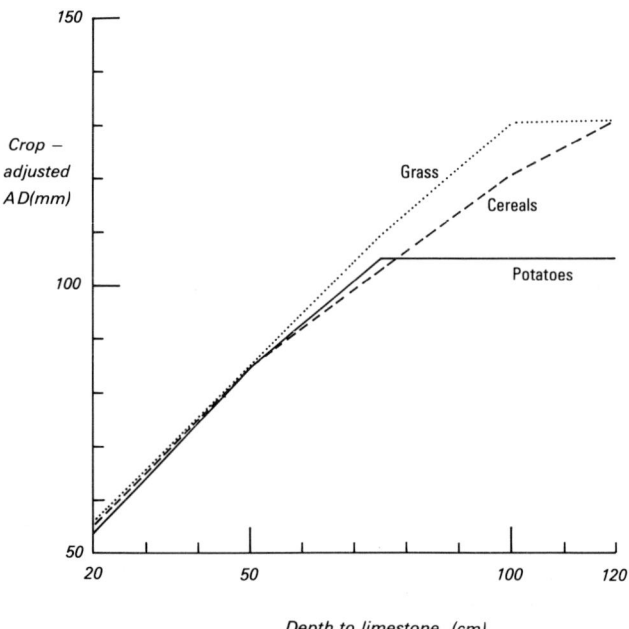

Figure 32. Soil depth and crop-adjusted profile available water in the East Keswick 3 association.

with different rooting depths. Potatoes (rooting depth 70 cm) grown in a soil 75 cm deep are not normally checked by drought while cereals which root to 120 cm in the same soil have only 75 per cent of the water normally available to them from a deeper profile. The crop-adjusted moisture deficit for grass is 100 mm at Caerwys but well below this elsewhere. For barley and potatoes at Caerwys it is 60 mm. Serious shortages of moisture can be expected in most years on the shallow Crwbin soils. Only in dry years will plants exhaust the moisture reserves of the deep soils (East Keswick, Newbiggin and Barkston series).

Cultivation and cropping
Fields with rock outcrops or a large proportion of Crwbin soils are of necessity under long-term grass or scrubland. The upper horizons of the deeper soils, while usually retaining adequate moisture, are easily worked. Rainfall is, however, high in west Wales and spring landwork is often restricted (Fig.33). Direct drilling techniques are suited to these soils and produce comparable yields to those of conventionally sown crops. In

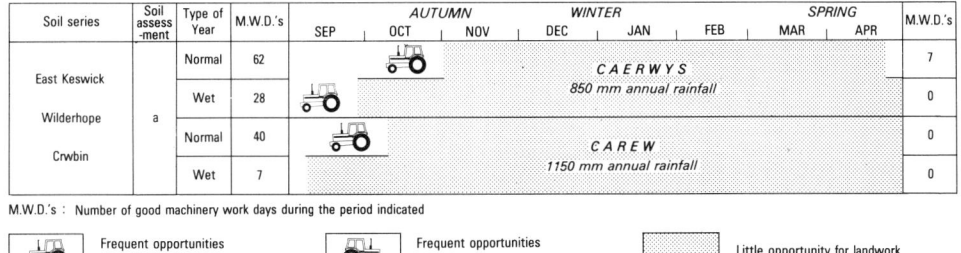

M.W.D.'s : Number of good machinery work days during the period indicated

Figure 33. The effects of soil and climate on landwork, East Keswick 3 association

Clwyd the soils are regularly below field capacity and can usually be cultivated during the spring. The west coast has a lower frost risk than inland and in west Dyfed and Anglesey early potatoes are grown. The February cultivations required for early potatoes damage the soil and reduce yields if the land is too wet, but, near the coast, full advantage is taken of the short rain-free periods when mild temperatures and strong winds assist in drying the soil rapidly. About half of this land in Dyfed is regularly cultivated but much is ley grassland. Barley is the main arable crop while potatoes are concentrated in the most favoured areas. Elsewhere in Wales grassland is more extensive, much of it reseeded as farms convert to silage; barley is the main cereal. In Clwyd where the soils are between 150 and 250 m O.D., the frost risk is greater and the few potatoes grown are maincrop. pH is mostly neutral or slightly acid in topsoils and generally increases with depth. In general there are no nutrient deficiencies or toxicities, although on Halkyn Mountain, Clwyd, the soils are contaminated with lead, zinc, cadmium and copper from mining and smelting, and locally crops are affected.

The soils are almost everywhere ideal for grassland with sufficient available water, acceptable poaching risk, good trafficability and a usable autumn flush of growth. In parts of Clwyd where the soils are shallow, yield is reduced by lack of moisture. There is little surface run-off from level ground during slurry spreading if the soils are uncompacted.

§ 38. ELLERBECK ASSOCIATION
541u

This association of mainly free draining soils is in very stony glaciofluvial or river terrace drift in Northern England, the Midlands and Wales. It contains the loamy gravelly Ellerbeck series, a typical brown earth, and the Baschurch series, a typical brown podzolic soil; also the less stony, coarse loamy Hall series, another typical brown earth, in which gravel occurs at between 40 and 80 cm depth (Appendix 2, p.315). The terrain is undulating or hummocky, on lower valley sides or bottoms, or almost level on outwash fans. The Ellerbeck series has a brown slightly or moderately stony surface horizon, changing abruptly at plough depth to a very stony sandy loam or clay loam, which overlies gravel with little fine earth. The Baschurch series is similar but the subsoil is enriched with translocated iron and aluminium. These porous soils allow roots to penetrate easily.

The association is extensive along the English border north of Shrewsbury in hummocky moraine. It has been described more fully by Jarvis R.A. *et al.* (1984) and by Lea and Thompson (1978). The less stony Wick (§ 93) and Hall series are dominant, with some Ellerbeck and Newport soils (§ 70). The Clifton series (§ 24) is limited to small pockets of reddish till. The gravels are of mixed origin and include rounded, often tabular sandstone, siltstone and greywacke with very hard rounded quartzites and igneous stones in a sandy matrix.

The principal soils are well drained (Wetness Class I) and readily absorb rainwater in winter. They are easily worked, although stones cause wear on implements and create difficulty when harvesting root crops. Cereals and ley grassland are common. Available water is limited and grass suffers drought in dry summers, but the land is valuable for winter grazing and can carry large numbers of stock with little damage to soil structure or the grass sward. The soils readily accept slurry except where the ground is steeply sloping.

§ 39. ENBORNE ASSOCIATION
811a

The Enborne association consists of stoneless moderately permeable, fine loamy and clayey alluvial soils on floodplains. Most are affected by seasonal flooding and fluctuating groundwater but there are permeable better-drained soils on levées. It is found throughout Northern England, the Midlands and to a small extent in Wales and in South

East England, at altitudes ranging from 10 to 120 m O.D. The most widespread soil, the Enborne series, is a fine loamy typical alluvial gley soil in non-calcareous riverine alluvium. The associated Trent series, fine loamy gleyic brown alluvial soils, is often found on levées. There are also clayey pelo-alluvial gley soils of the Fladbury series. The association is confined to valleys draining the South Wales coalfields, principally those of the Tawe, Neath, Ddawen and Ely.

Most Enborne soils are waterlogged for long periods in winter (Wetness Class V) but, where drained may be only seasonally waterlogged (Wetness Class IV). Trent soils are seasonally (Wetness Class III) and Fladbury soils severely waterlogged (Wetness Class V). There are large year-to-year differences in winter waterlogging depending upon rainfall in the respective catchments. Climate, soil wetness and the narrowness of the floodplains make the Enborne association unsuitable for arable cropping and it is mainly under permanent grass and rushy rough grazing, liable to flooding. Provided drainage measures are installed and an outfall maintained, the soils can support good grassland. Soils poach easily however and become inaccessible to stock at times in winter. The association has been fully described by Ragg *et al.* (1984).

§ 40. ESCRICK 1 ASSOCIATION
571p

This association contains deep well drained coarse loamy soils developed in reddish till and Head deposits. It is extensive in Herefordshire and south Shropshire and occupies 15 km^2 in eastern Powys and Gwent on gently undulating land. The till, deposited by Devensian ice in the Wye valley is often slightly calcareous at depth and is composed of local Devonian and Silurian rocks with a few far-travelled erratic stones. Many of the stones are soft siltstones and fine sandstones which readily weather into a reddish micaceous coarse loamy matrix with a large proportion of silt.

Coarse loamy Escrick soils, typical argillic brown earths, formerly described as Wootton series (Hodgson 1972, Palmer 1972), are predominant but fine loamy Newbiggin soils (p.320) are locally extensive. Stagnogleyic argillic brown earths, Nupend series (Palmer 1976), with their faintly mottled slowly permeable subsoils are common on level or gently sloping ground often associated with typical stagnogley soils, Vernolds series (p.325), on lower slopes and in depressions in the landscape. The till and Head deposits are generally thick but the underlying Devonian rocks occasionally protrude to give fine silty Bromyard soils (§ 22) developed in Devonian silty shale and siltstone.

A full description of the association, with a key to the main soil series is given by Ragg *et al.* (1984).

The largest occurrence of these soils in Wales is in the Wye valley between Three Cocks and Cusop. Other patches are found in small north-west to south-east tributary valleys of the Wye between Clyro and Llowes, and on the Herefordshire till plain which extends into Wales along the Arrow valley north-east of Michaelchurch-on-Arrow.

While the Escrick series is mainly well drained (Wetness Class I), it can suffer slight waterlogging locally (Wetness Class II). Nupend series with slowly permeable subsoil horizons is seasonally waterlogged (Wetness Class II and III) but with underdrainage can be improved to Wetness Class II. The slowly permeable Vernolds series is waterlogged for long periods during the winter (Wetness Class IV and V) although with underdrainage it is seasonally waterlogged (Wetness Class III). Escrick association readily absorbs winter rainfall except on steep slopes or on compacted arable land where there is some surface run-off when rain is heavy.

These deep valuable soils can be used for a wide range of crops and yield well under good management. Mixed farming is usual with cereals and some root crops.

Cultivations are easy and give a good tilth but heavy rain causes surface slaking because the soils contain large amounts of silt and fine sand. Recently ploughed or sparsely vegetated fields on moderate slopes are susceptible to rill and gully erosion especially where surfaces are slaked. Small retained water capacities of surface horizons and rapid permeability allow cultivations during rain-free periods in the winter and early spring.

About half the land is under ley or permanent grassland which yields well for much of the year. Good crops of hay or silage can be obtained. The small amounts of water held in the topsoils ensure that bearing strengths on long-term pastures are sufficient to allow grazing throughout most of the year. Only during the wettest winter periods therefore is there any risk of poaching. New leys are more at risk and increased care in the timing of grazings and stocking densities is then necessary.

§ 41. ESCRICK 2 ASSOCIATION
571q

The association consists of coarse and fine loamy brown soils developed in glaciofluvial drift. It has been mapped mainly in Yorkshire but is also found in Wales and inextensively in Lincolnshire and Durham. It occurs mainly on gently undulating outwash plains, on moraines with slopes up to 7 degrees and occasionally on steeper land. The main soils are coarse loamy typical argillic brown earths, Escrick series, occupying approximately half the area. The Bishampton series, fine loamy stagnogleyic argillic brown earths, and the Wick series, typical brown earths, each occupy about one-fifth. The Escrick and Wick series are generally found on the upper slopes of moraines, with the less well drained Bishampton soils on the gentler and lower slopes. Brief profile descriptions of the main series and a key to the component soils are given below.

The association occurs on moraines north of the Caldicot Levels, between Newport and Chepstow. Escrick and Wick soils are dominant but the drift is variable in thickness and scattered knolls occur, associated with shallower soils – mainly brown earths – but with some brown rankers and brown rendzinas where limestone is near the surface. Rougemont (Ragg et al. 1984) and Wighill (Hodge et al. 1984) are rare.

Escrick series
(Full description p.313)

0–20 cm — Ap
Dark brown, slightly stony sandy loam or sandy silt loam.

20–40 cm — Eb
Brown, slightly stony sandy loam or sandy silt loam; moderate medium subangular blocky structure.

40–60 cm — Bt
Reddish brown, slightly stony sandy loam or clay loam; medium angular blocky structure.

60–100 cm — BCu
Reddish brown, slightly stony or moderately stony sandy loam or clay loam; weak coarse angular blocky structure.

Bishampton series
(Full description p.310)

0–20 cm — Ap
Dark brown, stoneless or slightly stony sandy loam or sandy clay loam.

20–50 cm — Eb
Brown, slightly stony sandy clay loam or clay loam; moderate medium subangular blocky structure.

50–70 cm — Bt(g)
Yellowish brown, slightly mottled, slightly stony sandy clay loam or clay loam; moderate medium angular blocky structure.

70–100 cm — BCg
Greyish brown, mottled, moderately stony clay loam; weak medium prismatic structure; high packing density.

Wick series
(Full description Hollis 1978, p.57)

0–30 cm — Ap
Dark brown, slightly stony sandy loam or sandy silt loam.

30–60 cm — Bw
Brown, slightly stony sandy loam or sandy silt loam; moderate medium subangular blocky structure.

60–80 cm — Bw
Yellowish brown, slightly or moderately stony loamy sand or sandy loam; weak medium angular blocky or single grain structure.

80–120 cm — 2BCu
Brownish yellow, slightly or moderately stony sand or loamy sand; weak coarse angular blocky or single grain structure.

Key to component soil series

Unmottled soils; coarse loamy	1
Subsoils faintly mottled above 60 cm or distinctly mottled between 40 and 80 cm	3
1. Limestone gravel within 80 cm	Rougemont
Deeper soils	2
2. Reddish	ESCRICK
Brownish	WICK
3. Fine loamy	BISHAMPTON
Coarse loamy	Arrow or Wighill

Soil water regime
The dominant Escrick and Wick series are well drained (Wetness Class I) and readily absorb winter rainwater. The Bishampton series has a less permeable subsoil and is occasionally waterlogged (Wetness Class II). With a moisture deficit of more than 75 mm, potatoes benefit from irrigation.

Cultivation and cropping
The land is used mainly for dairying with some arable cropping, mainly of cereals, and long-term pasture. There are usually enough days in spring for cultivation although the soils remain drier for longer in autumn after harvesting cereals (Fig. 34).

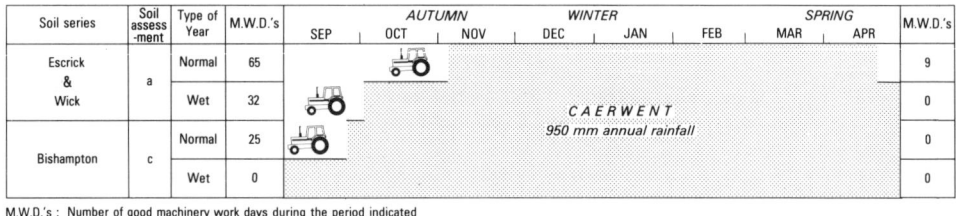

Soil series	Soil assess-ment	Type of Year	M.W.D.'s	AUTUMN			WINTER			SPRING		M.W.D.'s
				SEP	OCT	NOV	DEC	JAN	FEB	MAR	APR	
Escrick & Wick	a	Normal	65									9
		Wet	32				*CAERWENT* *950 mm annual rainfall*					0
Bishampton	c	Normal	25									0
		Wet	0									0

M.W.D.'s : Number of good machinery work days during the period indicated

Frequent opportunities for Autumn landwork   Frequent opportunities for Spring landwork   Little opportunity for landwork

Figure 34. The effects of soil and climate on landwork, Escrick 2 association

§ 42. EVERINGHAM ASSOCIATION
821a

This association is dominated by stoneless fine sandy permeable soils in aeolian sand which overlies clay in many places. Where undrained pale brown and grey subsoil colours indicate that many of the soils are waterlogged almost to the surface in winter. In some districts groundwater levels have been lowered by regional drainage. The association occurs widely in the Vale of York and is also found in west Wales. Altitude varies from sea level in Wales to 30 m O.D. near Selby.

The most common soil is the fine sandy Everingham series, which belongs to the typical sandy gley soils, with the Kexby series, gleyic brown sands, on slightly higher ground. The Holme Moor series of typical gley podzols is rarely found in Wales. Other soils include the Blackwood (§ 70) and Ollerton series (King 1977, p.65), Adventurers' (§ 8), Gilberdyke (Furness and King 1978) and Isleham series (Jarvis, R.A. *et al.* 1984). Brief profile descriptions of the main soils and a key to component series are given below.

Everingham series
(Full description p.313)

0–30 cm — Ap
Dark brown, stoneless loamy fine sand.

30–55 cm — Bg
Pale brown, mottled, stoneless loamy fine sand; very weak fine angular blocky structure.

55–100 cm — Cg
Light yellowish brown, mottled, stoneless loamy fine sand; single grain structure.

Kexby series
(Full description Furness and King 1978, p.72)

0–30 cm — Ap
Dark brown, stoneless loamy fine sand.

30–70 cm — Bw(g)
Brown, slightly mottled, stoneless loamy fine sand; weak subangular blocky or single grain structure.

70–90 cm — Cg
Very pale brown, mottled, stoneless fine sand; weak angular blocky or single grain structure.

Holme Moor series
(Full description Furness and King 1978, p.75)

0–30 cm — Ap/Ea
Very dark brown, stoneless loamy fine sand.

30–40 cm — Bh
Dark reddish brown, stoneless loamy fine sand; massive structure; weakly cemented.

40–100 cm — Bg
Very pale brown, mottled, stoneless fine sand; single grain structure.

The soils cover nearly 30 km², at Malltraeth Marsh in Anglesey and at intervals along Cardigan Bay and on the west side of Carmarthen Bay.

Key to component soil series

	Soils with brown or grey, distinct topsoil	1
	Soils with dark, humose or peaty topsoil	5
1.	Prominently mottled or greyish above 40 cm	2
	Subsoil faintly mottled above 60 cm or distinctly mottled between 40 and 80 cm	4
2.	Medium or coarse sandy; containing stones	BLACKWOOD
	Fine sandy; stoneless	3
3.	With bleached subsurface and dark humus-enriched subsoil pan	HOLME MOOR
	Without podzolic features	EVERINGHAM
4.	Fine sandy; stoneless	KEXBY
	Medium and coarse sandy; containing stones	Ollerton
5.	Peat thicker than 40 cm	Adventurers'
	Dark, humose or peaty topsoil less than 40 cm thick	6
6.	Fine sandy; stoneless	Gilberdyke
	Medium and coarse sandy; containing stones	Isleham

Soil water regime

In Wales, most of the soils are waterlogged almost to the surface for much of the year (Wetness Class V) and winter rain quickly leads to flooding. These permeable soils respond well to drainage by pipes and field ditches, provided that the water levels in the ditches can be lowered.

Cultivation and cropping

These easily worked stoneless soils are well suited to a wide range of arable and horticultural crops in the drier parts of England (Jarvis R.A. *et al.* 1984), but in west Wales much land remains undrained as rough pastures. Locally land drainage has been improved by ditching. There are also Sites of Special Scientific Interest, which contain diverse wetland vegetation and provide winter refuge for wildfowl.

§ 43. FFOREST ASSOCIATION
713c

This association consists predominantly of fine silty cambic stagnogley soils belonging to the Fforest series in reddish drift. It is found on gentle to moderate slopes, commonly footslopes and valley floors in areas of red sedimentary rocks. Fforest subsoils are slowly permeable and often contain a fragipan which, with the gentle relief, slows water movement so that the soils are waterlogged for much of the year (Wetness Classes IV to V) and excess winter rains run rapidly to the water courses. The subsurface is gleyed, pale grey and iron-depleted. This passes to a dense, strongly mottled horizon with prismatic structures which merges with compact, little altered drift within 1 m depth. Similar fine loamy soils of the Hollacombe series (Clayden and Hollis 1984) occur in places. On the wettest sites, usually boggy depressions, there are soils with humose or peaty tops over strongly gleyed slowly permeable subsoil and these are cambic stagnohumic gley soils belonging to Wenallt series (§ 91). Some well drained fine loamy typical brown earths of the Newbiggin series (Appendix 2, p.320) occur on strong slopes and convexities while stagnogleyic brown earths, Llangendeirne series, with a little gley mottling in their subsoils grade in their drainage characteristics towards the Fforest soils. There is more clay and silt in Fforest and Wenallt soils than in the drier Newbiggin and Llangendeirne soils because the rock fragments break down most completely where waterlogging is prolonged. Stoniness increases with depth in all these soils, but most are only slightly stony in the upper 40 cm, medium sized sandstones being most common.

Fforest series
(Full description Wright 1980, p.80)

0–20 cm — Ap
Dark brown, stoneless or slightly stony silty clay loam.

20–35 — Eg
Light brownish grey, mottled, stoneless or slightly stony silty clay loam; moderate medium prismatic structure.

35–60 cm — Bg
Reddish brown, mottled, slightly or moderately stony silty clay loam; moderate coarse prismatic structure.

60–100 cm — BCx(g)
Reddish brown, mottled slightly stony clay loam or silty clay loam; moderate medium platy structure; high packing density.

Wenallt series
(Full description Wright 1981, p.66)

0–20 cm — Oh
Very dark brown, stoneless humified peat.

20–30 cm — Eag
Light brownish grey, slightly stony silty clay loam; moderate coarse prismatic structure.

30–45 cm — Bg1
Grey, mottled, moderately stony silty clay loam or sandy silt loam; moderate coarse prismatic structure.

45–66 cm — Bg2
Reddish brown, mottled, moderately stony silty clay loam or clay loam; moderate coarse prismatic structure.

66–100 cm — BCg
Reddish brown, mottled, moderately stony silty clay loam or clay loam; massive structure.

Llangendeirne series
(Full description p.317)

0–25 cm — Ap
Dark reddish brown, slightly stony clay loam.

25–50 cm — Bw
Reddish brown, slightly stony clay loam; weak subangular blocky structure.

50–90 cm — BCx(g)
Reddish brown, slightly or moderately stony clay loam or sandy loam; massive structure; high packing density.

The association occupies patches of drift mainly over the Old Red Sandstone outcrop in Dyfed and Powys below 300 m O.D. It is most extensive in south-east Dyfed (Wright, 1981) where rainfall of more than 1,500 mm annually and altitudes over 180 m O.D. cause the Wenallt series to be the commonest associate soil. North of Ammanford reddish till has been spread over Carboniferous rocks by ice moving southwards into the Loughor valley and contamination of the till by Carboniferous sandstone has resulted in a larger proportion of the Hollacombe series than elsewhere. In the drier and lower parts of south-west Dyfed Llangendeirne and Newbiggin series are the only associates. The association includes a small area near Colwyn Bay of soils in reddish drift from Carboniferous basement rocks which were mapped by Ball (1960) as the Morfydd series. On Anglesey some soils previously identified as the Frogmoor series (Roberts, 1958) now form part of the association.

Key to component soil series

Soils prominently mottled or greyish above 40 cm	1
Other soils	3

1.	With dark, humose or peaty topsoil; fine loamy or fine silty	WENALLT
	With grey, distinct topsoil	2
2.	Fine silty	FFOREST
	Fine loamy	Hollacombe
3.	Subsoil faintly mottled above 60 cm or distinctly mottled between 40 and 80 cm; fine loamy	LLANGENDEIRNE
	Unmottled; fine loamy	Newbiggin

Land use

Most of the association is under permanent grassland some of which is rush-infested, and undrained soils are very easily poached. Underdrainage using permeable backfill followed by subsoiling or moling every four years facilitates pasture improvement but continued care is needed as there is still some risk of poaching even after drainage. Liver fluke is a hazard to stock on undrained Fforest soils. Arable cultivation is largely restricted to forage crops. Reserves of available water are usually adequate to ensure that Fforest soils are non-droughty in most years, although crops may be checked by drought where topsoils have become compacted and rooting is restricted. Small areas of the association are used for forestry, Sitka spruce being the preferred species. The sites are drained with open ditches and ploughed deeply to encourage root penetration and so reduce the risk of early windthrow.

§ 44. FLADBURY 1 ASSOCIATION
813b

The deep clayey alluvial soils of this association are widespread on flat valley floors of rivers draining catchments of Jurassic rocks in the Midlands and South West England. In Wales they are restricted to 5 km^2 of the floodplain of the Thaw and its tributaries which drain the Lias and Rhaetic shales of South Glamorgan. The association is fully decribed by Ragg et al. (1984).

Fladbury soils, pelo-alluvial gley soils, are clayey throughout and prominently mottled directly below the topsoil. The presence and proportions of the subsidiary Wyre

(Appendix 2, p.327) and Thames (Hazelden n.d.) soils vary along the rivers, reflecting changes in sedimentary regime, catchment lithology and influence of tributaries. Thames soils, pelo-calcareous alluvial gley soils, are similar to Fladbury series but calcareous and are most common where limestones are extensive in the catchment. Wyre soils, pelogleyic brown calcareous alluvial soils, occur on levées, or on other slightly elevated floodplain sites, and on narrow tracts of alluvium upstream. They have a brown subsoil without grey mottles above 40 cm depth.

In the winter a water-table is at shallow depth for long periods in many Thames and Fladbury soils (Wetness Class IV) and locally they suffer prolonged waterlogging (Wetness Class V) and perennial flooding. Thin peaty topsoils occur in some low-lying areas. On raised areas of the floodplain, where the waterlogging is less frequent, Wyre soils (Wetness Class II or III) are found.

The soils are predominantly under permanent grassland or long leys and rushes infest the wettest sites. They have a large retained water capacity and, because of the wet climate of south Wales, they lie wet for long periods and there are few days in an average year when the land can be grazed without some risk of poaching.

§ 45. FLADBURY 3 ASSOCIATION
813d

Clayey alluvial soils on river floodplains dominate this association which occurs mainly in South East and Northern England and locally in south Wales.

The Fladbury series of grey clayey pelo-alluvial gley soils, covers two-thirds or more of the ground and subsidiary silty, loamy or clayey soils the remainder. Thus the fine silty Conway series which belongs to the typical alluvial gley soils (§ 26) is a frequent associate, and the similar but fine loamy Enborne soils (Palmer 1982, p.116) occur in places. In the lower Usk valley north of Newport, Compton soils (Palmer 1982, p.121) in reddish alluvium are locally included and south of Llangorse lake on the Llynfi, clayey over peaty Midelney soils (Jones 1975, p.82) and peaty Adventurers' soils (§ 8) are common associates. Better-drained soils occur on narrow levées along the main river channels. A full account of the association is given by Jarvis, R.A. et al. (1984).

Where undrained the groundwater levels remain high into the growing season (Wetness Class V). In backswamps and other depressions where peat or humose topsoils have formed, soils are permanently waterlogged (Wetness Class VI). Fladbury soils are commonly slowly permeable, even within 40 cm depth, but the associated Conway and Enborne soils are moderately permeable. Winter and summer flooding are common and there is little scope for drainage improvement.

The land is mainly in permanent grass. Large reserves of available soil water supplemented by groundwater ensure growth throughout the season but there is severe risk of poaching which shortens the grazing period. Flooding further curtails winter grazing. Even in summer some of the land cannot be heavily grazed so stock rearing rather than dairying is the only practical use. In the Llangorse district, grass quality is poor on peaty soils and infestation with rushes is common. The natural phosphorus status of Fladbury soils is low but potassium levels tend to be high and magnesium levels moderate. Occasional liming is necessary, but there is a risk of manganese deficiency in grass if the soils are over-limed.

§ 46. FLINT ASSOCIATION
572 1

This association consists mainly of stagnogleyic argillic brown earths in thick reddish drift. The principal soils belong to the fine loamy over clayey Flint and the fine loamy Salwick series, both stagnogleyic argillic brown earths with slowly permeable subsoils which impede downward water movement. The association occurs in north Wales and is widespread in Midland and Northern England. The parent material is till or Head derived largely from Triassic mudstone but contains hard Triassic quartzite pebbles and also stones from the igneous formations of the Lake District, south-western Scotland and Ireland. The development of Flint and Salwick soils in drift which normally carries wetter soils is variously explained by steeper slopes, drier climate, permeable beds of sand and gravel within the drift at depths below 1 m or the presence of permeable bedrock within a few metres depth. As well as the Flint series, which occupies a third of the map unit, and the Salwick series, there are typical stagnogley soils belonging to the Salop (§ 79) and Clifton (§ 24) series.

Brief profile descriptions of the main series are given below.

Flint series
(Full description Jones 1975, p.70)

0–25 cm — Ap
Dark brown slightly stony clay loam.

25–60 cm — Eb(g)
Brown, slightly mottled, slightly or moderately stony clay loam; moderate medium subangular blocky structure.

60–100 cm — Btg
Reddish brown, mottled, slightly stony clay; strong coarse angular blocky or prismatic structure.

Salwick series
(Full description Jones 1975, p.59)

0–20 cm — Ap
Dark brown, slightly stony sandy loam or sandy clay loam.

20–35 cm — Eb(g)
Brown, slightly mottled, slightly stony sandy loam or clay loam; weak subangular blocky structure.

35–70 cm — Bt(g)
Reddish brown, slightly mottled, slightly stony clay loam; weak coarse prismatic structure.

70–100 cm — BCtg
Reddish brown, mottled, slightly stony clay loam; massive.

Salop series
(Full description Jones 1983, p.61)

0–25 cm — Ap
Very dark greyish brown, slightly stony clay loam.

25–45 cm — Eg
Brownish grey, mottled, slightly stony clay loam; moderate medium subangular blocky or prismatic structure.

45–100 cm — Btg
Yellowish red, mottled, slightly stony clay; moderate coarse prismatic structure.

100–120 cm — BCtg
Reddish brown, mottled, slightly stony clay; massive or coarse prismatic structure; sometimes with calcium carbonate concretions.

This association is mapped along the north coast of Wales and on Anglesey where it was previously mapped as Cottam and Flint series (Roberts 1958). On Anglesey, limestone bedrock is often within 2 to 3 m depth and slopes are up to 10 degrees, giving relatively well drained soils that ancillary Salop soils are rare. On the mainland, the association is found on gently sloping land along the coast, and here Salop series occupies small basins and level ground.

Key to component soil series

Mottled soils		1
Unmottled soils		4
1.	Subsoil faintly mottled above 60 cm or distinctly mottled between 40 and 80 cm	2
	Prominently mottled or greyish above 40 cm	3
2.	Fine loamy over clayey; reddish	FLINT
	Fine loamy; reddish	SALWICK
	Coarse loamy; brownish	Arrow
3.	Fine loamy over clayey	SALOP
	Fine loamy	Clifton
4.	Reddish; fine loamy	Newbiggin
	Brownish; coarse loamy	Wick

Soil water regime

Dense slowly permeable subsoils cause some degree of waterlogging in most soils. In wetter climates, drainage measures can alleviate conditions in surface horizons, but seasonal waterlogging remains (Wetness Class III). The soils absorb winter rainwater fairly rapidly but there is some surface run-off. There is generally enough water available for cereals, grass and potatoes so that droughtiness is nowhere severe although grass production is restricted by drought in some drier districts.

Cultivation and cropping

These soils provide good mixed farming land but, on Anglesey where the field capacity period is longer than 225 days, they are less suited to arable cropping. Here wetness restricts spring cultivation and autumn-sown crops are more reliable (Fig.35). Direct drilling is likely to produce comparable yields of autumn-sown crops but spring cereal yields will be lower than those sown conventionally.

Grassland on these soils is affected by drought along the mainland coast from Colwyn Bay to Abergele. On Anglesey this is less likely. Topsoils hold a moderate amount of water and there is an appreciable poaching risk in early and late season even in the drier districts. On level ground, the soils will accept moderate amounts of slurry in summer without risk but on slopes, especially under higher rainfall, surface run-off into streams restricts disposal to drier months. Although the deep subsoil is locally calcareous, the soils benefit from periodic liming and in the past marling has satisfied this need.

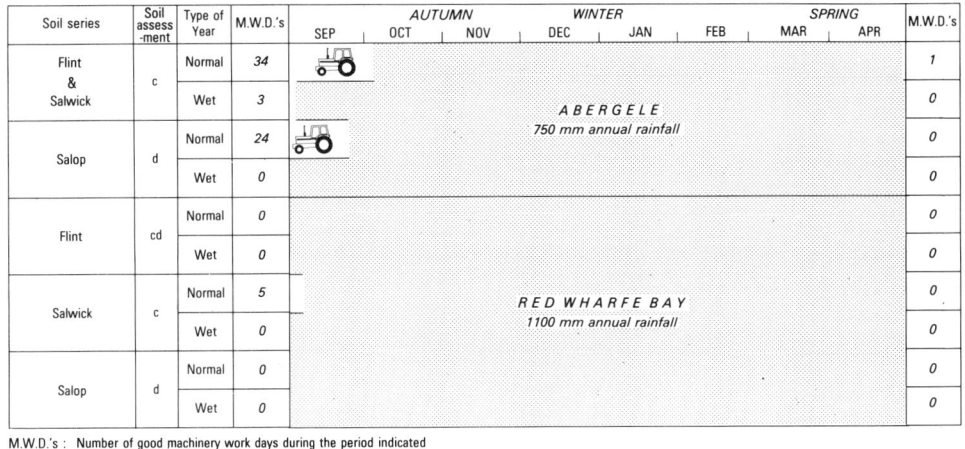

M.W.D.'s : Number of good machinery work days during the period indicated

Figure 35. The effects of soil and climate on landwork, Flint association

§ 47. FOGGATHORPE 1 ASSOCIATION
712h

This association is dominated by seasonally waterlogged clayey, often stoneless soils in till and glaciolacustrine clay. It occurs on flat or gently sloping ground, and is restricted to south-east Northumberland, Tyne and Wear, and Durham with smaller areas in basin sites in south-west Wales and near Clitheroe in Lancashire. The principal soils are the slowly permeable, stoneless Foggathorpe series on glaciolacustrine clay, and the Hallsworth series on clayey till. Slowly permeable fine loamy over clayey soils, Dunkeswick series, also occur but are usually of secondary importance. Brief profile descriptions of the main soils and a key to the component series are given below.

In Wales the association is dominated by poorly drained clayey stoneless soils on glaciolacustrine clay and slightly stony soils on till. It occurs mainly in flat or gently sloping basin sites around Cardigan and Whitland, between 15 and 120 m O.D. Some areas of more accentuated relief are the remains of pingos (Bradley 1980). These were ice-cored mounds formed in permafrost during the last glaciation, which collapsed on melting to leave basins surrounded by arcuate ridges. The soil pattern is locally complex although most of the ground is taken up by the two main series. The Hallsworth series, pelo-stagnogley soils, dominates in slightly undulating country where soils are on slightly stony till. The Foggathorpe series, pelo-stagnogley soils, is more widespread in the flatter areas where lacustrine clay overlies the till. The Dunkeswick series, typical stagnogley

Foggathorpe series
(Full description p.314)

0–20 cm — Ap
Very dark greyish brown, slightly mottled, stoneless clay or clay loam.

20–50 cm — Bg
Dark grey with many ochreous mottles, stoneless clay; moderate coarse prismatic structure.

50–90 cm — BCg
Grey with many brown mottles, stoneless clay; strong coarse prismatic structure.

90–120 cm — BCgk
Brown, mottled, stoneless silty clay; strong coarse prismatic, structure; calcareous laminae.

Hallsworth series
(Full description Bradley 1980, p.44)

0–20 cm — Apg
Dark greyish brown, slightly mottled, slightly stony clay loam or clay.

20–50 cm — Bg
Yellowish brown with many ochreous mottles, slightly stony clay; moderate coarse prismatic structure.

50–100 cm — BCg
Greyish brown, mottled, slightly stony clay; moderate coarse prismatic structure; high packing density.

Dunkeswick series
(Full description Hollis 1975, p.54)

0–20 cm — Ap
Very dark greyish brown, slightly stony clay loam or sandy clay loam.

20–40 cm — Eg
Greyish brown, mottled, slightly stony clay loam or sandy clay loam; weak medium subangular blocky structure.

40–60 cm — Btg
Yellowish brown with many grey mottles, slightly stony clay; moderate coarse prismatic structure.

60–100 cm — BCg
Grey, mottled, slightly or moderately stony clay; weak coarse prismatic or massive structure.

soils, occurs sporadically where fine loamy material more than 30 cm thick overlies the clay. It is most common where the association adjoins the Cegin association (§ 23). Elsewhere, particularly around the pingo ridges near Cardigan, occasional profiles of the fine silty Cegin series and coarse loamy over clayey Portington series (Furness and King 1978, p.48) also occur. Locally sandy strata have been recorded giving Blackwood series (§ 16).

Key to component soil series

Clayey soils	1
Soils with fine loamy or fine silty horizons	2
1. Stoneless	FOGGATHORPE
Containing stones	HALLSWORTH
2. Fine loamy over clayey	DUNKESWICK
Coarse loamy over clayey	Portington
Fine loamy	Brickfield
Fine silty	Cegin

Soil water regime

Nearly all the soils are slowly permeable so they suffer seasonal, sometimes severe, waterlogging. Drainage is essential, especially in Foggathorpe and Hallsworth soils (Wetness Class IV) if they are to be used other than in summer. Although the Dunkeswick series is less frequently waterlogged it is also in Wetness Class IV and land drainage is essential. Because of the slow permeability, excess winter rain runs off rapidly or remains on the gentle sloping or level land surfaces. Profile available water for grass in summer slightly exceeds the potential soil moisture deficit so growth is not often limited by drought.

Cultivation and cropping
The clayey texture and surface wetness of the Foggathorpe and Hallsworth series restrict cropping to grass and, here and there a little barley. Much of the land is in rough grazing, particularly on wetter ground where reseeding is difficult. Here the dense root mat, characteristic of old pastureland, reduces the risk of poaching. Potatoes and root crops can be grown but yield well in exceptional years only. Present farming reflects these limitations. Where there are more Dunkeswick soils the deeper fine loamy topsoil gives greater cropping flexibility. In south-west Wales there are few opportunities to work this land in suitable conditions in either spring or autumn. Cultivations for pasture reseeding are best done during summer, but care is needed to minimize structural damage which will inhibit germination. Although the soils are best suited to grassland they poach easily when wet. The land is not suited to repeated slurry applications as it is rarely dry enough to prevent severe wheel rutting and the rapid run-off would cause pollution of streams. The soils are only moderately fertile as they can be deficient in phosphorus; they are acid unless limed regularly. Unless drained, the slowly permeable clayey soils are poorly-suited to trees which tend to root shallowly and are thus vulnerable to the wind.

§ 48. FOGGATHORPE 2 ASSOCIATION
712 i

This association is dominated by slowly permeable clayey and fine loamy over clayey stoneless soils on glaciolacustrine clay. It is extensive in Northern England, from Northallerton through the Vale of York to the Doncaster district, and is also widespread in the western half of the Vale of Pickering. Elsewhere, it is found in north Nottinghamshire and near Welshpool in Wales. Here, the Foggathorpe series of pelo-stagnogley soils which are clayey and very strongly mottled cover about three-quarters of the association. Other soils in the association in Wales are the slowly permeable strongly mottled fine silty Cegin series (§ 23) on till and the silty Conway series (§ 26) on river alluvium.

Seasonal wetness is the main feature of the soils. Under-drainage is essential in Foggathorpe soils before the land can be brought into arable use. Foggathorpe and Cegin soils are slowly permeable and seasonally waterlogged (Wetness Class III and IV), even with drainage. Cegin soils have more permeable upper horizons than the other two. Excess winter rainwater is not readily absorbed and the soils lie wet in winter and early spring. However, in summer, much water is held in the clay at suctions too high for plants to use so in an average year the association is slightly droughty for most arable crops and moderately droughty for grass. In spring, grass grows well but the land is easily poached.

Slurry spreading is inadvisable in winter and spring, when there is severe risk of wheel rutting and stream pollution from surface run-off. It can be spread in summer however, when the ground is firm and this will also help to reduce the effects of droughtiness. A full description of similar soils is given by Jarvis, R.A. *et al.* (1984).

§ 49. GELLIGAER ASSOCIATION
654c

Coarse loamy ferric stagnopodzols (Gelligaer series) and ironpan stagnopodzols (Belmont series) are predominant above 350 m O.D. on the summit plateaux and ridges of ·the Pennant sandstone uplands in the south Wales coalfield (Plate 11). The soils are extremely acid with a peaty top about 20 cm thick but thinner on steep slopes. They have a grey, strongly leached, often very stony subsurface horizon with evidence of gleying underlain by a thin dark reddish zone of iron, aluminium and humus accumulation. In the Belmont series this takes the form of a thin ironpan no more than a few millimetres thick. Such ironpans are rarely continuous and in sections along forest roads can be seen to

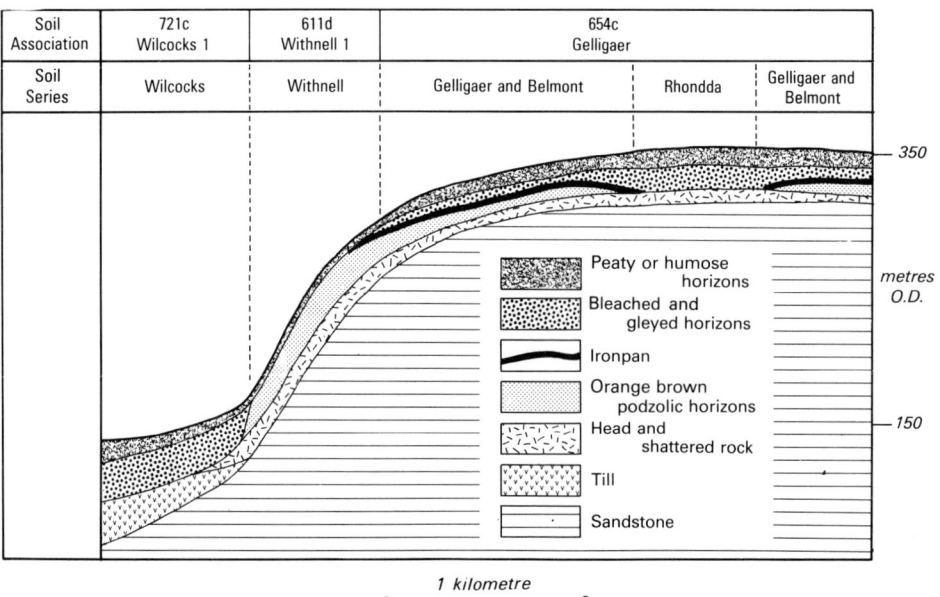

Figure 36. Topographic sequence of soils of the Pennant sandstone uplands.

disappear and reappear within a few metres. Weaker ironpans occur in the Gelligaer series in the form of egg-cup shaped coatings on prismatic peds in the subsurface gleyed horizon. The subsoils of Gelligaer and Belmont series are freely draining and brightly coloured like those of Withnell series (§ 76) on adjacent slopes. This, with archaeological evidence, suggests that stagnopodzols result from superficial modification of brown podzolic soils since Neolithic times (§ 5). Most of the soils overlie sandstone Head or bedrock between 40 cm and 80 cm depth, but some are shallower, in which case an ironpan may be developed in rock debris. The Rhondda series (stagnohumic gley soils) is also common and its relationship to the stagnopodzols is illustrated in Figure 36. These soils resemble stagnopodzols in having a peaty top over a gleyed horizon depleted of iron, but the underlying horizons are also gleyed with ochreous mottles, and they merge into rock debris at 50 to 80 cm depth. Where undrained, the soils are wet for most of the year (Wetness Class V) with rapid winter run-off. Soil wetness results from a combination of high rainfall, low evapotranspiration and gentle relief, and the sponge-like properties of their peaty tops although many of the subsoils are relatively permeable. Thus they differ in their water regime from Wilcocks series (§ 95) on adjacent lower ground, which has slowly permeable subsoils.

Gelligaer series
(Full description Carroll and Bendelow 1981, p.94)

0–10 cm — Oh or Ah
Black, humified peat or humose sandy loam; weak fine subangular blocky structure.

10–25 cm — Eag
Greyish brown, slightly stony sandy loam; weak fine angular blocky structure.

25–65 cm — Bs
Brown, moderately stony sandy loam; weak fine subangular blocky structure.

At 65 cm — R
Hard sandstone.

Belmont series
(Full description Jarvis, R.A. 1977, p.61)

0–15 cm — Oh
Black, humified peat; weak subangular blocky structure.

15–25 cm — Eag
Greyish brown, slightly stony sandy loam; weak medium angular blocky structure.

At 25 cm — Bf
Thin ironpan.

25–50 cm — Bs
Strong brown, moderately stony sandy loam or clay loam; weak medium subangular blocky structure.

At 50 cm — R or Cu
Hard sandstone or yellowish brown, very stony sandy loam.

Rhondda series
(Full description p.322)

0–20 cm — Oh
Black, humified peat.

20–30 cm — Eg
Greyish brown, slightly stony sandy loam; massive or weak coarse prismatic structure.

30–60 cm — Bg
Grey, mottled, moderately stony sandy loam or clay loam; weak coarse prismatic structure.

60–100 cm — Cu
Greyish brown, very stony sandy loam.

The Gelligaer series is most common overall but the proportions of Belmont and Rhondda series vary and either can be locally dominant. Gelligaer and Belmont soils are mainly found on convex, shedding sites while Rhondda soils generally lie on near-level or slightly concave slopes. The soil pattern is frequently complex, however, and soils with patchy ironpan development or gley mottling in the orange subsoil horizon are common. The association also includes patches of the Wilcocks series in impermeable drift and there is some peat (Crowdy series, § 28) usually in boggy depressions. On some narrow ridge tops there are typical podzols (Anglezarke series, § 12), which are drier than stagnopodzols and have a thin peaty top over a bleached horizon and a black humus pan about 3 cm thick, passing to a brightly coloured podzolic horizon and rock debris.

Key to component soil series

Peat thicker than 40 cm	Crowdy
Mineral soils	1

1. With bleached subsurface horizon, dark, humus-
    enriched or brightly coloured subsoil or ironpan   2
    Prominently mottled or greyish throughout; loamy   4

2. Unmottled bleached subsurface horizon and dark,
    humus-enriched subsoil; coarse loamy   Anglezarke
    With mottled, bleached subsurface horizon; loamy   3

3. With thin ironpan   BELMONT
    Without ironpan   GELLIGAER

4. Sandstone within 80 cm   RHONDDA
    Deeper soils   Wilcocks

The climate is wet, cold and very exposed so most of the land is under poor to moderate value rough grazing or forest. Bilberry heath and heather moor survive in places but mat-grass and purple moor-grass have become widespread after burning and grazing. The proportion of purple moor-grass varies considerably, often being abundant on the wetter Rhondda and Belmont series where the ironpan is well developed but only occasional on drier soils with thin surface peat. There has been little grassland improvement except on the margins of the association where Gelligaer soils are transformed by cultivation to typical or humic brown podzolic soils within a few years. The grassland would respond to the techniques of surface treatment developed at Pwllpeiran Experimental Husbandry Farm in mid-Wales on similar soils over shales (Hafren association, § 51). This involves flail mowing and burning to eliminate the mat of dead vegetation, followed by surface rotavation and reseeding, with large applications of lime, slag and nitrogen. There is, however, a great risk of poaching because of the wet climate and the water-holding properties of the peat.

To prepare sites for forestry, the soils are ploughed deeply to break any ironpan and mix the peat and upper mineral horizons. This improves aeration and encourages deep rooting while improving run-off along the furrows into deep drains. Sitka spruce is commonly planted as it withstands exposure and yields moderately well, although, on ground covered by heather, early growth is commonly checked and herbicide is needed to permit crop establishment. Alternatively Lodgepole pine can be planted because it tolerates competition from ericaceous species much better.

§ 50. GOLDSTONE ASSOCIATION
631e

This association of very acid very stony sandy and coarse loamy soils is developed on reddish Permo-Triassic and Devonian conglomerates and pebbly sandstones. It is most extensive in the Midlands and South West England but it is also mapped in Wales. The hardness of the underlying rocks makes them resistant to erosion and they form distinctive ridges or flat-topped hills. The association is fully described by Ragg *et al.* (1984).

The dominant soils belong to the Goldstone series, humo-ferric podzols, with extremely porous gravelly profiles over hard conglomerate. Sandstone beds occur within 80 cm of the surface locally and give reddish coarse loamy soils of the Eardiston (§ 34) and reddish fine loamy soils of the Milford series (§ 63). Sandy Delamere (Hollis 1978) and Redlodge soils (Corbett and Tatler 1970) have also been included.

All the soils are well drained (Wetness Class I), extremely porous and readily absorb winter rainfall, even on steep slopes. There is less water available to crops in the very stony Goldstone series than in the less stony Eardiston and Milford series.

The soils are found mostly on the Gower peninsula capping Cefn Bryn, Rhossili Down, Llanmadoc Hill and several other small hills, formed in conglomerates of Old Red Sandstone age. Goldstone is the main soil occurring mostly under heath with Eardiston and Milford series under grassland. The association has also been mapped in Tintern Forest where Goldstone, Delamere and Redlodge soils are co-dominant.

On the Gower, the land is used for rough grazing or recreation and part is owned by the National Trust. In the Tintern Forest, the soils are entirely under coniferous plantations.

§ 51. HAFREN ASSOCIATION
654a

This association consists mainly of soils with peaty surface horizons—stagnopodzols and stagnohumic gley soils on plateaux and steep valley sides. It covers some 1,300 km² of hill land mainly in Wales, with small areas in South West and Northern England. Hafren soils, loamy ferric stagnopodzols, are most widespread. Similar ironpan stagnopodzols belonging to the Hiraethog series are locally dominant. Wilcocks series, loamy stagnohumic gley soils in drift, and very stony (lithoskeletal) stagnopodzols are also common.

Most of the soils are developed in rock debris from Palaeozoic mudstones, shales, siltstones and occasionally slates. Altitudes range from about 300 m O.D. near the west coast of Wales to more than 900 m O.D. on the Carneddau range in Snowdonia, although further inland at Radnor Forest in east Wales the soils are found only above about 500 m O.D. Ironpans occur frequently and are most common in north Wales and Radnor Forest where the Hiraethog series dominates, whereas in the larger areas of central and south Wales Hafren series (Lea 1975) is usually more common. Together these series cover about half the association. In the mountains of north and mid-Wales however, high ridges and plateaux have similar soils in which organic matter was mixed with or washed into very stony shattered rock debris by frost heave or leaching. These were described as a dark brown subsoil phase of Hiraethog series by Rudeforth (1970). Stone stripes and polygons are found in some of these places (Ball and Goodier 1970). Cliffs, rock fields and scree are also features of this more rugged country. Cliffs like those at Graig Goch have an overriding influence on the use and management of the land despite their limited extent. Around Machynlleth and Corris the soils are developed on slaty rocks metamorphosed by the Cader Idris igneous activity. Shallow humic rankers, Skiddaw (Appendix, p.000) and Revidge series (§ 12), are frequent, especially in north and mid-Wales, where there are narrow rocky ridges. Basins, valleys and flushes have stagnohumic gley soils of the

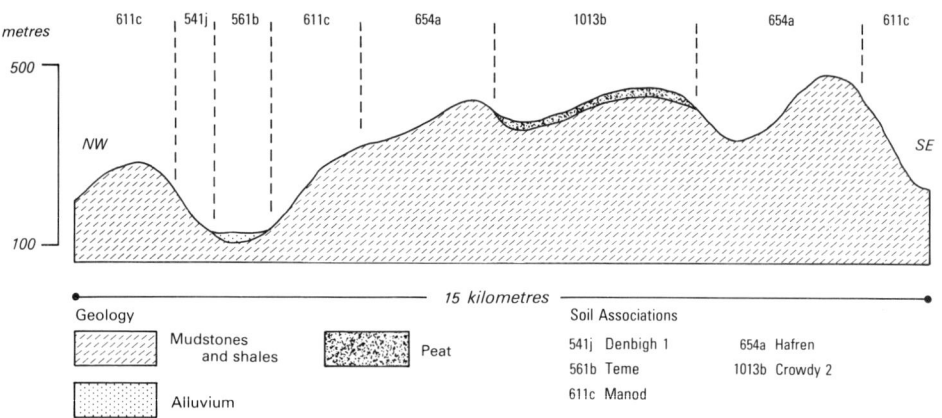

Fig 37. Relationship of soil associations on Palaeozoic mudstones, near Corwen, North Wales

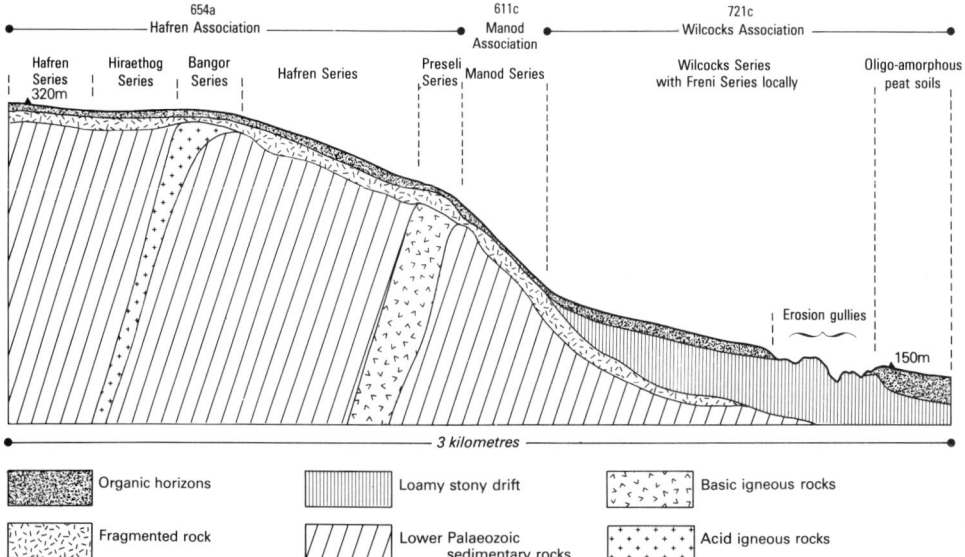

Figure 38. Soil series in the Hafren and Wilcocks associations in the Preseli Hills. Igneous rocks give Preseli and Bangor series locally. Manod association occurs on steep slopes and there are patches of raw oligo-amorphous peat in the larger basins

Wilcocks (§ 95), Mynydd (Clayden and Hollis 1984) and Rhondda series (§ 49), and also peat soils of the Crowdy (§ 28) and Floriston series (Clayden and Hollis 1984). Winter Hill soils (§ 14), fibrous peats, are included on some high plateaux. On valley sides at lower altitudes brown podzolic soils are frequent and include the Parc (§ 72) and Manod series (§ 60) (Fig.37, Plate 23). In the Preseli hills, igneous intrusions give rise to local occurrences of Bangor and Preseli series (Fig.38). A key to these soil series and brief descriptions of the main soils are given below.

Hafren series
(Full description p.315)

0–10 cm — Om or Ah
Dark reddish brown, stoneless semi-fibrous peat or slightly stony humose sandy silt loam or clay loam.

10–30 cm — Eg
Greyish brown, mottled, slightly stony clay loam or sandy silt loam.

30–45 cm — Bs
Strong brown, slightly or moderately stony sandy silt loam or clay loam.

45–60 cm — BCu
Yellowish brown, moderately or very stony sandy silt loam or clay loam.

At 60 cm — Cr
Silty shale *in situ.*

Hiraethog series
(Full description p.316)

0–20 cm — Oh
Black, stoneless humified peat; weak medium subangular blocky structure.

20–35 cm — Eg
Light brownish grey, slightly mottled, slightly stony clay loam; common organic coats.

At 35 cm — Bf
Thin continuous ironpan black above with root mat and red below.

36–60 cm — Bs
Strong brown, slightly stony clay loam or sandy silt loam; weak medium angular blocky structure.

60–80 cm — Cu
Dark greyish brown, very stony sandy silt loam or clay loam; massive structure.

Wilcocks series
(Full description Hollis 1975, p.75)

0–20 cm — Oh or Ah
Black, stoneless humified peat or humose clay loam.

20–50 cm — Bg
Light brownish grey, mottled, slightly stony clay loam or sandy clay loam; weak subangular blocky structure.

50–100 cm — BCg
Grey with many ochreous mottles, moderately stony clay loam; weak medium blocky or prismatic structure; high packing density.

The climate is cold and wet. Low summer temperature is the main influence on the development of stagnopodzols as distinct from brown podzolic soils on similar slopes and under similar rainfall. The accumulated temperatures range from about 800 to 1,500 day degrees centigrade and moisture deficits are generally less than 80 mm. Western districts have cooler summers and milder winters than in the east where the seasonal range of temperatures is greater.

Key to component soil series

Mineral soils	1
Peat thicker than 40 cm	7

1. Brown, distinct topsoil or thin, surface leaf mould; fine loamy — Manod
   Dark, humose or peaty topsoil — 2

2. With bleached subsurface horizon; fine loamy or
    fine silty   3
    Lacking bleached subsurface horizon and thin
    ironpan or brightly coloured subsoil   4

3. With thin ironpan   HIRAETHOG
    With brightly coloured subsoil; no ironpan   HAFREN

4. Rock within 30 cm; peaty   Skiddaw
    Deeper soils   5

5. Unmottled, with brightly coloured subsoil; fine
    loamy   Parc
    Prominently mottled or greyish above 40 cm; fine
    loamy or fine silty   6

6. Deeper than 80 cm   WILCOCKS
    Sandstone within 80 cm   Rhondda
    Siltstone, mudstone or slate within 80 cm   Mynydd

7. Fibrous peat containing remains of moss and
    cotton-grass   Winter Hill
    Amorphous peat   Crowdy

Soil water regime

The water regime of these soils is complicated by the presence of the contrasting horizons. Water is held in the surface horizons of Hafren and Hiraethog soils, the peat acting as a sponge so that they are seasonally waterlogged even though the subsoils drain freely (Wetness Class III or IV). Where strongly formed, the ironpan also impedes water movement. Wilcocks series, in slowly permeable thick drift on the lower ground, is naturally wetter (Wetness Class V). Rainwater passes rapidly into streams and rivers when the upper horizons are already saturated, particularly in winter.

Land use

Sheep grazing and forestry are the main uses of this land because of the wet peaty topsoils, steep slopes and climate. Much of the land is unenclosed, under semi-natural vegetation of moderate grazing value, mainly *Nardus*, with less valuable heather moor where grazing is light. A proportion of heather *(Calluna vulgaris)* is useful, however, as it is the only source of winter grazing. Wetter parts of the association on peats and stagnohumic gley soils provide poor grazing of heather moor, blanket bog communities, or *Molinia*, while narrow flushes and valley bottoms are dominated by bog-mosses and rushes. With adequate machinery, improvement is possible in many places, especially on the stagnopodzols where slope and soil pattern allow. Techniques developed on these soils at Pwllpeiran Experimental Husbandry Farm involve liming, flail mowing, shallow rotavation and seeding with cultivated grasses and wild white clover. This is followed by controlled grazing by sheep and cattle to curb regrowth of native species, and a programme of fertilizer application to maintain the new sward. Much of the land has been improved using these or similar techniques, but some is too far from roads and farm buildings for such methods to be feasible at present. Some of the lower parts have bent-fescue grassland with better grazing value than the more common *Nardus* grasslands.

Some 16 per cent of the association is under coniferous woodland, mostly Forestry Commission plantations. Before planting, wide deep furrows are drawn using specially adapted ploughs, then seedlings are planted on the upturned ridge of soil, thus keeping the young roots adequately aerated. Sitka spruce is best suited to the climate and soil, and yields well (§ 107) but Lodgepole pine competes better with heath vegetation and is grown above 600 m O.D. Some larch is grown on lower ground. In exposed places the problem of windthrow is exacerbated where there are waterlogged or shallow soils which prevent deep rooting. Crags and screes remain unplanted. Roads built for forestry also allow farmers access to remote land.

The semi-natural vegetation provides valuable refuges for native plants and animals. As agriculture advances, heather moors are becoming increasingly rare on an international scale, and despite low productivity they have a distinctive flora and fauna. Some places are of special interest for their birds and insects, notably part of the Cader Idris National Nature Reserve, and the association includes several designated Sites of Special Scientific Interest like those on the Berwyn hills. Much of this land and the neighbouring Crowdy or Winter Hill associations, is on water-gathering grounds of reservoirs. The soils do not absorb much winter rainwater after the peaty surface layers have become saturated, so flash-floods can occur on lower land soon after heavy rain.

The association provides relatively dry soils for footpaths. Cader Idris is particularly popular with walkers and campers, and local Outward Bound schools use it regularly. Many of the forest walks and nature trails laid out by the Forestry Commission pass through areas of the Hafren association.

§ 52. HALLSWORTH 1 ASSOCIATION
712d

This is an association mainly of pelo-stagnogley soils, belonging to Hallsworth and Tedburn series, which are slowly permeable clayey soils, over shales and shaly Head. Their topsoils are mottled dark grey or greyish brown and overlie strongly gleyed, grey and rusty mottled clay subsoils. The association is mapped in Devon and Cornwall on Carboniferous and Devonian shales and in mid-Wales near Llandrindod Wells on Ordovician shales. It commonly occupies gentle footslopes and basins, though in places it extends onto broad ridge crests. The two main soils are very similar, Tedburn series being distinguished from Hallsworth series only by having shale or shaly Head within 80 cm depth. Included in the association are small areas of Dale (Hollis 1975, p.62), Cegin (§ 23), Greyland (§ 23) and Onecote series (Hollis 1975, p.88). Short descriptions of the two leading soils follow.

Hallsworth series
(Full description Bradley 1980, p.44)

0–20 cm — Apg
Dark greyish brown, slightly mottled, slightly stony clay loam or clay.

20–50 cm — Bg
Yellowish brown with many ochreous mottles, slightly stony clay; moderate coarse prismatic structure.

50–100 cm — BCg
Greyish brown, mottled, slightly stony clay; moderate coarse prismatic structure; high packing density.

Tedburn series
(Full description Harrod 1981, p.69)

0–20 cm — Ahg
Dark grey, slightly mottled, slightly stony clay loam or clay.

20–50 cm — Bg
Grey with many ochreous mottles, slightly stony clay or silty clay; strong coarse prismatic structure.

50–100 cm — BCg or Cr
Grey with many ochreous mottles, very stony clay; weak coarse angular blocky or massive structure or shale *in situ*.

The association occurs in Powys over Ordovician Llanvirn shales and derived shaly Head at altitudes of 200 to 300 m O.D. The dominant Hallsworth soils are associated with fine loamy soils of the Greyland and Brickfield series. Shallower Dale and Tedburn soils are rare. Occasional outcrops of tuffs and tuffaceous shales of the Llandrindod Wells inlier give brown podzolic soils of the Malvern series, as at Llansantfraed-in-Elwell.

The climate is cold and wet with an average annual rainfall of about 1100 mm.

Key to component soil series

Dominantly clayey soils		1
Other soils		3
1.	With dark, humose topsoil	Onecote
	With grey, distinct topsoil	2
2.	Rock within 80 cm	TEDBURN
	Deeper soils	HALLSWORTH or Dale
3.	Fine loamy	Brickfield
	Fine silty	Cegin
	Fine loamy over clayey	Greyland

Soil water regime

Where undrained, these soils are usually waterlogged for long periods in the growing season (Wetness Class V) because their subsoils are slowly permeable. Drainage achieves some improvement (Wetness Class IV). Excess winter rainfall flows laterally over the saturated soil giving rapid run-off. Moisture deficits are small and droughtiness rarely restricts crop growth.

Cultivation and cropping

These soils are ill-suited to regular cultivation even where drained. In most years they are unworkable in spring or autumn because they retain water above their plastic limit at most times. Even summer cultivation for reseeding grassland can be delayed by wet weather. In drying conditions the soils remain friable and easily worked for only a short period, changing within a few days from wet and plastic to dry and very hard. Grassland yields well under careful management as growth is little restricted by droughtiness. However, wetness makes them very susceptible to poaching, not only in spring and autumn but also in summer during spells of wet weather. Where grazing is ill-timed pastures degenerate through poaching and compaction. The widespread rushy pastures are often associated with low density, uncontrolled grazing. Liver fluke can also affect stock where the land lies wet.

Some of the land is managed as coniferous woodland. Shallow cultivation and a skeleton drainage system is recommended prior to planting by Pyatt (1977), and application of phosphorus fertilizer prevents the usual check in growth after establishment. On exposed sites, mature trees commonly suffer windthrow as they root shallowly over the dense, coarsely structured and usually wet subsoil. Sitka spruce is widely planted as it is the species most able to thrive on these difficult soils. Elsewhere, unmanaged deciduous woodland, scrub, moor and heath have little economic value but provide game cover and refuges for wildlife.

§ 53. HEXWORTHY ASSOCIATION
651b

This association consists mostly of podzolic soils and extends over the granite outcrop of South West England from 510 m O.D. on Dartmoor to sea level on the Isles of Scilly. Smaller areas are found over granite on the Cheviot Hills and on microgranite and rhyolite in North Wales. In Devon and Cornwall most of the soils are coarse loamy and developed in gritty granitic Head though on the Isles of Scilly some of the component soils are in silty drift overlying granitic Head. In Wales bedrock is usually closer to the surface and much of the land is rocky and bouldery.

Over half the soils are ironpan stagnopodzols, Hexworthy series, (Harrod *et al.* 1976) and ferric stagnopodzols, Rough Tor series. Cambic stagnohumic gley soils of the Princetown series and humus-ironpan stagnopodzols, Trink series, are the main associates though there is considerable variation from place to place in the proportions of component series. Locally Laployd series (§ 55) Moor Gate series (§ 64), Bangor series (§ 14), Cucurrian series (Clayden and Hollis 1984) and Crowdy series (§ 28) occur. The Hexworthy series is described in § 64 and fully described in the Appendix, p.316. Short descriptions of the main associated soils are given below.

Trink series
(Full description p.324)

0–10 cm — Ah
Black, slightly stony humose
sandy silt loam or sandy loam.

10–25 cm — Ea
Greyish brown, moderately stony
sandy silt loam or sandy loam;
weak fine subangular blocky
structure.

25–35 cm — Bh
Black or dark reddish brown,
slightly stony sandy silt loam or
sandy loam; moderate fine
granular structure.

At 35 cm — Bf
Thin, hard undulating ironpan.

35–50 cm — Bs
Strong brown or reddish brown,
moderately stony sandy silt loam
or sandy loam; moderate fine
subangular blocky structure.

50–100 cm — BCx
Yellowish brown, extremely stony
sandy loam; moderate medium
platy structure; silt skins on
stones.

Rough Tor series
(Full description p.322)

0–20 cm — Oh
Black, stoneless humified peat.

20–30 cm — Ah/Ea
Black, slightly stony humose
coarse sandy silt loam or sandy
loam with bleached sand grains;
weak medium subangular blocky
structure.

30–35 cm — Eag/Bh
Dark grey, moderately stony
coarse sandy silt loam or sandy
loam; humus staining; weak
medium angular blocky structure.

35–45 cm — Bs(g)
Brown, slightly mottled, moderately stony coarse sandy silt loam
or sandy loam; moderate fine
angular blocky structure.

45–60 cm — Bs
Strong brown, moderately stony
coarse sandy silt loam or sandy
loam; moderate fine angular
blocky structure.

60–100 cm — BCu or BCx
Brown, very stony coarse sandy
loam or loamy sand; weak fine
angular blocky or strong medium
platy structure.

Princetown series
(Full description p.321)

0–20 cm — Oh
Black, humified peat.

20–35 cm — Ah
Very dark grey, moderately stony,
humose coarse sandy loam; weak
fine subangular blocky or massive
structure.

35–60 cm — Bg
Greyish brown, mottled, moderately stony coarse sandy loam;
weak coarse angular blocky or
prismatic structure.

60–100 cm — BCg or BCx
Brownish grey, mottled, very
stony coarse sandy loam or loamy
sand; single grain structure or
compact with platy structure.

There is much lateral variation over short distances, for instance the depth of the ironpan in Hexworthy soils fluctuates and they grade into the Trink series where organic matter has accumulated above the pan. In places organic matter masks the characteristics of eluvial horizons.

In North Wales, these soils occur over rhyolite on Garnedd-goch, Moel Tryfan, and the north slopes of Snowdon, on microgranite at Blaenau Ffestiniog and on the mixed acid and basic pyroclastic rocks of Arenig Fawr. The land is rocky and boulder-strewn and the soils are in stony sometimes skeletal material with bedrock normally within profile depth. Rock outcrops and shallow humic rankers are found on ridges while freely drained Moor Gate series cover steep slopes and screes. Humic gley soils of the Laployd and Fordham series occur on relatively flat ground with some deeper peat soils of the Crowdy series in basin and flush sites. Rough Tor series is widespread, most extensively on Garnedd-goch and near Blaenau Ffestiniog.

Key to component soil series

Soils with thin ironpan; coarse loamy		1
Soils without ironpan; coarse loamy		2

1.	With dark, humus-enriched subsoil	TRINK
	Without humus-enriched subsoil	HEXWORTHY

2.	With brightly coloured subsoil	3
	Prominently mottled or greyish above 40 cm;	
	without brightly coloured subsoil	4

3.	With bleached subsurface horizon	ROUGH TOR or
		Cucurrian
	Without bleached subsurface horizon	Moor Gate

4.	On hill crests and gentle slopes	PRINCETOWN
	In basins and flushes	Laployd

Land use

The land is open heather moorland with mire vegetation on wetter soils. The grazing is poor because of the low fertility of the acid soils, and the rocky mountainous terrain makes improvements uneconomic. None is afforested; deep ploughing to break the ironpan is largely precluded by rockiness and the risk of windthrow is considerable because the land is exposed. Lodgepole pine is the most suitable species but Sitka spruce is a better choice for the more sheltered, less rocky sites.

§ 54. HOLLINGTON ASSOCIATION
811c

Deep stoneless reddish fine silty and clayey alluvial soils of the Hollington association cover approximately 97 km² in the Welsh Borderland, Wales, South West and Northern regions. They suffer from winter and spring floods which severely restrict land use.

Fine silty Hollington series (typical alluvial gley soils; Palmer 1972) are dominant but drier profiles of the Mathon series (Palmer 1982) are common on the higher parts of the floodplains. Clayey pelo-alluvial gley soils, Compton series, formerly described as Woofferton series by Palmer (1972), often occur in low-lying places away from the rivers and are in places associated with wet silty, humose soils of the Dovey series (Clayden and Hollis 1984).

In Wales the association is limited to 4 km². It floors a small valley draining the Devonian outcrop into Milford Haven and in eastern Dyfed it occurs by the Gwendraeth Fach near Porthyrhyd. Near Trecastle the association is mapped on swampy alluvium and Dovey soils are common.

Hollington and Compton soils, with their pronounced grey and ochreous mottling are variably affected by groundwater (Wetness Class IV or V respectively). The less mottled Mathon soils are affected by groundwater for shorter periods (Wetness Class III).

Drainage improvement is difficult because there is little outfall and water backs up the drains when river levels rise.

The association is almost entirely under permanent grassland because of the risk of flooding and the difficulty of working the wet, fine-textured soils. Hollows and old meander channels, particularly those with Dovey soils, are waterlogged for long periods, often well into the grazing season. Available water is adequate for grass growth in most years and many fields are valued for their good hay crops or for fattening livestock. Reseeding has to be carefully planned as soil conditions are not suitable every year. Reseeded swards poach easily as their bearing strength is much less than that of permanent grassland. Careful timing of grazing and management of stocking densities are therefore necessary to prevent newly sown grass from quickly degenerating into poor quality rush-infested pasture.

A full description of the association, with a key to the recognition of the main soil series present, is given by Ragg et al. (1984).

§ 55. LAPLOYD ASSOCIATION
871a

This association consists predominantly of ground-water gley soils formed in drift derived mostly from acid igneous rocks with some slate. The soils occupy low-lying ground affected by groundwater mainly in South West England on the granite outcrops and very locally on Exmoor. In north Wales they are on rhyolite and microgranite. The main soils are coarse loamy humic gley soils of the Laployd series, characterized by a humose or peaty topsoil overlying mottled sandy silt loam or sandy loam which becomes very stony at shallow depth. The soils are permeable but are waterlogged most of the time unless artificially drained. Wetness is caused by a high groundwater-table or from flushes. Similar but deeper soils of the Fordham series and raw oligo-amorphous peat soils of the Crowdy series (§ 28) are the main associates with some thinner peat soils on rock classed as Hepste series (Clayden and Hollis 1984). Brief profile descriptions of the main series and a key to their identification are given below.

West of Snowdon these soils occupy boulder-strewn land receiving drainage water from higher ground. Here as well as the main soils are found stagnopodzols of the Hexworthy (§ 53) and Hafren series (§ 51) and stagnohumic gley soils of the Wilcocks series (§ 96). In Beddgelert Forest, drainage ditches reveal the local variability of the substrata which contain fragments and boulders of a range of igneous and metamorphic rocks. Fordham and Crowdy series predominate in some localities, with Laployd soils in debris from rocks upslope. East of Beddgelert the association is on the Llyn Dinas Breccia and Lower Rhyolitic Tuffs and where thin peaty topsoils overlie rock debris there are Bangor soils. Crowdy soils occupy flush sites with Laployd and Fordham series on steeper slopes. The steepest slopes have brown podzolic soils with stagnopodzols found on some convexities.

Laployd series
(Full description Clayden 1971, p.164)

0–15 cm — Oh or Ahg
Black or very dark grey, stoneless or slightly stony peat or humose sandy silt loam.

15–50 cm — Bg
Greyish brown, mottled, moderately stony coarse sandy loam or sandy silt loam; angular blocky structure.

50–100 cm — Cg
Greyish brown with many ochreous mottles, very stony coarse sandy loam or sandy silt loam; massive or single grain structure.

Fordham series
(Full description p.314)

0–20 cm — Oh or Ah
Black, humified peat or slightly stony, humose sandy loam.

20–50 cm — Bg
Light brownish grey, mottled, moderately stony sandy loam; weak medium subangular blocky structure.

50–100 cm — Cg
Light brownish grey, mottled, moderately stony sandy loam or loamy sand; weak coarse prismatic or massive structure.

Crowdy series
(Full description p.311)

0–10 cm — Om
Black, semi-fibrous peat, moderate granular structure.

10–20 cm — Oh1
Dark brown, stoneless humified peat; massive structure.

20–100 cm — Oh2
Black, stoneless humified peat; massive structure.

Near Blaenau Ffestiniog on microgranite, Laployd soils predominate, though much of the fine earth is probably derived from nearby slaty metamorphic rocks. The narrow floodplain with its alluvial gley soils at Tan-y-Grisiau is also included.

On Aran Benllyn in the upper catchments of several north-flowing streams Laployd soils are developed in debris from rhyolites of the Aran ridge. Hafren profiles are a component of this unit on convex slopes. The association is also mapped in the Rhinog Hills where Fordham soils are found in drift from Cambrian gritstone and sandstone.

Key to component soil series

Peaty or humose topsoils less than 40 cm thick; coarse loamy mottled subsoil	1
Amorphous peat thicker than 40 cm	2
1. Acid igneous rock within 80 cm	LAPLOYD
Deeper soils	FORDHAM
2. Peat thicker than 80 cm	CROWDY
Peat 40 to 80 cm thick	Hepste

Soil water regime

Undrained soils remain permanently wet (Wetness Class VI) from groundwater, the wetness aggravated by high rainfall in the uplands.

Land use

Undrained soils provide summer grazing of some value in dry seasons when the growth of grass on adjacent dry ground is checked by droughtiness. The soils provide wetland habitats for a wide variety of plants and animals. The semi-natural vegetation is often soft rush pasture or *Molinia* bog sometimes with tussocks of tufted hair-grass *(Deschampsia cespitosa)* or birch or willow scrub.

Some drainage improvement is achieved by open ditches where this is not prevented by boulders and rock. Most of the land is used for rough grazing which is of moderate or low value with large tracts of bog-moss wetland in hollows. The peaty surface is difficult to improve by cultivations and the land is often subsequently invaded by rushes. Improved land is easily poached as the better herbage is preferred by stock, so careful control of grazing is essential.

Forestry

The irregular surface of much of the land prevents efficient drainage and ridging for forestry. Sitka spruce withstands the conditions best but growth is patchy in the wettest parts. Weeds do not normally interfere with tree growth but groundwater restricts the depth of rooting and windthrow is likely.

§ 56. LUGWARDINE ASSOCIATION
561d

Reddish fine silty soils of the Lugwardine association are developed in riverine alluvium. They are extensive, covering 182 km², in the Welsh Borderland counties of Shropshire and Herefordshire and in south-east Wales. The soils are uniformly deep, stoneless, silty and permeable but many are subject to winter and spring floods which restrict their use. Typical brown alluvial soils of the Lugwardine and Walford series predominate with subordinate areas of soils affected by a seasonal groundwater-table represented by Mathon series, gleyic brown alluvial soils, and Hollington series, typical alluvial gley soils. Brief profile descriptions of the main series are given below.

The association is extensive in the valleys draining reddish Devonian rocks and related drifts, notably along the Wye around Monmouth and the Usk between Brecon and Usk. The freely drained fine silty Lugwardine and coarse loamy Walford series are usually found close to the streams, often on levées. Poorly drained Hollington soils occupy lower areas, commonly at the back of floodplains. Many of the valleys are narrow and the full sequence of soils is not always represented. In the Usk valley and its tributary near Tretower the Walford series predominates.

Occasionally small areas of river terraces are included where they are too small or narrow to map separately. They either flank the alluvium or occur as dissected remnants within the floodplain and carry stony coarse loamy typical brown earths (Newnham series, Appendix 2, p.320).

Lugwardine series
(Full description Palmer 1982, p.107–8)

0–20 cm — Ap
Dark reddish grey, stoneless silty clay loam.

20–70 cm — Bw
Reddish brown, stoneless silty clay loam; moderate medium subangular blocky structure.

70–100 cm — BCu
Reddish brown, stoneless silty clay loam; moderate medium prismatic structure.

Mathon series
(Full description Palmer 1982, p.113–4)

0–20 cm — Ah
Reddish brown, stoneless silty clay loam.

20–50 cm — Bw(g)
Reddish brown, slightly mottled, stoneless silty clay loam; moderate medium subangular blocky structure.

50–100 cm — Bg
Reddish brown, mottled, stoneless silty clay loam; moderate coarse prismatic structure.

Hollington series
(Full description p.316)

0–20 cm — Ahg
Dark reddish grey, mottled, stoneless silty clay loam.

20–50 cm — Bg1
Reddish grey with many ochreous mottles, stoneless silty clay loam; moderate medium prismatic structure.

50–100 cm — Bg2
Reddish brown with many ochreous mottles, stoneless silty clay loam; weak coarse prismatic structure with ferri-manganiferous concretions.

Walford series
(Full description p.325)

0–20 cm — Ap
Dark reddish brown, stoneless sandy silt loam.

20–70 cm — Bw
Reddish brown, stoneless sandy silt loam; moderate medium subangular blocky structure.

70–100 cm — BCu
Reddish brown, stoneless sandy silt loam; weak coarse subangular blocky or prismatic structure.

Key to component soil series

Unmottled soils	1
Mottled soils; fine silty	2

1. Fine silty — LUGWARDINE
   Coarse loamy — WALFORD

2. Subsoil faintly mottled above 60 cm or distinctly mottled between 40 and 80 cm — MATHON
   Prominently mottled or greyish above 40 cm — HOLLINGTON

Soil water regime

Short-lived winter and spring floods are common on this association unless there are protective embankments. Hollington soils are flooded longest because they are usually on the lowest ground.

The Lugwardine and Walford series are well drained (Wetness Class I) and usually unaffected by groundwater. Mathon and Hollington soils, although both permeable, suffer seasonal waterlogging from groundwater. The Mathon series is only occasionally waterlogged (Wetness Class II or III) but Hollington soils have more prolonged waterlogging (Wetness Class IV and V). If outfalls are satisfactory Hollington and Mathon soils will benefit from underdrainage and be improved to Wetness Class III and II respectively.

Cultivation and cropping
Lugwardine soils are mostly under permanent grassland because of the risk of flooding. Their deep profiles have large reserves of available water easily exploited by roots, so that grass yields well and the land is much valued for hay and silage. During dry spells growth continues longer than on adjacent higher land. There is little poaching risk on Lugwardine, Walford and Mathon soils. Hollington series, with its water retentive topsoil is often waterlogged well into the grazing season and therefore can usually only be safely stocked during the summer.

Cultivations are relatively easy and, where the flood risk allows, a wide range of crops can be grown but the typically high silt and fine sand contents render exposed soil susceptible to capping during heavy rain. In recent years there has been a distinct increase in arable cultivation, particularly for winter cereals. Where there is an appreciable risk of winter flooding crops are sown in spring to avoid rotting or damage by erosion.

Good yields are obtained from these naturally fertile soils although they tend to be slightly acid and require occasional dressings of lime.

§ 57. LYDCOTT ASSOCIATION
654b

The Lydcott association is mapped on hills above 300 m O.D. over the Old Red Sandstone in south Wales (Wright 1981) (Fig.39) and Devon (Hogan 1981) and over

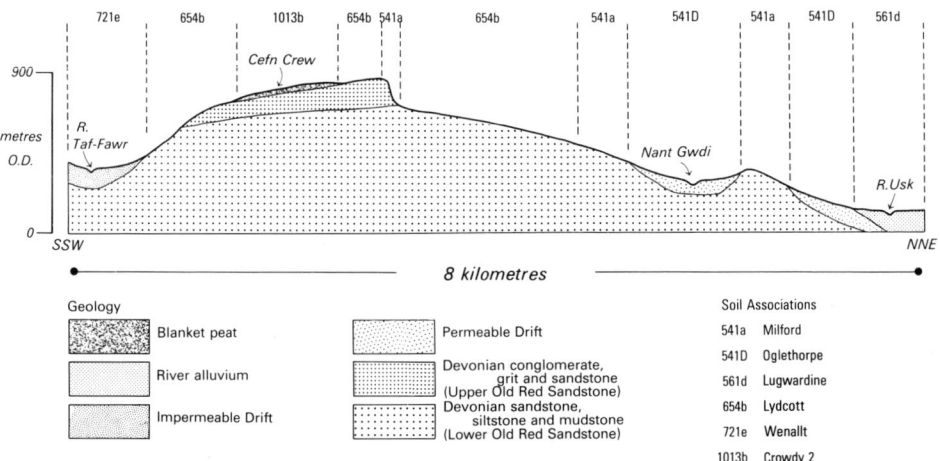

Figure 39. Section across the Brecon Beacons showing relationships between soil associations, relief and geology

Carboniferous rocks in the southern Pennines. Lydcott series, which belongs to the ferric stagnopodzols, predominates, with ancillary ironpan stagnopodzols, Burcombe series, and cambic stagnohumic gley soils, Beacon and Wenallt series.

The soils are extremely acid with peaty tops ranging in thickness from 5 to 40 cm, the thinner peat occurring on the steeper slopes. Below the peaty surface layer is a grey, strongly leached and often very stony horizon with evidence of gleying. At this depth some Lydcott profiles have weak irregular concentrations of iron coating prismatic peds, whereas Burcombe soils have an ironpan several millimetres thick at the base of the leached horizon. Ironpans are rarely continuous and disappear and reappear over distances of a few metres. The subsoils of Lydcott and Burcombe series are generally freely draining and often brightly coloured. They closely resemble the reddish subsoils of Milford and Whitcott series (§ 63) on adjacent steep slopes, suggesting that the stagnopodzols developed by superficial modifications of these soils following the spread of heath since Neolithic times (§ 5). Most of the soils overlie very stony sandstone Head or bedrock between 40 and 80 cm depth. Beacon series resembles the stagnopodzols in having a peaty topsoil over a gleyed subsurface horizon depleted of iron, but the underlying horizons are also gleyed, merging with rock debris at 40 to 80 cm depth. Wenallt series (§ 91) is similar to the Beacon series but is in thick drift.

Lydcott series
(Full description Wright 1981, p.82)

0-10 cm — Oh or Ah
Black, stoneless humified peat or slightly stony humose sandy silt loam or sandy loam.

10-30 cm — Eag
Pinkish grey, slightly stony sandy silt loam or sandy loam; weak coarse angular blocky or massive structure.

30-50 cm — Bs
Yellowish red, slightly or moderately stony sandy silt loam or clay loam; moderate fine angular blocky structure.

50-100 cm — Cu
Reddish brown, very stony sandy silt loam or clay loam; massive structure.

Burcombe series
(Full description Hogan 1981, p.88)

0-15 cm — Oh or Ap
Black, humified peat or slightly stony, humose sandy silt loam.

15-25 cm — Eag
Dark greyish brown, moderately stony sandy silt loam, clay loam or sandy loam; weak medium prismatic or angular blocky structure.

At 25 cm — Bf
Thin ironpan

25-45 cm — Bs
Yellowish red, moderately stony sandy silt loam or clay loam; moderate fine subangular blocky structure.

40-100 cm — BCu
Reddish brown, very or extremely stony sandy silt loam; massive or weak medium angular blocky structure.

Beacon series
(Full description p.309)

0-15 cm — Oh
Black, stoneless, humified peat.

15-25 cm — Eg
Light brownish grey, very slightly stony clay loam; weak coarse prismatic structure.

25-35 cm — Bg
Pinkish grey, mottled, moderately stony clay loam; moderate coarse prismatic structure.

35-50 cm — Bw(g)
Reddish brown, mottled, moderately stony sandy silt loam; moderate fine subangular blocky structure.

50-70 cm — Cu
Reddish brown, very stony sandy silt loam; massive structure.

At 70 cm — Cr
Shattered red sandstone.

The Lydcott series is the most common overall but proportions of Burcombe and Beacon series vary considerably and either may be dominant locally. Lydcott series occupies most of the steep slopes and convex sites, but it is mixed with Burcombe and Beacon series on gentle slopes and the soil pattern is frequently complex. Locally also there are small boggy depressions with thick peat (Crowdy series) among the wet Wenallt soils. Larkbarrow soils (Findlay *et al.* 1984) are rare in Wales.

Key to component soil series

Soils with bleached subsurface horizon and ironpan or brightly coloured subsoil	1
Soils prominently mottled or greyish throughout; loamy	3

1.	With thin ironpan; mottled subsurface horizon; loamy	BURCOMBE
	Without ironpan	2
2.	With mottled subsurface horizon; loamy	LYDCOTT
	Unmottled; coarse loamy	Larkbarrow
3.	Sandstone within 80 cm	BEACON
	Deeper soils	WENALLT

Soil water regime

The water regime of these soils is complicated by the presence of the contrasting horizons. Water is held in the surface horizons of Lydcott, Beacon and Burcombe soils, the peat acting as a sponge so that they are seasonally waterlogged even though the subsoils drain freely (Wetness Class III or IV). Where strongly formed, the ironpan also impedes water movement. Wenallt series in slowly permeable thick drift on the lower ground, is naturally wetter (Wetness Class V). Rainwater passes rapidly into streams and rivers when the upper horizons are already saturated, particularly in winter.

Land use and vegetation

The climate is wet, cold and exposed so the land is mostly under rough grazing with bilberry heath and moist heather moor although, in places, swards dominated by mat-grass and purple moor-grass have been established following grazing and burning. The proportion of purple moor-grass varies considerably but it is most common on Beacon and Burcombe soils, becoming rare on the drier Lydcott series where the peaty surface is thin. Relative grazing value is poor to moderate, the grassland providing better grazing than the heather.

Most of the land has never been cultivated but the soils would respond well to improvement using surface treatments developed at Pwllpeiran Experimental Husbandry Farm in mid-Wales on similar Hafren soils. This involves flail mowing and burning to eliminate the mat of dead vegetation followed by surface rotavation and reseeding, with large additions of lime, slag and nitrogen. The poaching risk is great, however, owing to the large retained water capacity of the peaty topsoils and the high rainfall.

For afforestation, sites are deep ploughed to mix the peat and upper mineral horizons and to break any ironpan. This improves aeration and encourages deep rooting, while facilitating movement of water along the furrows into deep drains. Sitka spruce is the preferred species but tree growth may be poor or checked where heather is present unless it is suppressed with herbicide. Scots pine competes better than Sitka spruce on the heathery land.

Much of the association is in the Brecon Beacons National Park (Plate 8) and there are a number of nature reserves and sites of Special Scientific Interest. Military training and hill-walking are other important activities. These often conflicting land uses have caused some dispute between those concerned with farming, forestry, conservation and recreation.

§ 58. MALHAM 2 ASSOCIATION
541p

This association is mapped where a variable thickness of silty aeolian drift overlies limestone in the Peak District of Derbyshire and Staffordshire and in the Vale of Glamorgan. The dominant soils are typical brown earths (Malham series) and brown rankers (Crwbin series).

Malham soils predominate on gently and moderately sloping ground and cover three-quarters of the land. They are shallow and moderately deep fine silty soils with a characteristic brown unmottled subsoil over hard limestone. Upper horizons are usually stoneless but the subsoil is stony immediately over limestone. The loamy shallow Crwbin soils (§ 30) occur on convex sites and steep slopes, often between rock outcrops. Both soils are well drained.

There are small pockets of typical paleo-argillic brown earths (Nordrach series, Appendix, p.320) on gentle slopes or level sites and there are some deep accumulations of silty drift, often in valley bottoms. Wetton series which has humose topsoil, occurs on the steepest sites or adjacent to rock outcrops.

Brief descriptions of Malham and Crwbin series are in § 30, and a key to identify the main soils is given below:

This association is mapped on Carboniferous limestone in south Gower and on the downlands of the Vale of Glamorgan where most was formerly mapped as Lulsgate series by Crampton (1972). The association also includes soils on Triassic limestone conglomerate where reddish fine loamy brown earths belonging to Wrington series (Findlay 1965) and reddish, clayey rankers of the Torbryan series (George 1978) are common. The Worcester series (§ 99) occurs locally where red mudstones outcrop with the conglomerate, as around Coychurch near Bridgend. In total the association covers 90 km².

Key to component soil series

Shallow soils, limestone within 30 cm	1
Deeper soils	2

1.	Fine loamy or fine silty	CRWBIN
	Clayey	Torbryan

2.	Limestone within 80 cm	3
	Deeper soils; reddish, clayey with subsoil faintly mottled above 60 cm or distinctly mottled between 40 and 80 cm	Worcester

3.	Fine silty; brownish	MALHAM
	Fine loamy; reddish	Wrington
	Fine silty over clayey; with reddish subsoil	Nordrach

Soil water regime

The dominant soils are porous and are underlain by fractured and well-jointed limestone which ensures good profile drainage (Wetness Class I). The land therefore absorbs winter rain easily and there are no streams.

Land use

Malham soils are easy to cultivate except where they contain many stones, which wear implements rapidly. The less stony Nordrach soils are easily worked. Crwbin soils are more difficult to cultivate because they are shallow, rocky and often steeply sloping. The climate, with 1,100 mm of rainfall annually, is relatively dry for Wales so there is some arable farming on these well drained soils and vegetables are grown on the Gower for the

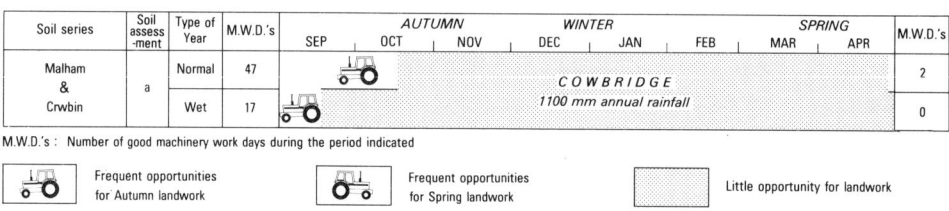

Figure 40. The effects of soil and climate on landwork, Malham 2 association

Swansea market. However, the period of winter landwork is short, and confined to autumn in most years (Fig.40). Reserves of available moisture in Malham and Nordrach soils are sufficient for most crops in most years, but the Crwbin series is moderately droughty for all crops and very droughty for grass. These soils provide good grassland, with little poaching risk, but summer growth is poor on Crwbin series.

§ 59. MALVERN ASSOCIATION
611a

The Malvern association is extensive in the Lake District, Snowdonia and the Cheviot Hills. It is also found on Bodmin Moor, Cornwall; near Alnwick, Northumberland; in upper Teesdale, Durham; and in central Wales and the Welsh Borderland. It extends over a wide range of altitude and climate, and occurs generally on steep bouldery slopes. Locally there are extensive crags and screes. The principal soils are typical brown podzolic soils in shallow drift over igneous rocks of varied acidity, the Malvern, Moretonhampstead and Davidstow series. Malvern and Davidstow soils are both developed on basic and intermediate igneous rocks; the former are very stony to the surface whereas the latter are often only slightly stony through to the subsoil. The Moretonhampstead series is similar to Davidstow soils but is developed on acid igneous rocks. Brief profile descriptions of the main soils and a key to the component series are given below.

Malvern series
(Full description Palmer 1976, p.100)

0–5 cm — Oh
Black, loamy peat.

5–25 cm — Ah
Brown, slightly stony sandy silt loam; moderate fine granular structure.

25–60 cm — Bs
Yellowish red, very stony sandy silt loam; moderate medium granular structure.

60–115 cm — BCu
Brown, extremely stony sandy loam; moderate angular fine angular blocky structure.

Moretonhampstead series
(Full description Staines 1976, p.99)

0–25 cm — Ap or Ah
Dark brown, slightly stony coarse sandy silty loam.

25–40 cm — Bs1
Brown, slightly or moderately stony coarse sandy silt loam or sandy loam; moderate fine subangular blocky or granular structure.

40–60 cm — Bs2
Strong brown, moderately stony coarse sandy silt loam or sandy loam; moderate fine subangular blocky or granular structure.

60–100 cm — BCu or BCx
Yellowish brown, very or extremely stony coarse sandy silt loam or sandy loam; weak fine subangular blocky structure or compact with platy structure.

Davidstow series
(Full description p.312)

0–20 cm — Ap
Dark brown, slightly stony sandy silt loam.

20–50 cm — Bs
Dark yellowish brown, moderately stony sandy silt loam; moderate fine subangular blocky structure.

50–65 cm — BCu
Olive brown, extremely stony sandy loam or sandy silt loam.

Small patches of shallow humose soils, the Bangor (§ 14) and Preseli series (Jarvis, R.A. et al. 1984), occur near crags and screes. In south-west Dyfed, shallow Preseli and Dunwell soils (Appendix 2, p.312) are common, as are the Bowden (Clayden and Hollis 1984) and Moor Gate series (§ 64). Locally there are inclusions of the Harthope series (Clayden and Hollis 1984) on outcrops of basic crystalline rocks. in Snowdonia the Preseli series is included with Bowden, Dunwell, Mayalls (Clayden and Hollis 1984) and Trusham series (Appendix 2, p.314). Iveshead and Gunnislake series are rare.

T.R.E. Thompson

Plate 22. The Lleyn Peninsula. Here hedges are cleared to enlarge small fields on the Malvern association against the backdrop of Aberdaron Bay and Bardsey Island. In the distance most soils are in the Brickfield 3 association, but the cultivated area on the cliffs is on well drained East Keswick soils.

Key to component soil series

	Soils with brightly coloured subsoil	1
	Other soils	5
1.	Over acid igneous rocks	2
	Over basic igneous rocks	3
2.	Coarse loamy	MORETONHAMPSTEAD
	Very stony, loamy	Iveshead
3.	Coarse loamy	DAVIDSTOW
	Very stony, loamy	4
4.	With brown, distinct topsoil	MALVERN
	With dark, humose topsoil	Bowden
5.	Shallow peaty soils, rock within 30 cm	6
	Deeper soils; with brown subsoil	7
6.	Over acid igneous rocks	Bangor
	Over basic igneous rocks	Preseli
7.	Over acid igneous rocks; coarse loamy	Gunnislake
	Over basic igneous rocks; fine loamy	Trusham

Land use

The soils are well drained and readily absorb winter rainwater except on the steepest slopes. On low ground the soils are in grass, with occasional barley crops on flatter land. The shallower soils are liable to drought but there are no poaching restrictions and, in western districts, there is a long autumn flush of grass, and grass continues to grow slowly for much of the winter. The land is used mainly for livestock rearing (Plate 22). At higher altitudes, in south-west Dyfed most of Snowdonia and around Builth Wells, bent-fescue grassland of good grazing value is common but bracken infestation occurs in places. In Snowdonia there is little other than rough grazing, with some oak woodland preserved on the steeper and more rocky land where cultivation is not feasible. The closely cropped bent-fescue grassland with an abundance of small herbs (Ratcliffe 1977) is grazed by sheep and has a large carrying capacity. Although this land has largely been untried for commercial forestry in Wales, planting with Sitka spruce and Douglas fir would be appropriate (Pyatt 1977). In exposed places trees suffer windthrow on the more shallow and stony soils.

§ 60. MANOD ASSOCIATION
611c

The Manod association consists mainly of free draining fine loamy soils over Palaeozoic mudstone, siltstone or slate. It is widespread in Wales and also occurs in the Midlands, Northern and South West England. Most of the land is above 200 m O.D. with more than 1,000 mm annual rainfall. Although some is level or gently sloping, much of it is steeper than 11 degrees (Plate 23). In the wetter western districts or where the rocks are particularly hard it reaches almost to sea level as in the Conwy valley and along parts of the west coast of Wales and Devon.

Typical brown podzolic soils, mainly Manod series, cover half the association and typical brown earths such as the Denbigh series (§ 31) one-fifth (Fig.41). The Powys soils, loamy rankers, are common where bedrock is close to the surface.

Brief profile descriptions of the main series are given below.

The Manod series has solid or shattered rock within 80 cm depth and is a permeable clay loam with dark topsoil over ochreous subsoil, usually with granular structure. Denbigh series differs in having a brownish subsoil with blocky structure; it usually occurs on gentler slopes and is more common on farm land. Powys soils, with solid or broken rock within 30 cm depth are most frequent in hilly country, on knolls and ridges or on slopes eroded by cultivation or solifluction.

In Wales fine silty Barton soils (§ 15) and fine silty brown podzolic soils similar to Manod series are more common than elsewhere. The fine loamy Meline series (Clayden and Hollis 1984) is a deeper soil in drift while the coarse loamy Withnell series (Crompton

Manod series
(Full description Bradley 1976, p.40)

0–25 cm — Ap
Dark brown, slightly stony clay loam.

25–60 cm — Bs
Strong brown, moderately stony clay loam; moderate medium granular or subangular blocky structure.

At 60 cm — R or Cu
Slaty mudstone *in situ* or extremely stony clay loam or sandy silt loam.

Denbigh series
(Full description p.312)

0–25 cm — Ap or Ah
Dark brown, slightly stony clay loam.

25–60 cm — Bw
Brown, slightly or moderately stony clay loam; moderate fine subangular blocky structure.

60–100 cm — BCu
Yellowish brown, very stony clay loam; moderate fine subangular blocky or massive structure or *in situ* slate or mudstone.

Powys series
(Full description Wright 1980, p.35)

0–25 cm — Ap or Ah
Dark brown, slightly stony clay loam or silty clay loam.

At 25 cm — Cr or Cu
Mudstone or slate *in situ* or extremely stony clay loam.

P.S. Wright

Plate 23. Mynydd Mallaen, between Tregaron and Llandovery, Dyfed. Hafren association covers the plateau with Manod association on steep valley sides under woodland and bracken.

1966) occurs on occasional sandstone bands. Many soils on slopes too steep or rocky to cultivate often have a thick surface mat of roots and plant remains. On the higher slopes Parc soils (§ 72) are found. Steep upland slopes often carry very stony profiles of the Banc series (Clayden and Hollis 1984). In localized wet spots over drift there are some stagnogleyic brown earths and cambic stagnogley soils. On high ground recently reclaimed from moorland, stagnopodzols and stagnohumic gley soils also appear.

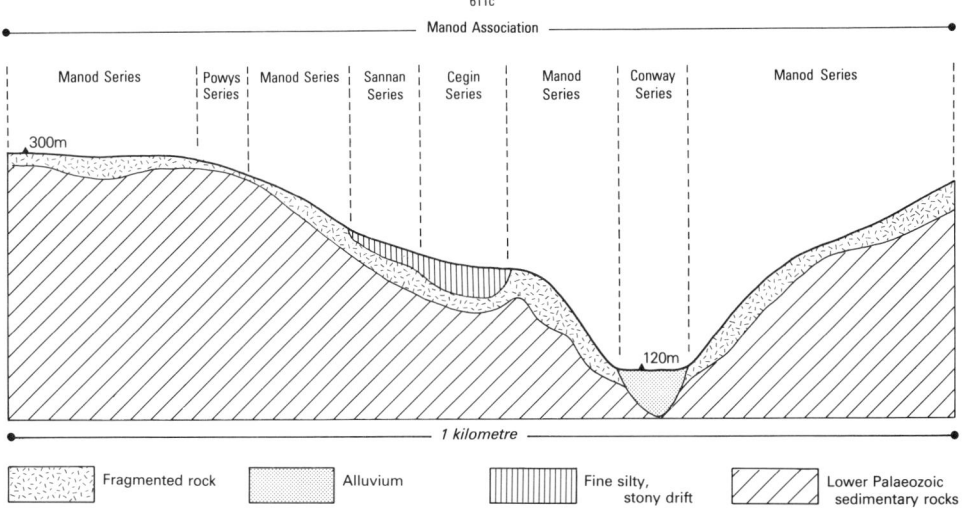

611c
Manod Association

Manod Series | Powys Series | Manod Series | Sannan Series | Cegin Series | Manod Series | Conway Series | Manod Series

300m

120m

1 kilometre

Fragmented rock · Alluvium · Fine silty, stony drift · Lower Palaeozoic sedimentary rocks

Figure 41. Distribution of soils in the Manod association. On low ground and gentler slopes Denbigh series tends to replace Manod series. At higher altitudes and in basin sites, Wilcocks series replaces Cegin series

Locally there are small areas of crystalline rocks. These give other minor component series such as Moretonhampstead series (§ 65) in Snowdonia, and Gunnislake (Clayden and Hollis 1984) and Trusham series (Appendix 2, p.324) in south-west Dyfed. Also included is a narrow fringe of reddish soils in drift from Carboniferous Basement Beds below the Eglwyseg rocks near Llangollen. Many of the valleys traversing the association are floored with strips of alluvium and discontinuous river terraces.

Key to component soil series

	Shallow soils, rock within 30 cm; fine loamy	POWYS
	Deeper soils	1
1.	Rock within 80 cm	2
	Deeper soils; brightly coloured subsoil, fine loamy	Meline
2.	With dark, humose or peaty topsoil; fine loamy	Parc
	With thin, surface leaf mould or brown, distinct topsoil	3
3.	With brightly coloured subsoil	4
	With brown subsoil	5
4.	Fine loamy	MANOD
	Coarse loamy	Withnell
5.	Fine loamy	DENBIGH
	Fine silty	Barton

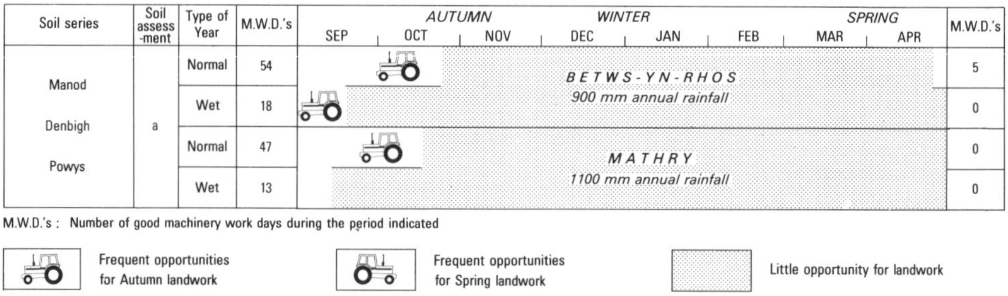

Soil series	Soil assess-ment	Type of Year	M.W.D.'s	AUTUMN			WINTER			SPRING		M.W.D.'s
				SEP	OCT	NOV	DEC	JAN	FEB	MAR	APR	
Manod	a	Normal	54		*BETWS-YN-RHOS*							5
		Wet	18		*900 mm annual rainfall*							0
Denbigh		Normal	47		*MATHRY*							0
Powys		Wet	13		*1100 mm annual rainfall*							0

M.W.D.'s : Number of good machinery work days during the period indicated

	Frequent opportunities for Autumn landwork		Frequent opportunities for Spring landwork		Little opportunity for landwork

Figure 42. The effects of soil and climate on landwork, Manod association

R. Hartnup

Plate 24. Manod association near Tafolog, Powys. Manod soils underlie the smooth bracken and gorse covered slopes. The grassland on the rounded crest has stagnopodzols of the Hafren association.

J.W. Lea

Plate 25. Forestry and grazing on the Manod association near World's End, Clwyd.

Soil water regime

The main soils are permeable and well drained (Wetness Class I) but as the climate over much of the association is wet, most of the soils remain moist throughout most years. Where average annual rainfall exceeds 1100 mm there is little or no drought restriction to grass growth on Manod and Denbigh soils. Grass on shallow very stony soils such as the Powys series can, however, suffer from drought. The soils readily absorb excess winter rainwater except on steep land or where bedrock is near the surface.

Cultivation and cropping

Most of the land is in permanent grass, leys or rough grazing. There is very little arable cropping but cereals are grown locally on gentle slopes, usually at low altitudes, and brassica and root crops are sometimes grown for feeding sheep and cattle. Some early potatoes are grown in south-west Dyfed. The land is well-suited for producing hay and silage where slopes are not too steep. The growing season is usually over 7 months (Smith 1976) and, in good years, 6 months are available for grazing in the drier parts without risking damage to the sward and soil structure (Fig.42). Locally, however, summer drought limits growth particularly where the soils are shallow. February cultivations, often required for early potatoes, are liable to damage the soil and reduce yields if the land is too wet, but full advantage is taken of the short rain-free periods when mild temperatures and strong winds assist in drying the soil rapidly near the coast.

The soils are naturally acid and deficient in phosphorus, but when adequately limed and fertilized grass grows well. Where grazing pressure is light, as on steep slopes marginal to moorland, the bent-fescue sward is infested with bracken and gorse (Plate 24). Care is needed to prevent erosion when cultivating or otherwise improving steeply sloping land. Where slopes are gentle, Manod and Denbigh soils are useful for dispersing slurry (Lea 1979), which is often spread prior to cultivation.

Other land use

Much of the land was forested before enclosure and some steep land has reverted to scrub woodland (Plate 25). Old oak woods and coniferous forests are widespread where agriculture is restricted by steep slopes (Plate 26). Woodlands occupy about one-fifth of the land, often in sheltered valleys. Sitka spruce is commonly planted and yields very well. Douglas fir is also productive in sheltered sites. Larch is often grown but yields are somewhat smaller. Grand fir and Western hemlock are suitable amenity species in the valleys. There is little risk of windthrow except where the soils are shallow and the site exposed. Many of the older deciduous woodlands are highly valued for recreation and as wildlife habitats.

J.W. Lea

Plate 26. View from the Horseshoe Pass south-west towards the Dee valley. Forestry and grazing land is on the Manod association. Far left there are the bare rocks and screes of the Crwbin association on the Eglwyseg escarpment.

§ 61. MIDDLETON ASSOCIATION
572b

Reddish fine silty and fine loamy seasonally wet soils on soft red Devonian silty shales and siltstones, are grouped as the Middleton association. The shales are micaceous and slightly calcareous where unweathered and contain a few thin interbedded argillaceous limestone bands and nodules. The rock has greenish grey blotches and streaks and thin lenses of micaceous sandstone in places. Surface wetness is often exacerbated by water seepage from the sandstones.

The association occupies about 100 km^2 in the Welsh Borderland and 11 km^2 in south-east Wales mainly on broad flat ridge crests or on gentle slopes in basin-like sites often around the heads of small valleys. Stagnogleyic argillic brown earths are dominant, the fine silty Middleton series (Appendix 2, p.318) being more widespread than the fine loamy Hodnet series (Clayden and Hollis 1984). Wet often rush infested patches contain Netchwood series (Appendix 2, p.319), typical stagnogley soils). Also included are coarse loamy, well drained soils over hard sandstone, the Eardiston series (§ 34), and Vernolds series (Appendix 2, p.325) in deep silty drift of flush sites.

A full description of the association, with a key to the main soil series, is given by Ragg et al. (1984).

Middleton and Hodnet series are seasonally waterlogged (Wetness Class III) but respond well to artificial drainage (Wetness Class II). Netchwood series is waterlogged for long periods during the winter (Wetness Class IV) and waterlogging may persist into the growing season (Wetness Class V). Surface wetness is caused by the slow permeability of subsoil horizons.

Land use is restricted to grassland and occasional cereal crops which on underdrained Middleton and Hodnet soils produce good yields. Substantial patches of wetter Netchwood soils within fields inhibit ploughing and such fields often contain rushes and other moisture loving plants. Poaching and compaction occur if the land is grazed early or late in the season so careful management is necessary.

§ 62. MIDELNEY ASSOCIATION
813a

The association is confined to flat lowlands where thin clayey river alluvium overlies peat. It is most extensive in central Somerset and in the Fens where rivers enter broad alluvial plains and is also found locally in narrower floodplains in Nottinghamshire, Staffordshire, Suffolk, Dorset and Gloucestershire. In Wales it is confined to a narrow belt of rushy land at the landward margin of the Caldicot levels in Gwent. The land is only a little above sea level and was frequently flooded in the past.

The main soils are pelo-alluvial gley soils of the Midelney series which have prominently mottled clayey upper horizons 40 to 50 cm thick over organic material.

Associated with these, are similar Wensum soils typical humic alluvial gley soils, which
have a humose surface horizon and tend to occur where there is unimproved grassland. In
Wales fine silty over peaty soils belonging to the Tregaron series (Rudeforth 1970) are also
found.

Slowly permeable horizons, high groundwater and a lengthy field capacity period result
in prolonged winter waterlogging (Wetness Class IV), so the land is mainly in permanent
grass. Potential yields of grass are large but there is a risk of poaching so access during the
growing season is limited. Partly for these reasons and because the land on Caldicot levels
is traversed by an embanked railway line which further affects drainage and access, the
pastures are weed infested and rather poorly used.

More detailed accounts of soils of this association are given by (Findlay *et al.* 1984).

§ 63. MILFORD ASSOCIATION
541a

The Milford association is mapped over 1,345 km² of the Old Red Sandstone outcrop in
south Wales, the Welsh Borderland, Devon and Somerset. It consists mainly of reddish
fine loamy typical brown earths on interbedded siltstone, sandstone and mudstone, on
gently sloping arable lowland in south-west Dyfed (Fig.43, Plate 27) to steep slopes up to
around 400 m O.D. In the Milford series the topsoil and subsoil are a reddish brown stony
clay loam with granular or fine subangular blocky structure over very stony Head or
bedrock at 40 to 80 cm depth. On steep slopes, especially above 150 m O.D. typical brown
podzolic soils (Whitcott series) are common. They are distinguished from the Milford
soils by a bright yellowish red subsoil, but are otherwise similar. Soils in thick drift occupy

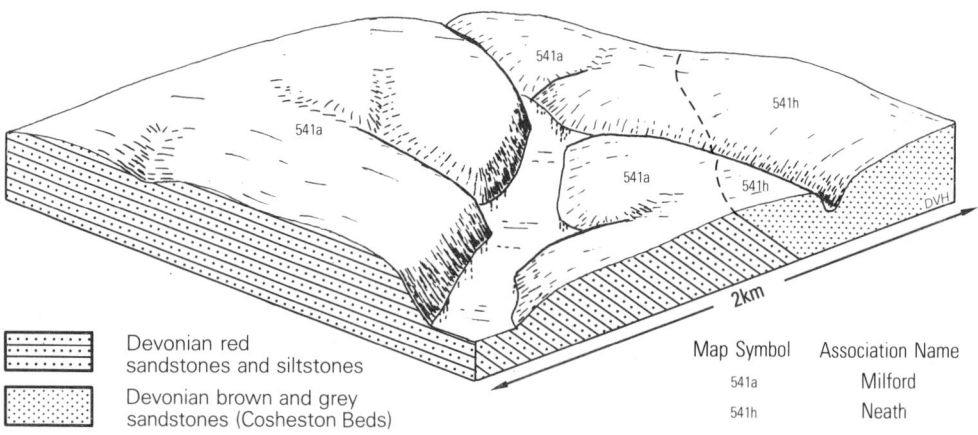

Figure 43. Soil associations near Honeyborough and Rosemarket, Dyfed

Plate 27. Milford association near Milford Haven. Arable fields form a patchwork with grassland. Oil refineries occupy much land in the middle distance.

a small part of the association usually on footslopes. These are either well drained typical brown earths (Newbiggin series, p.320) or stagnogleyic brown earths (Llangendeirne series, p.317) which are subject to periodic waterlogging and have some subsoil gleying. In small rushy depressions and flushes there are cambic stagnogley soils (Fforest series, § 43) but these are rare. Shallow brown rankers (Powys series, § 60) occur on some narrow ridges and plateaux. All the soils contain stones but most are only slightly stony in the upper 40 cm, small and medium sandstones being commonest.

The Milford association is the most extensive over the Old Red Sandstone outcrop of south Wales (Wright 1981). It includes land on the precipitous scarp of the Brecon Beacons where the proportions of Whitcott series and bare rock are greatest. Here too the shallow Newtondale soils (Carroll and Bendelow 1981, p.40) are found. Downslope, the association typically passes to the Fforest association (§ 43) of cambic stagnogley soils in drift, the boundary usually being a concave break of slope with a rapid transition to wet, rushy pasture. Upslope, above about 400 m O.D., Milford soils give way to the Lydcott association (§ 57) of stagnopodzols, the boundary usually marked by a change from bent-fescue pasture and bracken to bilberry heath or *Nardus* grassland.

Milford series
(Full description Wright 1980, p.73)

0–30 cm — Ap
Dark reddish brown, moderately or slightly stony clay loam.

30–70 cm — Bw
Reddish brown, moderately stony clay loam; moderate fine sub-angular blocky structure.

70–85 cm — Cu
Extremely stony, with reddish brown stone coatings; single grain structure.

Whitcott series
(Full description Wright 1981, p.44)

0–5 cm — Ah
Very dark grey, stoneless humose clay loam; moderate fine sub-angular blocky structure.

5–50 cm — Bw
Reddish brown, moderately stony clay loam; strong fine granular structure.

50–70 cm — Bs
Yellowish red, very stony clay loam; strong fine granular structure.

70–80 cm — Cu
Extremely stony with reddish brown loamy coatings.

Llangendeirne series
(Full description p.317)

0–25 cm — Ap
Dark reddish brown, slightly stony clay loam.

25–50 cm — Bw
Reddish brown, slightly stony clay loam; weak subangular blocky structure.

50–90 cm — BCx(g)
Reddish brown, slightly or moderately stony clay loam or sandy loam; massive structure; high packing density.

Key to component soil series

Shallow soils, rock within 30 cm; fine loamy		Newtondale or Powys
Deeper soils		1
1. Rock within 80 cm; fine loamy		2
Deeper soils; with reddish subsoil		3
2 Reddish		MILFORD or WHITCOTT
Brownish		Denbigh
3. Unmottled; fine loamy		NEWBIGGIN
Mottled		4
4. Subsoil faintly mottled above 60 cm or distinctly mottled between 40 and 80 cm; fine loamy		LLANGENDEIRNE
Prominently mottled or greyish above 40 cm; fine silty		Fforest

Soil water regime

The main Milford and Whitcott soils are permeable, well drained (Wetness Class I) and readily absorb winter rains.

Land use

The annual rainfall increases from around 1,000 mm in the lowlands of south-west Dyfed and Gwent to over 1,500 mm in the uplands of south-east Dyfed and south Powys. In the latter areas most land is permanent grassland and the farming system is based on livestock rearing with beef cows and hill sheep. Here arable crops are grown entirely for forage. In this difficult climate, Milford soils form the best land, being well-drained with a low poaching risk. Most of the grassland has been improved except on slopes greater than 15 degrees where there is usually bent-fescue pasture which is of good grazing value except

Table 21
Profile Available Water (A.P. mm), Crop-adjusted Mean Moisture Deficit (M.D. mm)
and Droughtiness Class for extensive crops—Milford Association

	Milford series	**Newbiggin series**
Location	Llanishen, Gwent	Llanishen, Gwent
Grid Ref.	SO480035	SO480035
Grass		
A.P.	120	130
M.D.	111	111
Droughtiness	slightly droughty	slightly droughty
Winter wheat		
A.P.	115	130
M.D.	76	76
Droughtiness	slightly droughty	non-droughty
Spring barley		
A.P.	115	130
M.D.	73	73
Droughtiness	slightly droughty	non-droughty
Potatoes		
A.P.	105	105
M.D.	71	71
Droughtiness	slightly droughty	slightly droughty

Soil series	Soil assess-ment	Type of Year	M.W.D.'s	SEP	OCT	NOV	DEC	JAN	FEB	MAR	APR	M.W.D.'s
Milford Whitcott Newbiggin	a	Normal	61									7
		Wet	30				*LLANISHEN*					0
Llangendeirne	cd	Normal	16				*950 mm annual rainfall*					0
		Wet	0									0
Milford Whitcott Newbiggin	a	Normal	47									0
		Wet	13				*MILFORD HAVEN*					0
Llangendeirne	cd	Normal	2				*1100 mm annual rainfall*					0
		Wet	0									0

M.W.D.'s : Number of good machinery work days during the period indicated

Frequent opportunities for Autumn landwork   Frequent opportunities for Spring landwork   Little opportunity for landwork

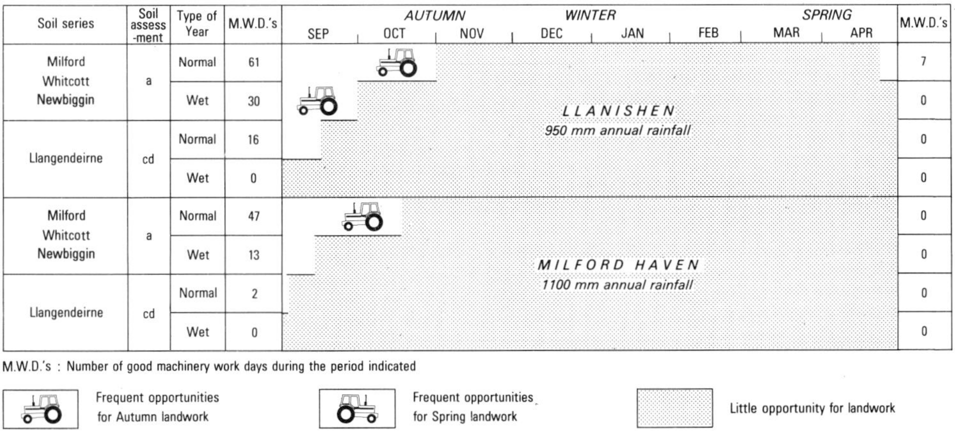

Figure 44. The effects of soil and climate on landwork, Milford association

C.C. Rudeforth

Plate 28. Milford association, cultivated for early potatoes, in south-west Dyfed. Broccoli is being harvested to the right of the lane.

where bracken infestation is heavy. The proportion of short-term grass and arable is greatest west of Carmarthen and in Gwent where the climate is drier and dairying the main activity. Some barley is grown. Where slope permits, Milford soils present few problems for cultivation and the only serious limitation is the wet climate, reflected in the long field capacity period of around 225 days. The winter work period is therefore short (Fig.44) with little or no opportunity for spring cultivation in most years, autumn providing the greater number of days when cultivation is possible without soil damage. Reserves of available water in most soil profiles are more than 100 mm, so that east of Milford Haven, where moisture deficits are less than 90 mm, droughtiness is not a problem in most years. Milford soils are droughty only in the lowlands of the extreme west and south-east where deficits are highest (Table 21) or where they are shallow. In the west, the mild climate favours early potato growing. February cultivations required for early potatoes damage the soil and reduce yields if the land is too wet, but near the coast, full advantage is taken of the short rain-free periods when mild temperatures and strong winds assist in drying the soil rapidly. Damage to soil structure by the early cultivations is lessened by using grass in rotation. Broccoli is also grown over winter (Plate 28). Small areas of the association, usually steep slopes, are under forestry or old sessile oak woodland.

§ 64. MOOR GATE ASSOCIATION
612b

The Moor Gate association consists mainly of podzolic soils developed over acid igneous rocks in South West England and in Wales. It is extensive on granite in Devon and Cornwall where the parent material is gravelly granitic Head (growan) with local additions of silty drift. In Wales the soils are on rhyolite and other acid igneous and metamorphic rocks, in south-west Dyfed and further north on the Lleyn Peninsula and around Cader Idris and Snowdon.

The association is widespread on the unenclosed moorland margins of Dartmoor, and at lower altitude in west Cornwall. In Wales, this land is generally rough grazing on rocky and bouldery steep slopes. Coarse loamy humic brown podzolic soils of the Moor Gate series (Clayden and Hollis 1984) have a dark humose topsoil overlying a finely structured permeable brown or brightly coloured subsoil of gritty sandy silt loam or gritty sandy

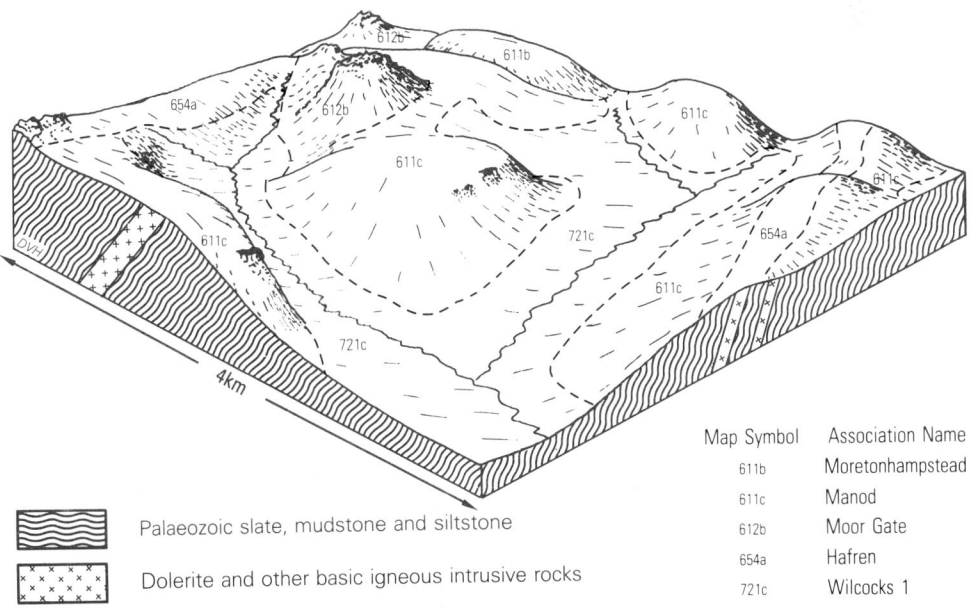

Map Symbol	Association Name
611b	Moretonhampstead
611c	Manod
612b	Moor Gate
654a	Hafren
721c	Wilcocks 1

Palaeozoic slate, mudstone and siltstone

Dolerite and other basic igneous intrusive rocks

Figure 45. Soil associations in the Preseli Hills south of Foeldrygarn

loam. These soils occupy a little more than half the land. The remainder is taken up by typical brown podzolic soils of the Moretonhampstead series (§ 65), and ironpan stagnopodzols of the Hexworthy series (§ 53), in which a bleached subsurface horizon overlies a thin ironpan. Very acid, permeable Cucurrian series, Bodafon series, (Clayden and Hollis 1984) and Trink series, (p.324) are locally important. Brief profile descriptions of the main series are given below.

Moor Gate series
(Full description p.319)

0–15 cm — Ah
Black, slightly stony humose
sandy silt loam; weak fine
subangular blocky structure.

15–25 cm — A/Bh
Dark reddish brown, slightly
stony humose sandy loam; weak
medium subangular blocky
structure.

25–50 cm — Bs
Strong brown, moderately stony
sandy loam; weak medium
subangular blocky structure.

50–100 cm — BCu or BCx
Brown, very stony sandy loam;
weak coarse angular blocky
structure or strong medium platy
structure in fragipan.

Moretonhampstead series
(Full description Staines 1976,
p.99)

0–25 cm — Ap or Ah
Dark brown, slightly stony coarse
sandy silt loam.

25–40 cm — Bs1
Brown, slightly or moderately
stony coarse sandy silt loam or
sandy loam; moderate fine
subangular blocky or granular
structure.

40–60 cm — Bs2
Strong brown, moderately stony
coarse sandy silt loam or sandy
loam; moderate fine subangular
blocky or granular structure.

60–100 cm — BCu or BCx
Yellowish brown, very or extremely stony coarse sandy silt
loam or sandy loam; weak fine
subangular blocky structure or
compact with platy structure.

Hexworthy series
(Full description p.316)

0–20 cm — Oh
Black, humified peat; weak
medium subangular blocky
structure.

20–30 cm — Ah/E
Black, slightly or moderately
stony, humose coarse sandy silt
loam or sandy loam with bleached
sand grains; weak fine subangular
blocky structure.

30–45 cm — Eag
Greyish brown, mottled, moderately stony coarse sandy silt loam
or sandy loam; weak fine
subangular blocky structure.

At 45 cm — Bf
Thin ironpan

45–60 cm — Bs
Strong brown, very stony coarse
sandy silt loam or sandy loam;
moderate fine subangular blocky
structure.

60–100 cm — BCu or BCx
Brown, very stony coarse sandy
loam or loamy sand; weak
angular blocky or strong platy
structure.

In north Wales, Bodafon series and freely drained coarse loamy humic cryptopodzols formerly included in the Ceiri series (Ball 1963), are often major components together with bedrock, boulders and scree. The soils in south-west Wales are closely associated with those of Hafren, Manod and Wilcocks associations (Fig.45).

Key to component soil series

Soils with brown, distinct topsoil or thin, surface leaf mould; coarse loamy	MORETONHAMPSTEAD
Soils with dark, humose topsoil; coarse loamy	1
1. Without bleached subsurface horizon	MOOR GATE
With bleached subsurface horizon	2
2. With thin ironpan	3
Without ironpan	4
3. With dark, humus-enriched subsoil	Trink
Without humus-enriched subsoil	HEXWORTHY
4. With dark, humus-enriched subsoil	Cucurrian
Without humus-enriched subsoil	Bodafon

Cultivation and cropping

Much of the land particularly in north Wales is under heather moor or *Nardus* grassland since rock and boulders discourage agricultural improvement. In parts of south-west Dyfed however, where rainfall is relatively low, there is fairly good quality grazing, though some grassland is deficient in essential trace elements such as copper, cobalt and selenium. On the islands of Skomer and Ramsey and on the mainland around St David's the rocky rough grazing land has considerable amenity value.

§ 65. MORETONHAMPSTEAD ASSOCIATION
611b

The Moretonhampstead association covers 550 km^2 and consists of well drained podzolic soils over acid igneous rocks. In South West England and the Lake District the parent material is a gritty, sometimes bouldery, granitic Head passing down into deeply weathered and weakly coherent granite. Hard rock is exposed in places as tors and buttresses and adjacent slopes are often strewn with granite blocks. In Wales the bedrock is more varied and, while acid igneous rocks such as rhyolite predominate, basic and intermediate rocks and intervening bands of sedimentary rock are also included. In north Wales rhyolite is the major parent rock and screes are common. Coarse loamy typical brown podzolic soils, Moretonhampstead series, usually occupy about two-thirds of the land, with similar but coarse silty soils. Humic brown podzolic soils of the Moor Gate series (§ 64) are important associates. Brief profile descriptions of the two main series are given below.

Moretonhampstead series
(Full description Staines 1976, p.99)

0–25 cm — Ap or Ah
Dark brown, slightly stony coarse sandy silt loam.

25–40 cm — Bs1
Brown, slightly or moderately stony coarse sandy silt loam or sandy loam; moderate fine subangular blocky or granular structure.

40–60 cm — Bs2
Strong brown, moderately stony coarse sandy silt loam or sandy loam; moderate fine subangular blocky or granular structure.

60–100 cm — BCu or BCx
Yellowish brown, very or extremely stony coarse sandy silt loam or sandy loam; weak fine subangular blocky structure or compact with platy structure.

Moor Gate series
(Full description p.319)

0–15 cm — Ah
Black, slightly stony humose sandy silt loam; weak fine subangular blocky structure.

15–25 cm — A/Bh
Dark reddish brown, slightly stony humose sandy loam; weak medium subangular blocky structure.

25–50 cm — Bs
Strong brown, moderately stony sandy loam; weak medium subangular blocky structure.

50–100 cm — BCu or BCx
Brown, very stony sandy loam; weak coarse angular blocky structure or strong medium platy structure in fragipan.

In Wales the various associated sedimentary and igneous rocks other than granite give predominantly well drained soils including Denbigh series (§ 31) over slates and mudstones, Trusham series (§ 87) on basic igneous rocks, and Meline (Clayden and Hollis 1984) and East Keswick series (§ 36) on drift. Brown earths of the Gunnislake series (Clayden and Hollis 1984) have also been included. Minor soils in north Wales include the Bangor series, with wetter soils in flushes.

Key to component soil series

	Soils over igneous rocks	1
	Other soils	5
1.	Over acid igneous rocks	2
	Over basic igneous rocks; fine loamy	Trusham
2.	With brown, distinct topsoil or thin, surface leaf mould; coarse loamy	3
	With dark, humose or peaty topsoil	4
3.	With brightly coloured subsoil	MORETONHAMPSTEAD
	With brown subsoil	Gunnislake
4.	Rock below 30 cm; coarse loamy	MOOR GATE
	Shallow soils, rock within 30 cm; loamy peat	Bangor
5.	Over slate or mudstone within 80 cm; fine loamy	Denbigh
	Deeper soils	6
6.	With brightly coloured subsoil; fine loamy	Meline
	With brown subsoil; fine loamy	East Keswick

Soil water regime

The soils are well drained (Wetness Class I) and in most circumstances they absorb winter rain with ease. On steep slopes there is some surface run-off. The soils are slightly droughty under grass in extreme south-west Dyfed and non-droughty for barley. Although harvested in June, early potatoes are commonly irrigated.

Cultivation and cropping

In the lowlands of south-west Dyfed arable and some horticultural crops are grown, whereas in the uplands grassland farming predominates. Where cultivation is not prohibited by slope or boulders, the soils are easily worked over a range of soil moisture contents and, although opportunities for landwork are few, the soils are commonly ploughed in winter as it allows maximum use of grass in autumn and limits weed growth during the mild winter weather. Specialized crops such as early potatoes and broccoli are grown in south-west Dyfed on ground relatively free from spring frosts. The commonest cereal crop is spring barley. Inland where rainfall is high, grassland and rough grazing of good and moderate value is widespread. On steep and rocky slopes where access is restricted bracken is common.

Forestry

The soils are well-suited for forestry in sheltered sites. Douglas fir and Japanese larch are favoured on the lower ground, while Sitka spruce succeeds well under higher rainfall. There is semi-natural oakwood in many steep-sided valleys providing rich wildlife habitats.

§ 66. NEATH ASSOCIATION
541h

Fine loamy brown soils over sandstone dominate the Neath association. Most of the soils derive from Carboniferous sandstone with local subordinate shales, but soils over Cambrian and Devonian sandstones are included in west Wales. The Neath soils, typical brown earths, are permeable and well drained, with brown, clay loam upper horizons containing sandstone and siltstone fragments. The stone content increases downwards, and the soils pass into rubbly Head or fragmented rock. Nercwys series (stagnogleyic brown earths) is a common associate soil which is differentiated by its mottled, less permeable, clay loam subsoil. Other ancillary soils vary from district to district. In Wales, the southern Pennines and the Bristol coalfield, wetter soils are less common than in the South West. Surface-water gleys occur in north Wales and the south-west Dyfed coalfield. Where shale and mudstone are interbedded, particularly on the Coal Measures, Denbigh (§ 31) and Barton series (§ 15) are intermixed with Neath soils, and, in Wales, Manod (§ 60) and Withnell series (§ 76) are also present. Brief descriptions of the three most extensive series follow.

Neath series
(Full description p.319)

0–20 cm — Ap
Dark brown, moderately stony clay loam.

20–50 cm — Bw
Yellowish brown, moderately stony clay loam; moderate medium angular blocky structure.

At 50 cm — R or Cu
Sandstone or brown, very stony clay loam or sandy loam.

Nercwys series
(Full description Rudeforth 1974, p.74)

0–25 cm — Ap
Dark brown, slightly stony clay loam.

25–50 cm — Bw
Dark yellowish brown, slightly stony clay loam; moderate medium subangular blocky structure.

50–100 cm — Bg
Yellowish brown, mottled, slightly stony clay loam; weak coarse angular blocky structure; high packing density.

Cherubeer series
(Full description p.310)

0–25 cm — Ap
Brown, slightly stony clay loam.

25–50 cm — Bw
Yellowish brown, moderately stony clay loam; weak fine blocky structure.

50–100 cm — Bg
Yellowish brown, mottled, moderately stony clay; massive structure.

In south Wales and Clwyd, the Neath association is mapped principally on the Coal Measures, where it occupies gentle or moderate slopes below 170 m O.D. Although the well drained Neath series predominates, some wetter soils are present in patches of drift.

In Clwyd these include Brickfield (§ 17) and Cegin series (§ 23), but in most of south Wales Nercwys series is the principal ancillary soil. On the south Dyfed coalfield there are more wet soils in the association. Well drained brown earths, brown podzolic soils and locally rankers are also present. They differ from Neath series in texture, parent material or soil depth. Shales are interbedded with the standstones in south Dyfed and carry Denbigh, Barton and Powys series (§ 31 and § 15). In Clwyd deeper brown soils in drift, with rocky layers below 80 cm depth, notably the fine loamy East Keswick series (§ 36), form an important element but shaly soils are largely absent. Typical brown podzolic soils are included on steep land. They are represented in Glamorgan mainly by the coarse loamy Withnell series (§ 76), but in Clwyd also by the fine loamy Manod series (§ 60). Small patches of Neath association are mapped over older rocks in south-west Dyfed near St David's, around St Ann's Head and near Pembroke. Here some coarse loamy brown earths of the Rivington series (§ 76) are included, while wetter associate soils are rare.

Key to component soil series

	Unmottled soils; rock within 80 cm	1
	Subsoils faintly mottled above 60 cm or distinctly mottled between 40 and 80 cm	3
1.	With brown subsoil; fine loamy	2
	With brightly coloured subsoil; over sandstone	4
2.	Over sandstone	NEATH
	Over mudstone or slate	Denbigh
3.	Fine loamy	NERCWYS
	Fine loamy over clayey	CHERUBEER
	Clayey	Halstow
4.	Coarse loamy	Withnell
	Fine loamy	Loxhore

Soil water regime

Neath series and the other ungleyed soils in this association have porous profiles over permeable substrata, so are rarely wet (Wetness Class I), although ill-timed use can reduce porosity causing occasional wetness (Wetness Class II). Nercwys and Cherubeer series, with their slowly permeable subsoils are seasonally wet (Wetness Class III), as are Halstow series and the included surface-water gley soils.

The soils readily absorb rain except on steep slopes and where there is a large proportion of wetter land. Hydrological studies on surface-water gleys in similar parent material in Devon (Harrod 1981) indicate that the underlying rubbly Head is permeable. In average years moisture reserves in the main soils are adequate. Even in the driest districts the soils are only slightly droughty for grass and non-droughty for other crops.

Cultivation and cropping

Grassland farming dominates with permanent pastures usually more numerous than leys. In west and north Wales cereals are grown locally on a small scale. In the mild, coastal part of Dyfed near St David's, Neath soils are used for early potatoes. With sufficient moisture reserves and favourable drainage, Neath series is well-suited to grassland, having little risk of damage by poaching or traffic. In dry districts, represented by Cresselly (Fig.46) Neath series is workable for long periods in autumns even in wet seasons, but there is usually less opportunity for landwork in spring. Because of the moderately high silt and fine sand contents, bare soil caps and slakes under rain impact.

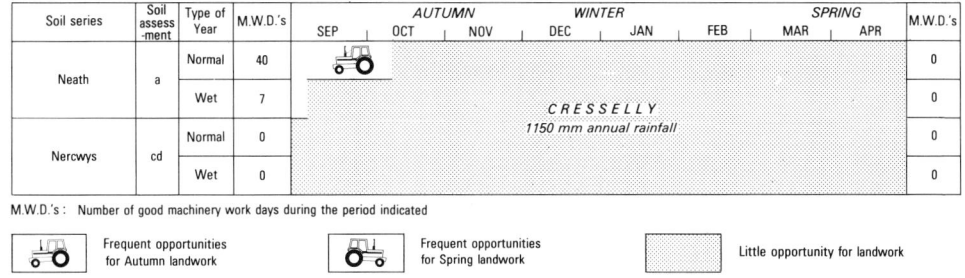

M.W.D.'s : Number of good machinery work days during the period indicated

Figure 46. The effects of soil and climate on landwork, Neath association

Other land use

Unmanaged oak woods occupy some steep slopes, but coniferous forestry is inextensive, only Douglas fir *(Pseudotsuga menziesii)* plantations at Penllergaer near Swansea, being of any size.

§ 67. NERCWYS ASSOCIATION
542

The Nercwys association consists of deep stony fine loamy soils in drift derived mainly from Carboniferous sandstones and shales. It is mapped in Clwyd, Devon, Derbyshire, North Yorkshire, Durham, Cleveland and Northumberland. The main soil is the Nercwys series of fine loamy stagnogleyic brown earths in drift with siliceous stones. It covers approximately half the land and usually occupies shedding sites or those where permeable bedrock occurs at between 1 and 2 m depth. The associated Brickfield series, fine loamy cambic stagnogley soils, occupies one-third of the area and is found mainly where the substratum is less permeable so soil conditions are wetter. The landscape is gently rolling but locally there are slopes up to 11 degrees. Brief profile descriptions are given of Nercwys series in § 67 and of Brickfield series in § 23.

In Clwyd, the association occurs south-west of Wrexham where the till, up to 30 m thick, mainly derived from the underlying Carboniferous sandstone and mudstone, includes a little material brought in by Irish Sea ice. Both the Nercwys and Brickfield soils contain lenses of coarse loamy or sandy material formed by weathering of stones. Coarse loamy Wick soils (§ 93) occur on isolated deposits of glaciofluvial material. Further north, the association occupies slopes running down to the Dee at Holywell, where Nercwys and Neath series have been mapped by Thompson (1978). In Radnor there are Nercwys soils near Gladestry.

Key to component soil series

	Unmottled; coarse loamy, sandstone within 80 cm	Rivington
	Mottled	1
1.	Subsoil faintly mottled above 60 cm or distinctly mottled between 40 and 80 cm	2
	Prominently mottled or greyish above 40 cm	4
2.	Coarse loamy	Arrow
	Fine loamy over clayey	Kearby
	Fine loamy	3
3.	Brownish subsoil	NERCWYS
	Reddish subsoil	Salwick
4.	Fine loamy	BRICKFIELD
	Fine loamy over clayey	Dunkeswick

Soil water regime

Waterlogging in winter months, above the slowly permeable subsoil places Nercwys soils in Wetness Class II or III, depending on the thickness of permeable upper horizons and the existence of artificial drainage. While Brickfield soils are in Wetness Class IV, or V if undrained, Neath profiles are permeable (Wetness Class I). Nercwys soils can be slightly droughty for grass but not for cereals or potatoes.

Cultivation and cropping

Dairying is the main source of income, the soils being mostly in grass on short and long term rotations. Poaching risk restricts winter use, particularly on the wettest soils. Cereal crops, mostly barley, are most common where either a coarse loamy topsoil affords better conditions for cultivation or the climate is drier. Autumn cultivations can be made over several weeks but the soils are mostly too wet in spring, when occasional days following dry spells will afford the only opportunities for landwork if compaction is to be avoided. There is some woodland, oak being the main species.

§ 68. NEWCHURCH 2 ASSOCIATION

814 c

This association is extensive on stoneless clayey marine alluvium in Humberside, Lincolnshire, Norfolk, Essex, Kent, Sussex, Somerset, Gloucestershire, Avon and south

Wales. The Newchurch series, which belongs to the pelo-calcareous alluvial gley soils, dominates the association which also contains non-calcareous clayey Wallasea series and, locally, calcareous silty Agney (Appendix 2, p.309), Romney and Blacktoft series and silty over clayey Stockwith series (Reeve and Thomasson 1981). Brief profile descriptions of Newchurch and Wallasea series are given in § 89.

These soils were formerly mapped by Crampton (1972) as the Wentlloog series and cover some 67 km² from the Caldicot Levels in Gwent along the Severn to Cardiff. The Newchurch series is dominant with some Wallasea but few other soils.

Key to component soil series

	Subsoils non-calcareous above 40 cm; clayey	WALLASEA
	Subsoils calcareous above 40 cm	1
1.	Prominently mottled or greyish above 40 cm	2
	Subsoil faintly mottled above 60 cm or distinctly mottled between 40 and 80 cm	3
2.	Clayey	NEWCHURCH
	Fine silty	Agney
	Silty over clayey	Stockwith
3.	Coarse silty	Romney
	Fine silty	Blacktoft

Soil water regime

The land is very low-lying and protected from high spring tides by sea-defences. Most of the soils are at least moderately permeable but field drainage systems and pump drainage are necessary for efficient groundwater control. These soils are usually waterlogged for long periods in winter (Wetness Class IV) particularly where field drainage systems have not been installed and where the arterial system depends on gravity drainage. The soils are non-droughty for arable crops and grass.

Cultivations and cropping

Though there is some evidence that the Romans grew cereals here after reclaiming the land from the Severn estuary, it is now mainly used for grazing. The land poaches easily in the autumn and spring, especially where undrained. Field drainage systems are necessary for successful arable use. Recently some land has been drained and about one-tenth is again in arable use. Even after drainage there is serious risk of structural damage during cultivations which need to be timed carefully to coincide with drier soil conditions. Potassium and magnesium levels are generally satisfactory but phosphorus levels are often low when poor-quality grassland is brought into arable use. Manganese deficiency sometimes occurs on Newchurch soils.

§ 69. NEWNHAM ASSOCIATION
541w

The Newnham association includes reddish and occasional brownish, mainly well drained soils in river terrace deposits and associated drift. These versatile loamy soils often rest on gravel at relatively shallow depth and occur on level terraces along the main valleys in the Midlands and South West England, and locally in Powys. The terrace gravels, composed mainly of siltstones, sandstones and hard Bunter quartzites, have a sandy loam matrix. The association is dominated by the reddish coarse loamy typical brown earths, Newnham series (Palmer 1972), over non-calcareous gravel. The associated fine loamy Huntworth series (Findlay 1965) and coarse loamy Rushwick series (Palmer 1982) are both typical argillic brown earths developed in the same parent material. Similar deeper but brownish porous soils of the Hopsford series (Whitfield and Beard 1977), seasonally affected by groundwater, are locally extensive. Most of the soils are slightly stony in the surface horizon but stoniness increases with depth. Silt contents, approaching 50 per cent in the upper horizons, decline in the gravelly subsoils. A full description of the association, with a key to the main soil series is given by Ragg *et al.* (1984).

Almost 13 km² are mapped in Powys in the Wye and Usk valleys. The reddish Newnham soils predominate but the depth to gravel is variable and deeper reddish coarse loamy Oglethorpe soils – formerly Castleton series (Crampton 1972) – are common.

Newnham soils on permeable gravels are well drained (Wetness Class I). Hopsford soils, although affected by fluctuating groundwater, are relatively porous (Wetness Class I or II depending on provision of underdrainage). The Newnham association occupies level ground and readily accepts winter rainfall, unless structure has been weakened by repeated or ill-timed cultivations.

The well drained coarse loamy soils and level land encourage cultivation. Cereals are widespread and good yields can be obtained from a wide range of crops, though subsoil compaction and subsequent surface wetness become limitations if the soils are worked too soon after rain. Permeability can be restored easily, however, by subsoiling or deep cultivation. The large proportion of sand and silt make the soil susceptible to slaking particularly where organic matter is depleted. Seedlings emergence is checked if, after sowing, heavy rain is followed by dry weather.

§ 70. NEWPORT 1 ASSOCIATION
551d

The Newport association consists mainly of well drained medium sandy soils formed in glaciofluvial and river terrace deposits and, in a few places, stoneless, aeolian sand. Although widely distributed in England and Wales, the association is rarely extensive, but in many wetter districts, it provides the only well drained, easily-worked soils suitable for arable use. Relief varies from gently to moderately sloping.

The principal soil is the Newport series which belongs to the typical brown sands. Other main soils include the well drained coarse loamy Wick series, typical brown earths, and the medium or coarse sandy Blackwood soils, typical sandy gley soils, which are affected by seasonal groundwater where undrained. The Rudge series, stagnogleyic brown sands, common in some areas, resembles the Ollerton series, except that the subsoil mottling results from a slowly permeable, fine or coarse loamy, subsoil horizon rather than from fluctuating groundwater. Brief profile descriptions of the Newport, Blackwood and Rudge series are given below, and of the Wick series in § 80.

Newport series
(Full description Jones 1975, p.38)

0–25 cm — Ap
Dark brown, slightly stony sandy loam or loamy sand.

25–55 cm — Bw
Brown, slightly stony loamy sand or sand; weak fine subangular blocky structure.

55–120 cm — Cu
Yellowish red or brownish yellow, slightly stony sand; single grain structure.

Blackwood series
(Full description Hollis 1978, p.101)

0–20 cm — Ap
Very dark greyish brown, slightly stony or stoneless loamy sand.

20–35 cm — Bg1
Pale brown, mottled, slightly stony loamy sand; weak medium and coarse subangular blocky structure.

35–90 cm — Bg2
Light brownish grey, mottled, slightly stony loamy sand; weak medium subangular blocky or single grain structure.

90–100 cm — Cg
Greyish brown, mottled, stoneless, slightly or moderately stony sand; single grain structure.

Rudge series
(Full description Hollis 1978, p.74)

0–30 cm — Ap
Dark brown, slightly stony sandy loam.

30–55 cm — Eb(g)
Yellowish red, slightly mottled, slightly stony loamy sand; very weak medium subangular blocky structure.

55–85 cm — Btg
Reddish brown, mottled, slightly stony loamy sand; weak coarse prismatic structure.

85–105 cm — 2Btg
Reddish brown, mottled, slightly stony sandy clay loam or sandy loam; weak coarse prismatic structure.

In Wales the association occurs on glaciofluvial drift in the Lleyn Peninsula, near Cardigan and on river terrace deposits around Wrexham (Fig.47). Relief varies from gently to moderately sloping and is often hummocky. Rabbit warrens marked by eroded patches of loose sand commonly occur in grassland on sloping ground. The land ranges from 6 m to about 200 m O.D. Exposure is a limitation near the west coast.

The typical soil pattern is determined by relief and the depth to groundwater. Newport or, where the drift is loamy, Wick series is widespread on the dry upper slopes and crests. Seasonal groundwater gives strongly mottled Blackwood soils in the lowest sites. The Rudge series is locally important at intermediate positions. Other soils include Ollerton series (King 1977) where groundwater only marginally affects the soil profile, and Salwick (§ 80) and Clifton (§ 24) series, on lenses of reddish till. Quorndon series (Appendix 2, p.321) also occurs on seasonally waterlogged coarse loamy drift. Where the land is in heath or has been reclaimed only recently, Crannymoor series (Furness 1971), characterized by an ochreous subsoil layer, is likely to be found.

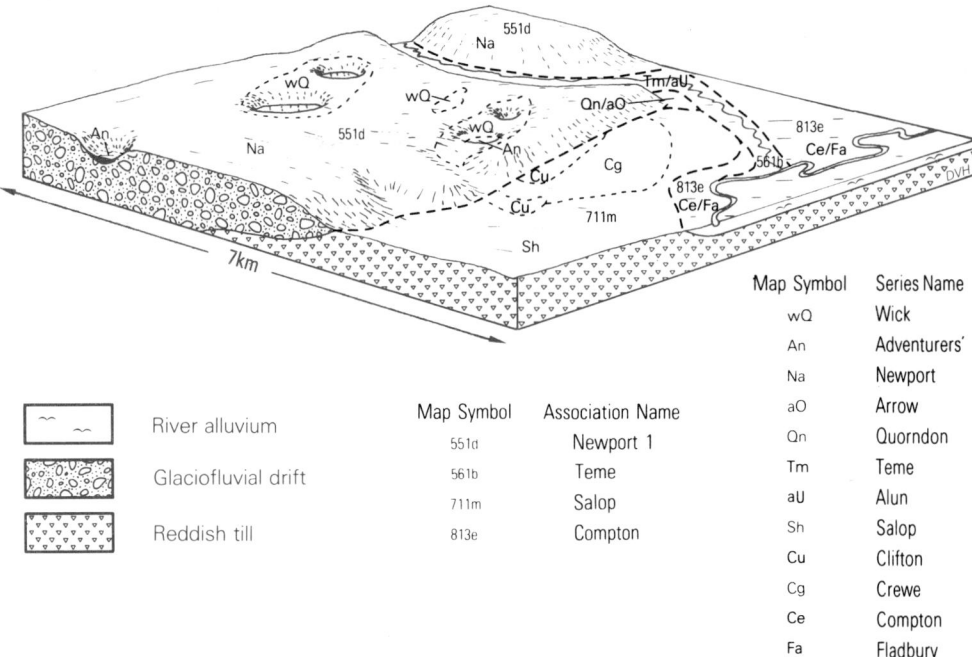

Figure 47. Soils and landscape north-east of Wrexham

Key to component soil series

Unmottled soils	1
Mottled soils; sandy	2

1. Sandy — **NEWPORT**
   Coarse loamy — **WICK**

2. Subsoil faintly mottled above 60 cm or distinctly — **RUDGE** or
   mottled between 40 and 80 cm — Ollerton
   Prominently mottled or greyish above 40 cm — **BLACKWOOD**

Soil water regime

Free drainage and summer droughtiness are the outstanding features of the association, particularly around Wrexham. Drainage measures are not necessary except for Blackwood soils, where field ditches often suffice to remove excess water yet leave groundwater within rooting depth. The Newport and most of the Ollerton soils are well drained (Wetness Class I). Undrained Ollerton soils however, suffer from occasional waterlogging (Wetness Class II), especially where rainfall is high. Waterlogging in the Blackwood series depends on drainage and climate, varying from occasionally to seasonally waterlogged (Wetness Class II or III). Most component soils are permeable and absorb winter rainwater without run-off but soil water reserves for demanding crops like grass and potatoes are often less than the soil moisture deficit. The Newport and Wick series are slightly or even moderately droughty under most arable crops and can be very droughty under grass in the drier districts for example around Wrexham (Table 22). Droughtiness is uncommon on the Ollerton series and rare on the Blackwood series because groundwater is available.

Table 22
Profile Available Water (A.P. mm), Crop-adjusted Mean Moisture Deficit (M.D. mm)
and Droughtiness Class for extensive crops—Newport 1 Association

	Newport series		Wick series	
Location	Wrexham	Monington	Wrexham	Monington
Grid Ref.	SJ350520	SN130440	SJ350520	SN130440
Grass				
A.P.	85	85	130	130
M.D.	138	70	138	70
Droughtiness	very droughty	slightly droughty	moderately droughty	non-droughty
Winter wheat				
A.P.	95	95	140	140
M.D.	94	50	94	50
Droughtiness	slightly droughty	slightly droughty	slightly droughty	non-droughty
Spring barley				
A.P.	95	95	140	140
M.D.	89	50	89	50
Droughtiness	slightly droughty	slightly droughty	non-droughty	non-droughty
Potatoes				
A.P.	70	70	100	100
M.D.	92	30	92	30
Droughtiness	moderately droughty	slightly droughty	slightly droughty	non-droughty

Cultivation and cropping

The land is easily worked, and in wetter parts of Wales provides the best soils for arable use. Cereals, potatoes and root crops are grown. The effects of soil properties and climate on land work in two contrasting areas of Wales are summarized in Figure 48. The soils can be cultivated in the winter months. Continuous arable use reduces reserves of organic matter and weakens topsoil structure so compaction and subsequent surface wetness result if the ground is worked under unsuitable conditions although these can readily be corrected by subsoiling. Surface capping can occur after heavy rain, especially if the seedbed is fine.

Grass yields vary widely. In the drier Wrexham district, grass grows well in spring and there is little risk of poaching, but summer yields are small because of drought. In wetter

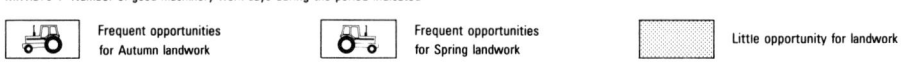

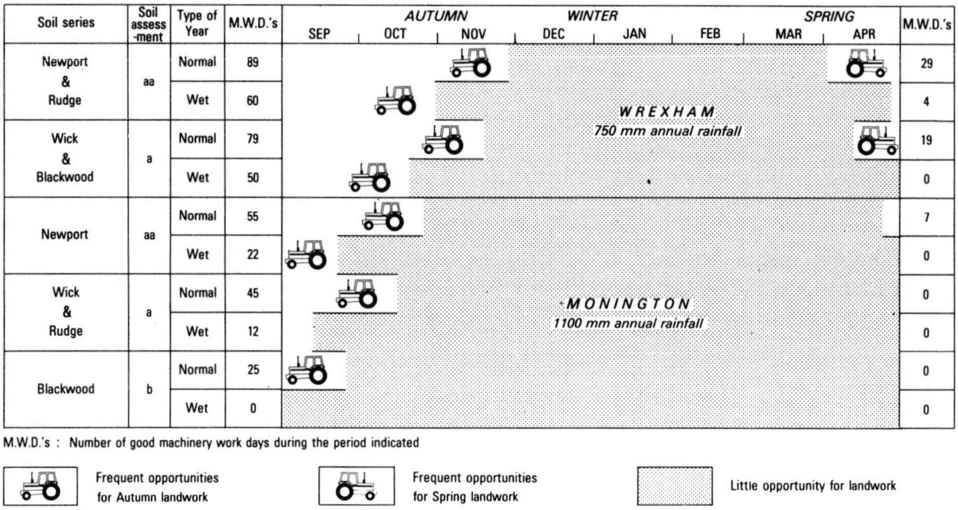

Soil series	Soil assess -ment	Type of Year	M.W.D.'s	AUTUMN			WINTER			SPRING		M.W.D.'s
				SEP	OCT	NOV	DEC	JAN	FEB	MAR	APR	
Newport & Rudge	aa	Normal	89									29
		Wet	60									4
Wick & Blackwood	a	Normal	79									19
		Wet	50									0
Newport	aa	Normal	55									7
		Wet	22									0
Wick & Rudge	a	Normal	45									0
		Wet	12									0
Blackwood	b	Normal	25									0
		Wet	0									0

WREXHAM 750 mm annual rainfall

MONINGTON 1100 mm annual rainfall

M.W.D.'s : Number of good machinery work days during the period indicated

Frequent opportunities for Autumn landwork　　　Frequent opportunities for Spring landwork　　　Little opportunity for landwork

Figure 48. The effects of soil and climate on landwork, Newport 1 association

districts, freedom from poaching and ample growth throughout most seasons, make the land well-suited to pasture. Slurry spreading is possible for most of the year except on the wetter Blackwood series as the soils are able to accept heavy applications without excessive wetness and surface run-off. The soils are naturally acid and infertile but with lime, suitable fertilizers and irrigation the soils give good yields of a wide range of crops. Around Wrexham, there is large-scale fruit and vegetable cropping, particularly where irrigation is available. These sandy soils give clean crops which can be harvested without difficulty, well into the autumn. Potatoes, carrots, beetroot, brussels sprouts, peas, lettuce, onions, strawberries and raspberries are grown and yield well. Near towns, or in holiday districts, the land is suitable for pick-your-own fruit and vegetable growing. The association provides dry and firm ground so is also well-suited for garden centres, golf courses, playing fields and parks.

§ 71. OGLETHORPE ASSOCIATION
541D

The three main soil series in the Oglethorpe association are all typical brown earths in reddish drift. It is mapped in Wales around Brecon and west of Chepstow and in Lancashire, Northumberland and Cumbria. The land is generally undulating and the drift is mostly till, but there are some gravelly glaciofluvial deposits. The Oglethorpe soils are coarse loamy but similar fine loamy Newbiggin soils are also common. Topsoils and subsoils are generally reddish brown with fine subangular blocky structure and common or many medium and large sandstones increasing in abundance with depth. Where subsoils are gravelly within 80 cm depth the Newnham series is found and this is a common associate. Brief descriptions of main series are given below.

Oglethorpe series
(Full description Crompton and Matthews 1970, p.95)

0–25 cm — Ap
Dark reddish brown, slightly stony sandy loam.

25–50 cm — Bw
Reddish brown, moderately stony sandy loam; weak fine subangular blocky structure.

50–100 cm — BCu
Reddish brown, moderately stony sandy loam; single grain structure.

Newnham series
(Full description p.320)

0–20 cm — Ap
Reddish brown, stoneless or slightly stony sandy loam or clay loam.

20–50 cm — Bw
Reddish brown, moderately stony sandy loam or sandy silt loam; weak fine subangular blocky structure.

50–70 cm — 2Cu1
Reddish brown, very stony sandy loam; massive structure.

70–100 cm — 2Cu2
Reddish brown, very stony loamy sand; single grain structure.

Newbiggin series
(Full description p.320)

0–25 cm — Ap
Reddish brown, slightly stony clay loam.

25–50 cm — Bw
Yellowish red, slightly stony clay loam; moderate medium subangular blocky structure.

50–100 cm — BCu
Reddish brown, moderately stony clay loam; moderate coarse subangular blocky or prismatic structure.

The association is most extensive south of Brecon between the Usk and the Brecon Beacons (Fig.39, Plate 29) at heights from 100 to 300 m O.D. Newnham series is most common in the main valleys where there are hummocky glaciofluvial deposits and river

P.S. Wright

Plate 29. Pasture land in the Usk valley between Sennybridge and Brecon. The Oglethorpe association is on low undulating ground with the Milford association on the background hills to the north.

terraces. Cambic stagnogley soils, Hollacombe series (Clayden and Hollis 1984), are rare associates in rushy or wooded depressions.

Key to component soil series

Coarse loamy soils		1
Fine loamy soils		2
1.	No gravel within 80 cm	OGLETHORPE
	Gravel within 80 cm	NEWNHAM
2.	Unmottled	NEWBIGGIN
	Prominently mottled or greyish above 40 cm	Hollacombe

Soil water regime
The soils are permeable, naturally well drained (Wetness Class I) and readily absorb winter rainwater.

Land use
The annual rainfall ranges from 1,000 mm in Gwent to about 1,500 mm on the flanks of the Brecon Beacons, corresponding to field capacity periods ranging from 215 days to more than 270 days. With little risk of poaching because of the coarse textures these are valuable grassland soils, used for dairying at lower elevations and livestock rearing on higher ground. Some cereals are grown in Gwent, but otherwise there is little arable apart from forage crops because cultivation is restricted by the climate and widespread presence of large stones in the topsoil. Soil depth ensures good reserves of available water, usually more than 125 mm, so the soils are not droughty in most years when soil moisture deficits are around 80 mm. Many tree species grow well and enhance the scenic value of the land in the Brecon Beacons National Park.

§ 72. PARC ASSOCIATION
612a

This association consists mainly of fine loamy humic brown podzolic soils on mudstone, shale and sandstone (Parc series) with similar fine silty soils. It is mapped on gentle to moderate slopes on plateaux at 130 to 300 m O.D. in south-west Wales and on steep slopes of north-west Dartmoor where slates are also a parent material (Fig.49). Many Parc soils appear to have formed through cultivating stagnopodzols (Hafren and Hiraethog series, § 51) which occur within the association on uncultivated land such as commons.

Parc soils are permeable, well drained (Wetness Class I) and readily accept winter rain. They have dark, humose topsoils about 25 cm thick over brightly-coloured subsoils, passing to loose stony Head or deeply shattered shale or mudstone locally with sandstone. The associated Hafren series is similar but has a bleached and gleyed horizon directly

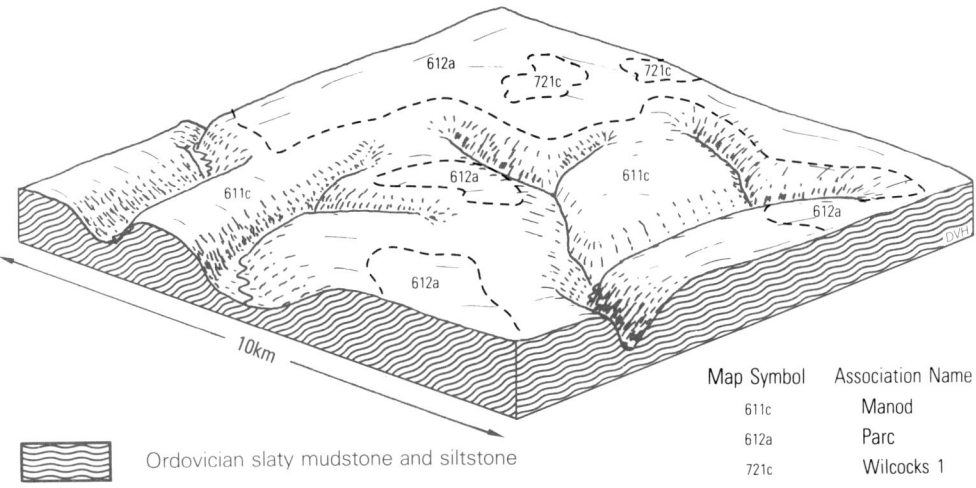

Figure 49. Soil associations on Ordovician rocks around Cynwyl Elfed, Dyfed

Map Symbol	Association Name
611c	Manod
612a	Parc
721c	Wilcocks 1

Ordovician slaty mudstone and siltstone

beneath the humose topsoil, while the Hiraethog series also has a thin ironpan. Cambic stagnohumic gley soils (Wilcocks series, § 95) in slowly permeable drift occupy wet depressions with rushes and purple moor-grass, and have a peaty top over a strongly gleyed, compacted subsoil. Stones in all the soils are mostly shale, mudstone and sandstone but in places large angular quartz boulders, of residual rather than glacial origin are scattered over the surface of fields or along the field boundaries to where they have been moved.

Parc series
(Full description Thompson 1978, p.36)

0–20 cm — Ap
Very dark greyish brown, slightly stony humose clay loam.

20–65 cm — Bs
Strong brown, moderately stony clay loam; weak fine granular structure.

65–100 cm — Cu
Yellowish brown, very stony sandy silt loam; massive structure.

Wilcocks series
(Full description Hollis 1975, p.75)

0–20 cm — Oh or Ah
Black, stoneless humified peat or humose clay loam.

20–50 cm — Bg
Light brownish grey, mottled, slightly stony clay loam or sandy clay loam; weak subangular blocky structure.

50–100 cm — BCg
Grey with many ochreous mottles, moderately stony clay loam; weak medium blocky or prismatic structure; high packing density.

Hafren series
(Full description p.315)

0–10 cm — Om or Ah
Dark reddish brown, stoneless semi-fibrous peat or slightly stony humose sandy silt loam or clay loam.

10–30 cm — Eg
Greyish brown, mottled, slightly stony clay loam or sandy silt loam.

30–45 cm — Bs
Strong brown, slightly or moderately stony sandy silt loam or clay loam.

45–60 cm — BCu
Yellowish brown, moderately or very stony sandy silt loam or clay loam.

At 60 cm — Cr
Silty shale *in situ.*

Some Parc subsoils show grey mottling superficially like that associated with wetness but structure faces are generally aerobic and the colours are thought to be related to podzolization. Dipwell measurements confirm that such profiles are not significantly wetter than ungleyed soils. The association includes soils similar to Parc series but developed in pockets of till or on sandstone. The Wilcocks series is a common associate on the higher land east of the Preseli hills but it disappears almost entirely elsewhere.

Key to component soil series

	Soils without brightly coloured subsoil; mottled throughout, fine loamy or fine silty	WILCOCKS
	Soils with brightly coloured subsoil	1
1	Without bleached subsurface; fine loamy	PARC
	With bleached subsurface; fine loamy or fine silty	2
2	With thin ironpan	Hiraethog
	Without ironpan	HAFREN

Land use

The climate is generally wet with about 1,500 mm rainfall and a field capacity period usually more than 250 days per year. Grass is therefore the main crop but swards are readily damaged by poaching because the humose topsoils retain much water. The grazing period is restricted and careful management is required. The proportions of ley grassland and cereals increase westwards on lower ground where rainfall is less and there are sufficent days in autumn and spring which are dry enough to allow cultivations without damage to the soil. On undergrazed commonland there is heather moor and small areas are afforested.

§ 73. PINDER ASSOCIATION
711q

The Pinder association consists mainly of slowly permeable brownish fine loamy and fine silty over clayey soils and is found in Shropshire and Powys. It is developed in locally-derived loamy drift containing Palaeozoic siltstones and sandstones and some far-travelled Welsh greywackes and igneous stones. It occurs between 75 and 300 m O.D. on the undulating Shropshire till plain but is confined to land below 150 m O.D. in Wales. The main soils are typical stagnogleys notably the fine loamy Pinder series and the fine silty over clayey Prolleymoor series (p.321). Less mottled soils, belonging to the Bishampton series (p.310), are found mainly on steeper slopes and are classified with the stagnogleyic argillic brown earths. There are smaller inclusions of Hamperley (Clayden and Hollis 1984), Cegin (§ 23), Stanway, Baschurch (p.300) and Brickfield series (§ 17). A full description of the association, with a key to the main soil series is given by Ragg *et al.* (1984).

The association is restricted to eastern Powys around the northern and eastern slopes of the Breidden Hills. Pinder series is usually dominant and Cegin series also present. The boundary between cambic stagnogley soils in moist western districts and typical stagnogley soils on the lowland plains is unclear and gradual. This land lies on that boundary.

Pinder, Bishampton and Cegin series are seasonally waterlogged (Wetness Class IV or III). All three series respond well to artificial drainage following which Pinder and Cegin soils are often Wetness Class III and Bishampton series Wetness Class II. Surface waterlogging results from dense slowly permeable subsoils and slow surface run-off from relatively flat land.

Dairying and livestock rearing are the main enterprises and most of the land is under permanent grass. The growing season is long and grass growth is seldom restricted by drought though utilization of early and late growth is often checked by wet topsoil conditions. When the soils are wet, grazing or silage harvesting causes soil compaction with a consequent deterioration in the sward and soil drainage. Careful grazing management is necessary, particularly on reseeded pastures.

§ 74. REVIDGE ASSOCIATION
311a

Shallow peaty soils of the Revidge series on gritstone and sandstone are interspersed with deeper peat soils of the Crowdy and Winter Hill series in this association. The soils are very acid, infertile and often wet, largely because rainfall is high and summer

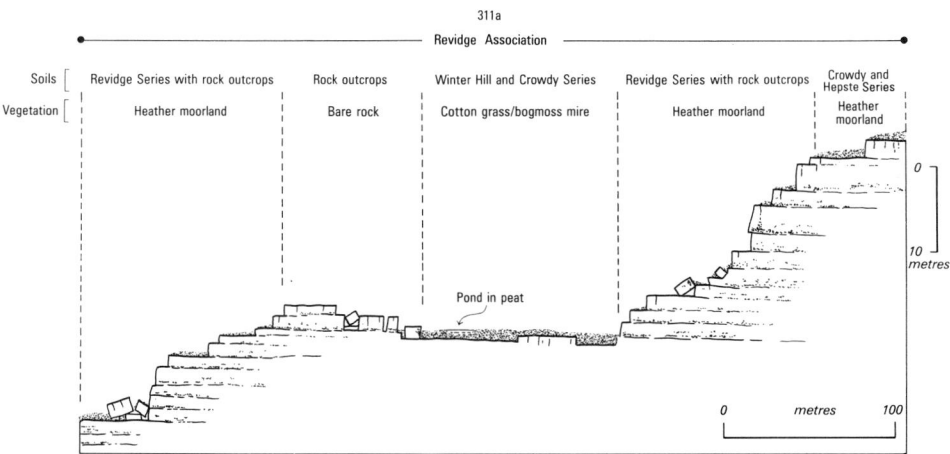

Figure 50. Cross-section of Revidge Association showing soil series and vegetation

temperatures low. The terrain is generally rough and mountainous with heather moor and extensive bare rock and scree. The association dominates the Rhinog hills north of Barmouth, where it rises to 700 m O.D. and receives an annual average rainfall of 2,000 mm. Rankers occupy up to half the land, peat soils about a fifth and bare rock a quarter to a third (Fig.50). There are small areas of stagnopodzols, including Belmont (Jarvis 1977), Hiraethog and Hafren series (§ 51) and coarse loamy brown podzolic soils belonging to the Withnell series (§ 98) occupy bouldery lower slopes under local remnants of deciduous woodland.

Revidge series
(Full description Carroll and Bendelow 1981, p.31)

0–20 cm — Oh or Ah
Dark reddish brown to black, humified peat or dark reddish brown, stoneless humose sandy loam; moderate fine blocky structure.

At 20 cm — R or Cu
Hard sandstone or grit or extremely stony sandy loam.

Crowdy series
(Full description p.311)

0–10 cm — Om
Black, semi-fibrous peat, moderate granular structure.

10–20 cm — Oh1
Dark brown, stoneless humified peat; massive structure.

20–100 cm — Oh2
Black, stoneless humified peat; massive structure.

Winter Hill series
(Full description Carroll *et al.* 1979, p.88)

0–10 cm — Om1
Black, semi-fibrous peat.

10–40 cm — Om2
Dark reddish brown, semi-fibrous or amorphous *Eriophorum-Sphagnum* peat; moderate coarse platy structure.

40–70 cm — Om3
Dark reddish brown, semi-fibrous *Eriophorum-Sphagnum* peat; weak coarse platy structure.

70–120 cm — Om4
Dark reddish grey, semi-fibrous *Eriophorum-Sphagnum* peat; massive structure.

Most Revidge soils have organic rather than humose surface horizons. There is usually a thin surface layer of heather litter, which overlies the main horizon of well humified amorphous peat, which shows a blocky structure when moist or dry. This peat normally lies directly on bedrock or unweathered coarse scree and has a concentration of heather roots. Crowdy series which consists of deeper, acid, amorphous peats often rests on bedrock within 1 m depth. The shallower Crowdy profiles dry out occasionally encouraging humification and the development of a weak blocky structure. The included Winter Hill soils are dominantly fibrous, seldom if ever becoming dry. They may be solely rain fed or flushed with laterally moving water, but, because of the poor nutrient status of the underlying rock, the soil is very acid. Winter Hill soils are composed almost entirely of *Sphagnum* and cotton-grass remains. They rarely have rock within 1 m depth.

Peat development is partly related to rainfall but differences in hardness, bedding and jointing of the rock account for much of the soil variation. In the Rhinog hills, the rock is extremely hard, yielding little fine earth. Where rock bedding is thickest, the land is roughly stepped with Revidge soils on the lip of each ledge above a sheer rock face (Fig.50). Peat thickens towards the backs of the ledges to give Crowdy soils. Winter Hill series occupies intricate patterns of mires where *Eriophorum* and *Sphagnum* have filled waterlogged hollows with fibrous peat. It also covers mountain tops to form blanket peat where rainfall is greatest. Bare rock is most extensive on Rhinog Fawr but to the south, where the rocks are more thinly bedded, terracing is less marked and peat soils confined to the mires. Shaly drift occupies pockets around the periphery of the gritstone outcrop. Stagnopodzols of the Belmont series occur where loamy drift covers the sandstone, and Hiraethog and Hafren series are found where the drift rests on shales. Stagnohumic gley soils are minor constituents in flush sites. On the lowest, often wooded, but boulder-strewn slopes, well drained Withnell soils have characteristically brightly coloured iron-enriched subsoils. Outcrops of shale on the ridge linking Diffwys to Llawlech carry loamy podzolic and stagnogleyic rankers, these have thin peaty mats over loamy horizons resting on shale at very shallow depth.

Key to component soil series

	Peat, rock within 30 cm	1
	Thicker peat	2
	Mineral soils	5
1.	Over gritstone or sandstone	REVIDGE
	Over slate, mudstone or siltstone	Skiddaw
	Over acid igneous rock	Bangor
2.	Rock within 80 cm	Hepste
	No rock within 80 cm	3
3.	Amorphous peat	CROWDY
	Fibrous peat	4
4.	With moss and cotton-grass remains	WINTER HILL
	With grass and sedge remains	Floriston
5.	With thin, surface leaf mould or brown, distinct topsoil over brightly coloured subsoil; coarse loamy	Withnell
	With peaty topsoil over bleached subsurface horizon	6
6.	Without ironpan; fine silty or fine loamy	Hafren
	With thin ironpan; coarse loamy	Belmont

Land use

The shallow peaty soils, rockiness and steep slopes preclude agricultural improvement. The cool climate and exposure on the higher land are further severe limitations to productivity. In the Rhinog hills these factors and the poor soil fertility are reflected in the small numbers of sheep that graze there during the summer. The vegetation has little value for grazing and improvement is possible only in a few peripheral areas. Surface horizons are often wet, poach easily and cannot support machinery or heavy stocking without serious damage. The soils are easily eroded, and footpaths are soon worn to rock.

Afforestation is hampered by the rugged terrain which prevents drainage improvement and ridging for planting. Exposure and shallow waterlogged soils favour Lodgepole pine *(Pinus contorta)* but windthrow, particularly on western slopes, is an enduring limitation. The soils are almost continually wet and their vegetation has rarely if ever been burnt. They therefore support some of the most luxuriant heather and atlantic bryophyte communities in Britain. The patches of peat soils carry bog-moss *(Sphagnum* spp.*)* with varying amounts of deer grass *(Trichophorum cespitosum)*, purple moor-grass *(Molinia caerulea)* and cotton-grass *(Eriophorum* spp.*)*. Heather communities on the high peaks are the closest to dwarf montane heath in Wales and are accompanied by grassland of very low productivity. Woodland, which may once have covered the entire Rhinog hills, is now restricted to steep lower slopes. In places the terrain is scenically attractive and provides a valuable amenity for walkers.

§ 75. RHEIDOL ASSOCIATION
541v

This association consists mainly of well drained permeable fine loamy brown earths over gravels. It is mapped on river terraces and glaciofluvial outwash deposits and covers 87 km² mainly in narrow strips along rivers and in larger areas around river confluences. It is most extensive in Dyfed along the middle reaches of the Teifi and Tywi. Locally there are kettle holes, small enclosed depressions, some of which contain small lakes surrounded by ground-water gley soils.

The Rheidol series which belongs to the typical brown earths is most extensive with the typical brown podzolic Nefern series distinguished by its brighter ochreous subsoils. In hollows, gleyic brown earths of the Hopsford series occur with fine loamy wetter soils related to the Rockland series (Clayden and Hollis 1984). Deeper soils include the East Keswick (§ 36), Meline and Wigton Moor series (Appendix 2, p.326). Southminster soils (Sturdy 1976) are rare.

Brief descriptions of the main series are given below.

The terraces along the larger rivers lie up to 15 m above the floodplain from which they are usually separated by a steep bluff. Elsewhere, especially in the upper reaches of rivers the terraces are more recent so they grade gently into the alluvium. In places small strips of alluvial Conway (§ 26) and Teme series (§ 86) are included.

Key to component soil series

Unmottled soils; fine loamy	1
Mottled soils; fine loamy	4
1. Gravel within 80 cm	2
Deeper soils	3

2. With brown subsoil   RHEIDOL
    With brightly coloured subsoil   NEFERN

3. With brown subsoil   East Keswick
    With brightly coloured subsoil   Meline

4. Subsoil faintly mottled above 60 cm or distinctly
    mottled between 40 and 80 cm   HOPSFORD
    Prominently mottled or greyish above 40 cm   5

5. Gravel within 80 cm   Southminster
    Deeper soils   Wigton Moor

Rheidol series
(Full description p.322)

0–15 cm — Ap
Dark brown, slightly stony clay loam.

15–60 cm — Bw
Brown, moderately stony clay loam; weak fine subangular blocky structure.

60–80 cm — Cu
Yellowish brown, very or extremely stony sandy loam; single grain structure.

Nefern series
(Full description Bradley 1980, p.63)

0–15 cm — Ap
Dark brown, slightly stony clay loam.

15–30 cm — Bw
Dark yellowish brown, moderately stony clay loam; moderate medium subangular blocky structure.

30–50 cm — Bs
Strong brown, moderately stony clay loam; moderate fine subangular blocky.

50–75 cm — Cu
Loamy gravel with occasional very stony infillings from horizon above.

Hopsford series
(Full description Whitfield and Beard 1977, p.50)

0–25 cm — Ap
Dark brown, slightly stony clay loam.

25–55 cm — Bw(g)
Dark yellowish brown, slightly mottled, slightly stony clay loam; moderate medium subangular blocky structure.

55–75 cm — Bg
Brown, mottled, slightly stony clay loam; weak medium subangular blocky structure.

75–100 cm — BCg
Greyish brown with many ochreous mottles, moderately or slightly stony sandy clay loam; weak fine subangular blocky structure.

Land use

The land is usually more sheltered than that of surrounding slopes but some occurs near the coast (Plate 30). It is locally much valued, supporting arable crops such as winter barley as well as providing ley grassland and good permanent pastures. Near Welshpool summer temperatures are higher than elsewhere so there are more arable crops, although grass is dominant almost everywhere. In many districts, this level land with its well drained soils is the best available. The soils readily absorb winter rain so that waterlogging is rare and there is little risk of poaching. The soils are mainly non-droughty, but are moderately droughty in the Severn valley near Welshpool where moisture deficits are highest.

The association frequently occurs at the bridging points of rivers or at confluences where villages and towns have developed, or is occupied by old parkland, where the presence of hollows and kettle holes adds to the landscape value. The land is generally suitable for playing fields and camp sites where topsoils are not too stony.

Plate 30. Coastal lowlands near Llanon, Dyfed. The flat well drained arable land of the Rheidol association, is bounded by a plateau mainly with soils of the Denbigh 1 association. The Cambrian mountains with Plynlimon form the skyline.

§ 76. RIVINGTON 2 ASSOCIATION
541g

This association of loamy brown earths and brown podzolic soils over Palaeozoic, predominantly Carboniferous, sandstones and shales occurs on moderate to steep valley sides, hills and ridges throughout the Pennines and also in west Wales, north Devon and west Somerset. The land is mostly below 450 m O.D. The two soils which dominate the association, the Rivington series, typical brown earths, and the Withnell series, typical brown podzolic soils, are both coarse loamy, well drained and overlie hard sandstone within 80 cm depth. The subsidiary Heapey series, fine loamy stagnogleyic brown earths, overlies shale and is occasionally waterlogged. Brief profile descriptions of these series are given below.

Rivington series
(Full description Reeve 1975, p.86)

0–20 cm — Ap
Dark greyish brown, slightly stony sandy loam or sandy silt loam.

20–50 cm — Bw
Yellowish brown, slightly or moderately stony sandy loam or sandy silt loam; weak medium subangular blocky structure.

At 50 cm — R or Cu
Hard or soft sandstone or extremely stony sandy loam.

Withnell series
(Full description Crompton 1966, p.79)

0–15 cm — Ah
Dark brown, slightly stony sandy loam or sandy silt loam.

15–50 cm — Bs
Strong brown, moderately stony sandy loam or sandy silt loam; moderate fine subangular blocky or granular structure.

At 50 cm — R or Cu
Hard sandstone or extremely stony sandy loam.

Heapey series
(Full description Carroll, Hartnup and Jarvis 1979, p.62)

0–20 cm — Ap
Dark greyish brown, slightly stony clay loam.

20–45 cm — Bw
Yellowish brown, slightly stony clay loam; weak fine blocky or subangular blocky structure.

45–60 cm — Bg
Yellowish brown, slightly mottled, moderately stony clay loam; moderate coarse blocky or platy structure.

At 60 cm — Cr or BCg
Platy siltstone or mudstone or extremely stony mottled silt loam or clay loam; high packing density.

The association in Wales is over Cambrian Sandstones between Harlech and Barmouth, and lies mainly below 200 m O.D. The small amount of included wetter land usually has coarse loamy Arrow series (§ 93), and where acid peat has accumulated, pockets of Floriston series (Clayden and Hollis 1984). Other soils of minor occurrence including Melbourne series (Whitfield and Beard 1980) are listed in the key.

Key to component soil series

	Unmottled soils	1
	Subsoils faintly mottled above 60 cm or distinctly mottled between 40 and 80 cm	7
1.	Shallow soils, sandstone within 30 cm; loamy	Newtondale
	Deeper soils, rock within 80 cm	2
2.	Over mudstone; fine loamy	Denbigh
	Over sandstone	3
3.	With bleached subsurface horizon; loamy	6
	Without bleached subsurface horizon	4
4.	Sandy	Bridgnorth
	Coarse loamy	5
5.	With brown subsoil	RIVINGTON
	With brightly coloured subsoil	WITHNELL
6.	With thin ironpan	Belmont
	With dark, humus-enriched subsoil; no ironpan	Anglezarke

7.	Fine loamy; mudstone within 80 cm	HEAPEY
	Coarse loamy	8
8.	Sandstone within 80 cm	Melbourne
	Deeper soils	Arrow

Soil water regime

The major soils are well drained (Wetness Class I). Excess winter rainwater generally passes rapidly downwards through the permeable subsoil, although there is some run-off on steep slopes.

Land use

As much of this association occurs in districts with more than 1,000 mm rainfall where the field capacity period exceeds 200 days, there is little opportunity for arable cropping. Most of the land is under grass, and used for livestock rearing. Some slopes are too steep for the maintenance of good pasture and erosion can be serious if steep slopes are ploughed. Elsewhere the association is well suited for intensively used grassland. Grass yields are good and there is ample growth throughout the season. Most sites have adequate moisture in normal years. Soil conditions rarely limit the access of stock and farm machinery and there is little risk of poaching. Reseeding is easy and slurry is readily absorbed. The soils tend to become acid and require frequent liming to prevent the formation of a peaty surface mat and reversion to poor pasture. The soils are well suited to timber production and can grow a wide range of species. Good yields can be had from Sitka spruce, Norway spruce, Japanese or hybrid larch and Douglas fir. There is some deciduous woodland, particularly of oak.

§ 77. ROCKCLIFFE ASSOCIATION
811d

The association occurs on reclaimed stoneless marine alluvium in north-west England, Wales and Lincolnshire. It contains coarse and fine silty typical alluvial gley soils, Rockcliffe and Tanvats series, and coarse silty gleyic brown alluvial soils, Snargate series. The main soils have a large silt or very fine sand content and are strongly mottled in the subsoil. Rockcliffe and Snargate soils generally become sandier with depth, although in some places they have fine silty or alternating silty and sandy subsoil layers.

   The soils, which are fully described by Jarvis R.A. *et al.* (1984), occur on the north-west Wales coast near Conwy, Aberffraw, Llandwrog, Pwllheli and Porthmadog. Rockcliffe and Tanvats series and their calcareous equivalents, the Wisbech (§ 97) and Agney (§ 9) series are the most common soils with Blankney soils (Clayden and Hollis 1984) in the wettest places. The Snargate series is inextensive and found mainly at Porthmadog where the best land in the association is found. At Pwllheli, there are limited areas of peat interstratified with the alluvium.

The main soils are seasonally waterlogged (Wetness Class IV) on the better land at Porthmadog but severely waterlogged (Wetness Class V) elsewhere. The ditches are drained by gravity as the land is not extensive enough to warrant pumped schemes. Effective drainage in the better areas could greatly reduce waterlogging (Wetness Class II) and increase the potential of the land to produce grass and carry stock.

The land is generally used for summer grazing. Manganese toxicity has been recorded at Llandwrog. The land provides a valuable winter feeding ground for many species of birds.

§ 78. ROWTON ASSOCIATION
571A

This association of yellowish brown silty soils, is extensive in the Welsh Borderland notably in south Shropshire, north Herefordshire and east Powys. It is found on gently undulating glaciofluvial terraces and till in the valleys of the Clun, Teme, Onny and Lugg. The terrace deposits and till are of similar composition and difficult to distinguish from each other (Hodgson 1972, Mackney and Burnham 1966). The included stones are mainly local in origin, generally siltstones, fine grained sandstones and shales with occasional Welsh greywackes and igneous stones.

On the terraces, the soils usually overlie gravel at less than 1 metre depth and are either silty typical argillic brown earths, Rowton series, or less commonly, fine loamy typical brown earths, Rheidol series (§ 75). The soils in till have slowly permeable subsoils, and are represented by fine silty stagnogleyic argillic brown earths, Hamperley series (Hodgson 1972) and the fine silty over clayey typical stagnogleys, Prolleymoor series (p.321). There are small inclusions of East Keswick (§ 36), Sannan (Lea 1975), Pinsley (Hodgson 1972) and Ludford series (Heaven 1978). A full description of the association, with a key to the main soil series is given by Ragg et al. (1984).

Rowton and Rheidol series dominate especially on gently undulating gravelly terrace deposits along the Teme and Lugg. These deposits contain patches of till and, away from the rivers, merge into mounded till country with occasional lenses of glaciofluvial outwash gravels. Hamperley and Sannan series are extensive on the compact tills with Prolleymoor soils generally restricted to hollows.

Narrow tracts of alluvium, adjacent to the streams have been included in the association and Rowton soils often merge gradually into alluvial soils as they become deeper and less stony.

Rowton and Rheidol series are well drained (Wetness Class I) because they overlie permeable gravels. The Hamperley soils are seasonally waterlogged (Wetness Class III) but on steep slopes, particularly under permanent grass where run-off is substantial, or where they are artificially drained they have a drier water regime.

This association provides good farming land. The level ground typical of the Rowton and Rheidol series is often used for cereals and, occasionally, roots whilst permanent pasture is more common on Prolleymoor series and to a lesser extent Hamperley and Sannan soils. Rowton soils are easy to cultivate but the large silt content makes them susceptible to capping on recently ploughed or sparsely vegetated fields.

Grass swards on Hamperley, Sannan and Prolleymoor soils poach and compact readily and therefore deteriorate as weed species colonize the bare patches. On Rowton soils there is little poaching risk though there is some risk from drought during dry summer spells.

§ 79. SALOP ASSOCIATION
711m

This association consists mainly of stagnogley soils with slowly permeable subsoil in reddish drift mostly derived from Permo-Triassic rocks. There is a small proportion of stagnogleyic argillic brown earths. As there is little run-off on the level or gently sloping land these slowly permeable soils are seasonally waterlogged. The association occupies large areas in Midland and Northern England and occurs on the narrow lowland along the coast of north Wales. The Salop series, fine loamy over clayey typical stagnogley soils, occupies one to two-thirds of the ground, Clifton series, similar but fine loamy throughout, is generally a minor associate but in Cheshire covers about a quarter. The clayey Crewe series, pelo-stagnogley soils, is included on level land. Coarse loamy over clayey Rufford soils occur locally where there are glaciofluvial deposits nearby. Stagnogleyic brown earths belonging to Flint series cover up to one-eighth of the ground, mainly on the steeper slopes. Salop, Flint and Crewe series are described briefly below, and Clifton series in § 24.

Salop series
(Full description Jones 1983, p.61)

0–25 cm — Ap
Very dark greyish brown, slightly stony clay loam.

25–45 cm — Eg
Brownish grey, mottled, slightly stony clay loam; moderate medium subangular blocky or prismatic structure.

45–100 cm — Btg
Yellowish red, mottled, slightly stony clay; moderate coarse prismatic structure.

100–120 cm — BCtg
Reddish brown, mottled, slightly stony clay; massive or coarse prismatic structure; sometimes with calcium carbonate concretions.

Flint series
(Full description Jones 1975, p.70)

0–25 cm — Ap
Dark brown slightly stony clay loam.

25–60 cm — Eb(g)
Brown, slightly mottled, slightly or moderately stony clay loam; moderate medium subangular blocky structure.

60–100 cm — Btg
Reddish brown, mottled, slightly stony clay; strong coarse angular blocky or prismatic structure.

Crewe series
(Full description Jones 1975, p.76)

0–20 cm — Ap
Very dark greyish brown, stoneless clay loam or clay.

20–50 cm — Bg
Strong brown, mottled, stoneless clay; moderate coarse angular blocky structure.

50–100 cm — BCg
Reddish brown, mottled, stoneless clay; strong coarse prismatic structure

Along the north coast of Wales these soils are found where reddish Devensian drift is sufficiently thick to impede drainage (Fig.51). There is a small area at Beaumaris on Anglesey but the largest extent is in the Vale of Clwyd and along the border with Cheshire and Shropshire where rigg and furrow and water-filled marl pits are common features of the landscape. The proportion of Clifton and Salop soils is determined by the depth of fine loamy drift over the reddish clay. East of Wrexham (Lea and Thompson 1978) there are fewer profiles of the Clifton series but Crewe replaces Salop series in the lowest and most level parts, probably in glaciolacustrine deposits. In the Vale of Clwyd and on Anglesey in particular, the proportion of Flint profiles is greater and Crewe soils are rare. Clifton profiles are most common at Hawarden where these soils adjoin the Clifton association to the north.

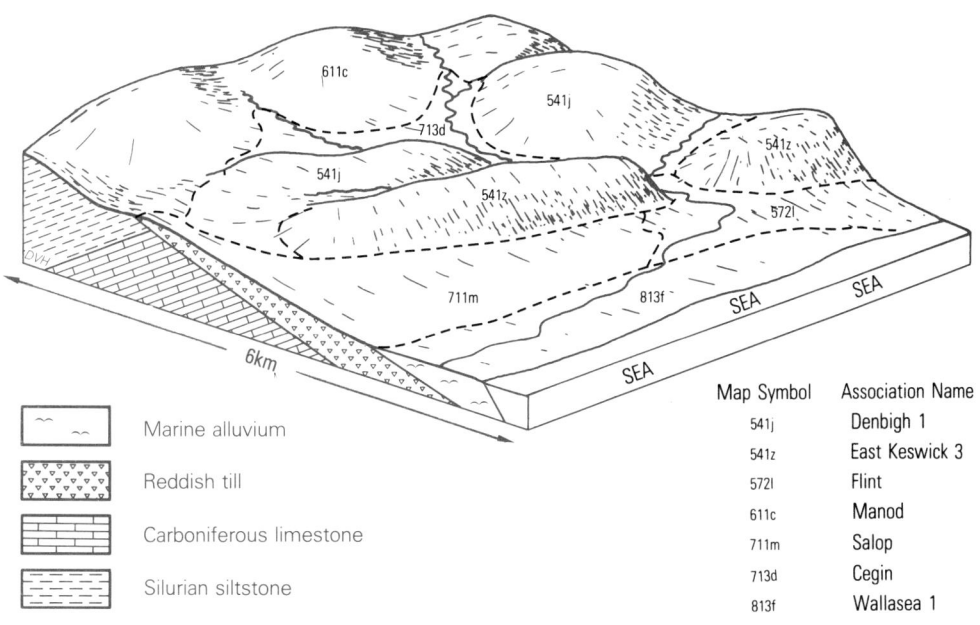

Map Symbol	Association Name
541j	Denbigh 1
541z	East Keswick 3
572l	Flint
611c	Manod
711m	Salop
713d	Cegin
813f	Wallasea 1

Marine alluvium

Reddish till

Carboniferous limestone

Silurian siltstone

Figure 51. Soil associations near Abergele

Key to component soil series

Soils prominently mottled or greyish above 40 cm	1
Subsoils faintly mottled above 60 cm or distinctly mottled between 40 and 80 cm	3

1.	Fine loamy	CLIFTON
	With clayey horizons	2

2. Fine loamy over clayey   SALOP
    Clayey   CREWE
    Coarse loamy over clayey   Rufford

3. Fine loamy over clayey   FLINT
    Fine loamy   Salwick

Soil water regime

Most of the soils when undrained are waterlogged for long periods in winter (Wetness Class IV). They can be improved to Wetness Class III with underdrainage especially in the drier eastern districts. Flint soils are better drained (mostly Wetness Class III) depending on climate and the efficiency of drainage measures. Surface waterlogging results from the combination of slowly permeable subsoil and slow surface run-off from relatively flat land. Crops may suffer from slight seasonal drought and grass from moderate drought in the driest districts.

Cultivation and cropping

These soils are traditionally used for grass production and form the basis of the dairy industry in Cheshire and Shropshire. In mid-Lancashire, rainfall prevents regular cultivation but elsewhere cropping is more mixed with a variety of cereals and fodder crops between leys. The land is generally difficult to work and timing of cultivations is critical especially on the wetter, heavier soils. With suitable drainage and regular subsoiling there are adequate machinery work days (Fig.52) in the autumn on all but clayey Crewe soils.

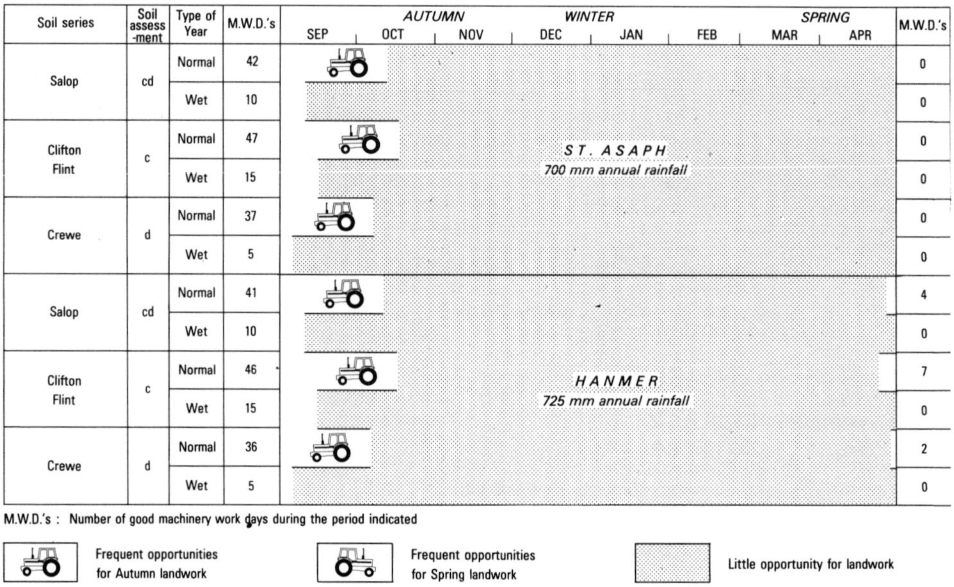

Soil series	Soil assess-ment	Type of Year	M.W.D.'s	AUTUMN			WINTER			SPRING		M.W.D.'s
				SEP	OCT	NOV	DEC	JAN	FEB	MAR	APR	
Salop	cd	Normal	42									0
		Wet	10									0
Clifton Flint	c	Normal	47				*ST. ASAPH*					0
		Wet	15				*700 mm annual rainfall*					0
Crewe	d	Normal	37									0
		Wet	5									0
Salop	cd	Normal	41									4
		Wet	10									0
Clifton Flint	c	Normal	46				*HANMER*					7
		Wet	15				*725 mm annual rainfall*					0
Crewe	d	Normal	36									2
		Wet	5									0

M.W.D.'s : Number of good machinery work days during the period indicated

Frequent opportunities for Autumn landwork

Frequent opportunities for Spring landwork

Little opportunity for landwork

Figure 52. The effects of soil and climate on landwork, Salop association

Grassland suitability varies with locality. In the west, potential grass yields are large because drought seldom restricts growth, and there is a valuable autumn flush. However, grazing and silage harvesting on wet soils lead to poaching and compaction with subsequent deterioration of grass growth and soil drainage. In the east, moisture stress restricts growth in mid- and late season, and in some years there is little autumn flush although the longer grazing period compensates for this to some extent. Overall, winter wetness restricts grazing to summer as the soils are easily damaged by untimely stocking. Slurry is stored in winter because spreading is impracticable while the land is wet.

Surface horizons become slightly acid despite calcium-rich subsoils so occasional liming is required.

Other land use

Although oak *(Quercus robur)* and holly *(Ilex aquifolium)* are the commonest woodland and hedgerow trees on these soils, most native trees thrive. The marl pits associated with these support valuable base-rich wetland communities (Day *et al.* 1982) and older pastures, particularly if undrained, can develop a distinctive base-rich vegetation.

The soils are ill-suited for use as recreation grounds, camping sites or sports fields. Although often level, the ground is too wet for winter use. Providing the clay subsoil is continuous without sandy pockets it forms an impermeable lining for pool and reservoir construction. The soils are abnormally corrosive however and buried ironwork should be protected (Argent and Furness 1979).

§ 80. SALWICK ASSOCIATION
572m

This association, developed in reddish till and glaciofluvial drift, consists of fine loamy soils with slight seasonal waterlogging, and well drained coarse loamy soils. It occurs sporadically throughout the Midlands and Northern England, and locally in south Wales and Clwyd. The land is mainly gently or moderately sloping, often forming broad ridges rising above low ground which carries surface-water gley soils.

The dominant Salwick series of stagnogleyic argillic brown earths, has coarse loamy upper horizons overlying dense fine loamy reddish till. The subsidiary Wick soils, typical brown earths, are developed in coarse loamy glaciofluvial drift and are very porous. In places, similar but fine loamy glaciofluvial drift gives gleyic brown earths of the Hopsford series. Where coarse loamy horizons greater than 40 cm thick overlie the till Nupend soils (Palmer 1976) are found; these were formerly mapped as a sandy loam phase of the Salwick by Hollis (1978). Gleyic brown earths belonging to the Arrow series (§ 93) developed in coarse loamy drift are also included. Clifton soils (§ 24) are found on low-lying, wet ground. Brief profile descriptions of the main component series are given below.

Salwick series
(Full description Jones 1975, p.59)

0–20 cm — Ap
Dark brown, slightly stony sandy loam or sandy clay loam.

20–35 cm — Eb(g)
Brown, slightly mottled, slightly stony sandy loam or clay loam; weak subangular blocky structure.

35–70 cm — Bt(g)
Reddish brown, slightly mottled, slightly stony clay loam; weak coarse prismatic structure.

70–100 cm — BCtg
Reddish brown, mottled, slightly stony clay loam; massive.

Wick series
(Full description Hollis 1978, p.57)

0–30 cm — Ap
Dark brown, slightly stony sandy loam or sandy silt loam.

30–60 cm — Bw
Brown, slightly stony sandy loam or sandy silt loam; moderate medium subangular blocky structure.

60–80 cm — Bw
Yellowish brown, slightly or moderately stony loamy sand or sandy loam; weak medium angular blocky or single grain structure.

80–120 cm — 2BCu
Brownish yellow, slightly or moderately stony sand or loamy sand; weak coarse angular blocky or single grain structure.

Hopsford series
(Full description Whitfield and Beard 1977, p.50)

0–25 cm — Ap
Dark brown, slightly stony clay loam.

25–55 cm — Bw(g)
Dark yellowish brown, slightly mottled, slightly stony clay loam; moderate medium subangular blocky structure.

55–75 cm — Bg
Brown, mottled, slightly stony clay loam; weak medium subangular blocky structure.

75–100 cm — BCg
Greyish brown with many ochreous mottles, moderately or slightly stony sandy clay loam; weak fine subangular blocky structure.

The association occupies undulating terrain with some kame and kettle topography on heterogeneous morainic drift. It commonly flanks the main valleys. Salwick series is the most extensive soil in reddish drift containing Devonian mudstone and sandstone. Wick soils which are brownish, coarser textured and more stony are developed in drift derived from Carboniferous rocks. The complex nature of the drift is such that other soils are widespread, notably reddish typical brown earths of the Newbiggin and Oglethorpe series (§ 71).

Key to component soil series

	Mottled soils	1
	Unmottled soils; brown, coarse loamy	WICK
1.	Subsoil faintly mottled above 60 cm or distinctly mottled between 40 and 80 cm	2
	Prominently mottled or greyish above 40 cm	5
2.	Reddish	3
	Brownish	4
3.	Fine loamy	SALWICK
	Coarse loamy	Nupend
4.	Fine loamy	HOPSFORD
	Coarse loamy	Arrow
5.	Coarse loamy; greyish	Quorndon
	Fine loamy; reddish	Clifton

Soil water regime

Salwick and Hopsford soils suffer from seasonal waterlogging (Wetness Class III) although drainage can be improved to Wetness Class II and Class I respectively. The well drained coarse loamy Wick soils are naturally well drained (Wetness Class I). Overall the soils readily absorb winter rainwater.

Cultivations and cropping

The association is used for mixed farming with grass and cereals but with the emphasis on dairying. There is also some horticulture, especially near Cardiff around Michaelstone.

§ 81. SANDWICH ASSOCIATION
361

The Sandwich association consists of deep calcareous and non-calcareous sandy soils on sand dunes, marine shingle and related beach deposits (Fig.53). The dunes, which rise to over 25 m O.D., are unstable in places. The association occurs in all regions and contains a wide variety of soils. It is exposed everywhere to sea winds.

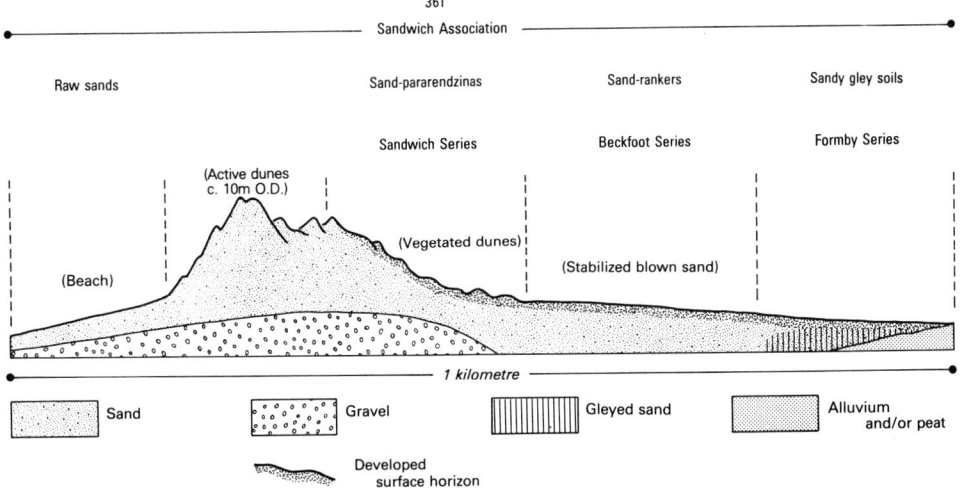

Figure 53. Distribution of soil series in the Sandwich association

Soil development on the dunes largely depends on the vegetation and the degree to which this has stabilized the dune system. The most extensive soil is the Sandwich series, typical sand pararendzinas, found on fixed dunes, usually over 100 years old and often considerably older. Here the vegetation is dominated by herbs, brambles, grasses and deciduous scrub. Younger, unstable dunes, particularly those nearest the sea, are unvegetated or only thinly colonized by marram grass *(Ammophila arenaria)* and lymegrass *(Leymus arenarius)*. Their shape constantly alters as a result of wind erosion and deposition and raw sands predominate there. Soils of the Beckfoot series, typical sand-rankers, occur on stabilized non-calcareous dunes. Wind erosion ceases only when the land surface nearly reaches groundwater level, at which stage small hollows (slacks) are formed. Larger elongated depressions (swales) result from the enclosure of lengths of beach by a newly-formed line of dunes. Typical sandy gley soils, Formby series, occur in wet hollows among the dunes. As sand is often blown on to fixed dunes, buried topsoils are widespread and it is common for several to occur in one profile where they form distinctive organic-rich layers.

Many of the dune systems have developed on the lee side of shingle bars or spits formed by tidal or other currents. There is great variety in their length, shape and age, and they have often been breached and re-formed in historical times. Brief profile descriptions of the main series and a key to the component series are given below.

Sandwich series
(Full description p.323)

0–5 cm — Ah
Dark brown, stoneless fine sand; very weak fine subangular blocky structure.

5–120 cm — Cu
Light brownish grey, stoneless fine sand; single grain structure; slightly calcareous.

"Raw sand"
(Full description Hall and Folland 1967, p.45)

0–90 cm — Cu
Light yellowish brown, stoneless sand; single grain structure; calcareous.

Beckfoot series
(Full description p.310)

0–10 cm — Ah
Dark reddish brown, stoneless sand; single grain structure.

10–30 cm — BCu
Dark brown, stoneless sand; single grain structure.

30-120cm
Dark yellowish brown, stoneless sand, single grain structure

Formby series
(Full description Hall and Folland 1967, p.46)

0–30 cm — Ap
Very dark greyish brown, stoneless medium loamy sand.

30–55 cm — Bg
Greyish brown, mottled, stoneless medium sand; single grain structure.

55–120 cm — Cg
Brown, mottled, stoneless medium sand; single grain structure.

Apart from Talacre Warren on the north-eastern tip of Wales, sand dunes are in three groups. On the south-west coast of Anglesey there are dunes at Newborough, Valley and Aberffraw. Ranwell (1959) has described the history and vegetation of Newborough Warren, considering it to be Boreal or earlier in origin. The sand at Valley extends more than one kilometre inland and, although construction of the airfield has destroyed most of the dunes, there are large expanses of stabilized dunes and wet slacks. There is a protecting seaward wall of high dunes colonized by marram grass, many of which are suffering erosion as a consequence of human trampling. Around Cardigan Bay there are narrower dune systems. At Morfa Harlech, dunes protect extensive alluvial and sandy gley soils. These dunes also support marram grass but some form more stable herb-rich fescue grassland. Willow scrub grows in slacks and parts have been afforested. Similar dunes lie seaward of the airfield at Morfa Dyffryn and further south at Tywyn and Ynyslas. On the south coast the largest system is at Pembrey (19 km²) flanking the Tywi estuary. There are also extensive dunes at Freshwater West, Pendine, Kenfig Burrows and Merthyr Mawr, with smaller areas elsewhere.

Most of the dunes have sufficient shell material to be calcareous although older dunes become acid through leaching. Raw sands, on unstable dunes, and the Sandwich series are the main soils, both on dunes and in dry slacks. Wet slacks have sandy gley soils. Coniferous plantations cause more rapid leaching of carbonates than natural vegetation, and the Beckfoot series is more commonly formed under them. Slightly podzolized soils are found under old plantations at Pembrey and Harlech.

Key to component soil series

	Vegetated soils	1
	Sparsely vegetated soils	3
1.	Prominently mottled or greyish above 40 cm	FORMBY
	Unmottled	2
2.	Calcareous	SANDWICH
	Non-calcareous	BECKFOOT
3.	Sandy	RAW SANDS
	Gravelly	Raw alluvial soils

Land use

The land is rarely used for agriculture because the soils are droughty, and unstable when ploughed. Topography, exposure and local salinity also help to limit land use to recreation (mainly golf courses) and amenity use, nature reserves or military training. There is some rough grazing in the moist hollows and on stabilized dunes. Because there is little clay and organic matter, even in the surface horizons, the soils are easily eroded. Disturbance along paths and tracks through the dunes, usually to the beaches beyond, increases vulnerability to erosion, which is common during strong winds. Blown sand sometimes reaches agricultural land or forms new, mobile dunes further inland. Attempts at stabilization by replanting with marram grass or laying brushwood are only successful where they are comprehensive and fenced off from the public. Walkways to the beach, fenced on either side, are sometimes necessary, supplemented by planking in the worst cases.

Plantations, both pure and in mixed stands, of Scots pine *(Pinus sylvestris)*, Corsican pine *(Pinus nigra* var. *maritima)* and Austrian pine *(Pinus nigra* var. *austriaca)* have been established with some success on fixed dunes by the Forestry Commission and private landowners. The effect of exposure is well demonstrated by the severe wind-pruning and premature death of trees on the seaward side of the plantations.

§ 82. SKIDDAW ASSOCIATION
311b

The Skiddaw association is extensive in Cumbria, on the northern and north-western hills of the Lake District, on high land above the upper Lune Valley and on Black Combe, in the extreme south-west. It is also found in small areas in north Wales. It is generally above 450 m O.D. on moderate to steep slopes, craggy in places, underlain by Palaeozoic slaty mudstone and siltstone. The highest land is on Skiddaw at over 915 m O.D. All districts are cold, wet and exposed. Humic rankers, Skiddaw series, predominate but on gentler slopes there are ironpan stagnopodzols, Hiraethog series, and raw oligo-fibrous peat soils, Winter Hill series. Brief profile descriptions of the main series and a key to the component soils are given below.

Skiddaw series
(Full description p.323)

1-0 cm — L & F
Root mat of sheep's fescue, heather and lichens.

0–10 cm — Oh
Black, humified peat (H10); moderate fine granular structure.

At 10 cm — R
Greywacke or siltstone.

Hiraethog series
(Full description p.316)

0–20 cm — Oh
Black, stoneless humified peat; weak medium subangular blocky structure.

20–35 cm — Eg
Light brownish grey, slightly mottled, slightly stony clay loam; common organic coats.

At 35 cm — Bf
Thin continuous ironpan black above with root mat and red below.

36–60 cm — Bs
Strong brown, slightly stony clay loam or sandy silt loam; weak medium angular blocky structure.

60–80 cm — Cu
Dark greyish brown, very stony sandy silt loam or clay loam; massive structure.

Winter Hill series
(Full description Carroll *et al.* 1979, p.88)

0–10 cm — Om1
Black, semi-fibrous peat.

10–40 cm — Om2
Dark reddish brown, semi-fibrous or amorphous *Eriophorum-Sphagnum* peat; moderate coarse platy structure.

40–70 cm — Om3
Dark reddish brown, semi-fibrous *Eriophorum-Sphagnum* peat; weak coarse platy structure.

70–120 cm — Om4
Dark reddish grey, semi-fibrous *Eriophorum-Sphagnum* peat; massive structure.

The association includes areas of deeper soils of the Manod (§ 60). Hafren and Hiraethog series (§ 51). In hillside flushes, the Wilcocks series (§ 96) and occasionally the Longmoss series (Kilgour 1979, p.106) are found.

Key to component soil series

Peat thicker than 40 cm		WINTER HILL
Other soils		1
1. Shallow soils, rock within 30 cm		2
Deeper soils		3

2. Peaty   SKIDDAW
    Brown, distinct topsoil; fine loamy   Powys

3. Brown, distinct topsoil; fine loamy, unmottled
    with brightly coloured subsoil   Manod
    Peaty topsoil; loamy   4

4. With bleached subsurface horizon over unmottled
    subsoil; fine loamy or fine silty   5
    Without podzolic features; prominently mottled or
    greyish above 40 cm, loamy   Wilcocks

5. With thin ironpan   HIRAETHOG
    Without ironpan   Hafren

Soil water regime

The shallow Skiddaw soils absorb little winter rain and in the wet climate lie waterlogged
for much of the winter. Shallowness and small available water capacity make the main
soils droughty in dry weather, however. Although ironpan stagnopodzols are deeper, the
ironpan and other layers reduce percolation. Deep peat has formed in very wet conditions
where run-off is slow.

Land use

The land has little agricultural value because of the shallow wet soils and rainy climate.
Most is under *Nardus*, grazed by a few sheep, but heather is becoming dominant on the
Hiraethog series. Although the vegetation is of moderate grazing value its improvement
would not be economic because of remoteness, persistent wetness, which generally cannot
be alleviated, and the risk of soil poaching. Some areas are used for forestry but the
shallow rooting depth causes vulnerability to windthrow. Most of the land is good
walking country and is included in the Snowdonia National Park.

§ 83. STANWAY ASSOCIATION
711a

The Stanway association consists of slowly permeable, seasonally waterlogged fine silty
soils developed over soft shales and siltstones. It is found mainly in Shropshire and
Herefordshire, but extends into Powys around Corndon Hill and Presteigne, covering
some 15 km². The easily eroded parent rocks give moderate to gentle valley slopes and
most of the land lies between 150 and 200 m O.D.

The association is dominated by fine silty, typical stagnogley soils of the Stanway series
(Appendix 2, p.323) and similar but fine silty over clayey soils of the Martock series
(Findlay 1976). Stanway soils are most common, but both series are closely interrelated
and often difficult to separate.

Most footslopes have a thin veneer of Head though this is usually silty and almost
indistinguishable from the underlying shales and siltstones. Fine loamy patches occur

downslope from local outcrops of fine-grained sandstones or igneous rocks. Here soils similar to the Stanway series, but with fine loamy upper horizons (Palmer 1976, pp. 91–93) are found along with Bardsey series, a fine loamy over clayey stagnogley soil. Fine silty stagnogleyic argillic brown earths of the Yeld series (Palmer 1976) occur on moderate to steep slopes and are less mottled than other soils in the association. Around Corndon Hill, prominently mottled fine silty over clayey Prolleymoor soils (Appendix 2, p.321) occur on patches of local drift. Detailed descriptions of the component soil series are given in Ragg *et al.* (1984).

The soils vary widely in base status, ranging from neutral to moderately acid according to the nature of the parent rocks. Calcareous shales and siltstones are common within the Silurian strata giving relatively base-rich profiles. Where limestones occur within or upslope of the association, flushes of base-rich soil water have produced patches of dominantly neutral soils. Elsewhere, however, non-calcareous rocks predominate and more acid soils are developed. Subsoils in the association are slowly permeable and impede the downward percolation of excess surface water causing periodic waterlogging in upper horizons. Stanway, Martock and Bardsey soils are all seasonally waterlogged for prolonged periods (Wetness Class IV), but Yeld soils which are developed on moderate or steep slopes shed much of their excess water by surface run-off and are seasonally waterlogged for slightly shorter periods especially in the driest areas around Presteigne (Wetness Class III). Because there is much gently or moderately sloping land and lower subsoils are slowly permeable, most soil water moves laterally through the topsoils and upper subsoils. Generally the association sheds excess winter rain rapidly.

Almost all the land is under permanent grass but some cereals are grown in the Camlad valley south of Corndon Hill. The grassland is used mainly for stock rearing, predominantly sheep. Intensive cattle rearing needs careful management to avoid poaching or damage by farm machinery. The large amounts of water retained in the fine silty topsoils and the frequent occurrence of subsoil waterlogging make soils poach easily. Risk is most severe around Corndon Hill but with careful management good grass yields are possible, although some of the growth early and late in the season cannot be fully used.

§ 84. STON EASTON ASSOCIATION
571a

The Ston Easton association is extensive over Lower Lias limestone in the Vale of Glamorgan (Crampton 1972) and is also found in the eastern Mendips (Findlay 1965) and the Forest of Dean.

It consists mainly of fine silty over clayey typical argillic brown earths on limestone (Ston Easton series). Thin shale partings are usually present in the limestone sequence. Profiles are stony and seldom deeper than 60 cm and are decalcified in the fine earth. The upper profile consists of a greyish brown topsoil over a yellowish brown horizon of silty clay loam. Below is a clayey, yellowish brown or strong brown argillic horizon which rests on and often infills cracks in the limestone bedrock. In places, often where the soils are deeper than 60 cm the clayey subsoil is reddish and these soils are classed as paleo-argillic brown earths (Nordrach series). Tetbury soils are also present; they, like the Ston Easton series, are typical argillic brown earths but less silty. Argillic pelosols of the Hornton series which are common locally were originally mapped by Crampton (1972) as Sigingstone series. Some profiles have no clay increase with depth and these typical brown earths belong to the fine silty Malham and fine loamy Waltham series. Inclusions of stony calcareous soils less than 30 cm deep are brown rendzinas belonging to the Sherborne series; these were mapped as Somerton series by Crampton (1972) and Findlay (1965). Similar, but decalcified clayey brown rankers of the Torbryan series (Clayden and Hollis 1984) occur rarely.

Ston Easton series
(Full description Crampton 1972, p.23)

0–20 cm — Ap
Dark greyish brown, stoneless or slightly stony silty clay loam.

20–40 cm — Eb
Yellowish brown, slightly stony silty clay loam; medium angular blocky structure.

40–60 cm — Bt
Yellowish brown, moderately stony silty clay; medium angular blocky structure.

At 60 cm — R
Limestone.

Nordrach series
(Full description p.320)

0–20 cm — Ap
Dark brown, very slightly stony silty clay loam.

20–40 cm — Eb
Brown, very slightly silty clay loam; fine subangular blocky structure.

40–75 cm — Bt
Yellowish red, stoneless or slightly stony clay; fine or medium angular blocky structure.

At 75 cm — R
Hard jointed limestone

Hornton series
(Full description Reeve 1978, p.86)

0–20 cm — Ap
Dark greyish brown, stoneless or slightly stony silty clay or clay.

20–55 cm — Bw
Yellowish brown, slightly mottled, stoneless or slightly stony silty clay or clay; strong medium angular blocky structure.

55–100 cm — Btg
Greyish brown, mottled, stoneless clay; strong coarse angular blocky structure.

At 100 cm — Cr
Interbedded limestone and shale.

Sherborne series
(Full description Burton 1981, p.60)

0–25 cm — Ap
Dark brown, stoneless or slightly stony silty clay.

25–30 cm — Bw
Yellowish brown, slightly or moderately stony clay; moderate medium subangular blocky structure; calcareous.

At 30 cm — Cr
Limestone

In Wales the parent material is of mixed origin, the soils being only partly derived from the strata below. The soils include stones that indicate a contribution from Irish Sea drift. Mineralogical studies (Crampton 1961) suggest some addition of sand blown from coastal dunes. There is also variation in the distribution of soils within the Vale of Glamorgan. Profiles are generally shallower and more clayey in the east, where the Sherborne series is common on ridge crests, while the Nordrach series occurs throughout the Vale as isolated caps on the plateaux.

Key to component soil series

Shallow soils, limestone within 30 cm; clayey	1
Deeper soils, limestone within 80 cm	2

1. Calcareous topsoil   SHERBORNE
   Non-calcareous topsoil   Torbryan

2. With reddish subsoil; fine silty over clayey   NORDRACH
   With brownish subsoil   3

3. Fine silty over clayey   STON EASTON
   Fine loamy over clayey   TETBURY
   Clayey   HORNTON
   Fine silty   Malham
   Fine loamy   Waltham

Soil water regime

All the soils of the association except Hornton series are permeable and well drained (Wetness Class I) and readily absorb winter rain.

Cultivations and cropping

The Vale of Glamorgan with about 1,000 mm of rainfall annually is one of the driest parts of Wales and Ston Easton soils are naturally fertile and highly valued for agriculture. Inland they are used for mixed farming with an emphasis on dairying. More corn, mainly winter wheat and spring barley, is grown in the drier coastal areas. Return to field capacity is in early October in most years and field capacity ends in early May. Figure 54 indicates

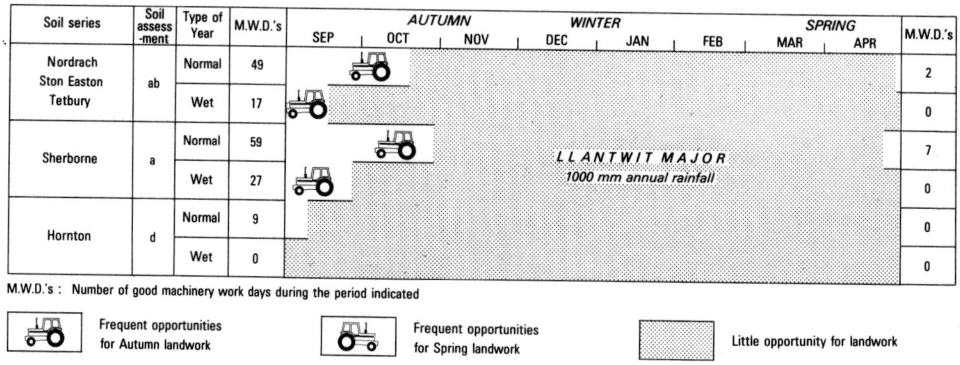

Soil series	Soil assess-ment	Type of Year	M.W.D.'s	AUTUMN			WINTER			SPRING		M.W.D.'s
				SEP	OCT	NOV	DEC	JAN	FEB	MAR	APR	
Nordrach Ston Easton Tetbury	ab	Normal	49									2
		Wet	17									0
Sherborne	a	Normal	59									7
		Wet	27									0
Hornton	d	Normal	9									0
		Wet	0									0

M.W.D.'s : Number of good machinery work days during the period indicated

Frequent opportunities for Autumn landwork   Frequent opportunities for Spring landwork   Little opportunity for landwork

LLANTWIT MAJOR
1000 mm annual rainfall

Figure 54. The effects of soil and climate on landwork, Ston Easton association

that on the Ston Easton and other well drained soils the greatest opportunities for cultivation are in the autumn. On the wetter Hornton series only autumn cultivation is advisable. Maximum soil moisture deficits of about 100 mm indicate that all the soils are to some degree droughty for grass (Table 23) particularly those which are shallow. For cereals drought restriction in average years is slight. There is little risk of poaching on grassland except on Hornton series.

Table 23
Profile Available Water (A.P. mm), Crop-adjusted Mean Moisture Deficit (M.D. mm) and Droughtiness Class for extensive crops—Ston Easton Association

	Ston Easton series	Nordrach and Tetbury series	Hornton series
Location	Llantwit Major	Llantwit Major	Llantwit Major
Grid Ref.	SS970680	SS970680	SS970680
Grass			
A.P.	100	125	115
M.D.	104	104	104
Droughtiness	moderately droughty	slightly droughty	slightly droughty
Winter wheat			
A.P.	100	115	110
M.D.	74	74	74
Droughtiness	slightly droughty	slightly droughty	slightly droughty
Spring barley			
A.P.	100	115	110
M.D.	71	71	71
Droughtiness	slightly droughty	slightly droughty	slightly droughty

§ 85. TANVATS ASSOCIATION
811e

The Tanvats association consists of stoneless, silty, silty over clayey and clayey soils developed in marine alluvium. Tanvats, Rockcliffe and Pepperthorpe series, typical alluvial gley soils and Wallasea series, pelo-alluvial gley soils are dominant. Calcareous Agney (Appendix 2, p.309) and Stockwith (Reeve and Thomasson 1981, p.115), Newchurch and sandy Everingham (Appendix 2, p.313) and Loggans series (Clayden and Hollis, 1984) are locally present. The association occurs between 3 and 6 m O.D. on flat low-lying land around the Wash and in smaller estuaries in Wales.

Brief profile descriptions of three of the main soils are given below, and of Wallasea series in § 89.

Tanvats series
(Full description p.324)

0–25 cm — Ahg
Dark brown, mottled, stoneless
silty clay loam.

25–60 cm — Bg1
Greyish brown with many
ochreous mottles, stoneless silty
clay loam; weak to moderate
medium subangular blocky
structure.

60–80 cm — Bg2
Greyish brown with many
ochreous mottles, stoneless silty
clay loam; moderate coarse
angular blocky structure.

80–100 cm — BCg
Brown with many grey mottles,
stoneless silty clay or silty clay
loam; weak medium angular
blocky structure with relic
stratifications.

Pepperthorpe series
(Full description p.321)

0–30 cm — Ap
Dark brown, stoneless silty clay
loam.

30–50 cm — Bg
Brown, mottled, stoneless silty
clay loam; moderate coarse
subangular blocky structure.

50–75 cm — 2Bg
Reddish grey, mottled, stoneless
silty clay; moderate coarse
angular blocky structure; slightly
calcareous.

75–100 cm — 2BCg
Reddish grey, mottled, stoneless
silty clay; weak coarse angular
blocky structure; slightly calcareous.

Rockcliffe series
(Full description Kilgour 1979,
p.90)

0–20 cm — Apg
Dark brown, mottled, stoneless
silt loam.

20–60 cm — Bg
Greyish brown with many
ochreous mottles, stoneless silt
loam; weak medium angular
blocky structure.

60–100 cm — Cg
Greyish brown with many
ochreous mottles, stoneless silt
loam; strong medium platy
structure.

Locally in north-east Wales, the association consists almost entirely of Tanvats series, but
Wallasea soils are more common in south Wales. It includes small patches of peaty soils
around the Dovey estuary. Where there are nearby coastal sand dunes, the calcareous
Loggans and non-calcareous Everingham soils also occur.

Key to component soil series

Subsoils non-calcareous above 40 cm		1
Subsoils calcareous above 40 cm		2
1.	Fine silty	TANVATS
	Coarse silty	ROCKCLIFFE
	Clayey	WALLASEA
	Fine silty over clayey	PEPPERTHORPE
2.	Fine silty	Agney
	Clayey	Newchurch
	Silty over clayey	Stockwith

Soil water regime

The soils are mainly waterlogged for long periods in winter (Wetness Class IV) but
neglected land is severely waterlogged (Wetness Class V). In drier localities in north-east
Wales where there is efficient field drainage and groundwater control, the soils are only
seasonally waterlogged (Wetness Class III).

Tanvats and Wallasea soils are non-droughty for all crops in south Wales but Tanvats
soils are slightly droughty for most arable crops and moderately droughty for grass in
north-east Wales.

Cultivation and cropping

Where these soils are drained there is ample opportunity for autumn cultivations in north-east Wales. In wetter south and west Wales, however, there are normally few days in autumn and spring when the soils can be worked without risk of structural damage. The soils are mainly under long term grassland or rough grazing, the latter being especially common near industrial complexes. In the north-east, where there are few Wallasea soils, there is some potential for horticulture or field vegetables though groundwater needs to be controlled. The risk of structural damage during spring cultivations is, however, considerable.

§ 86. TEME ASSOCIATION
561b

The Teme association occurs on river alluvium in Wales, Northern, Midland and South West England. Deep permeable fine silty typical brown alluvial soils, Teme series, dominate the broader floodplains, covering more than half the land. Wetter fine silty typical alluvial gley soils, Conway series (§ 26) and gleyic brown alluvial soils, Clwyd series are found mainly along valley margins and within abandoned meanders (Fig.55).

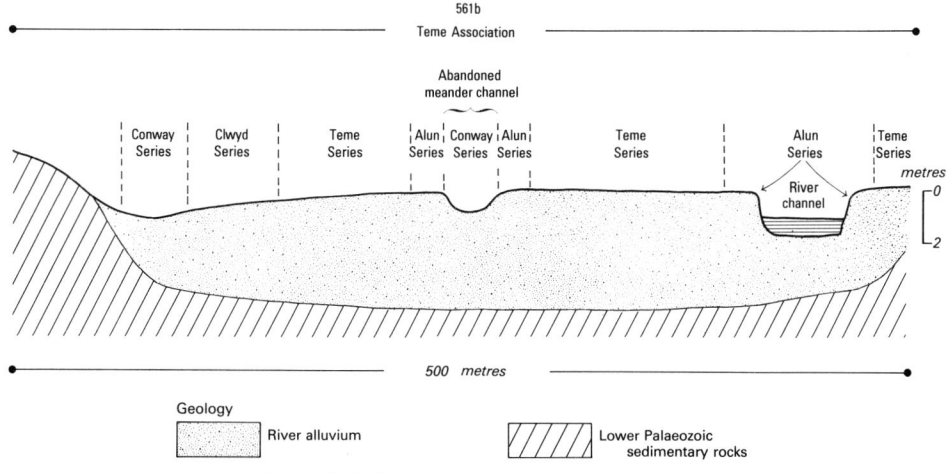

Figure 55. Soil series within the Teme association

There are well drained coarse loamy typical brown alluvial soils of the Alun series near the river or fringing abandoned channels often on the inside of small meanders. Most of the soils are well drained because stream levels are as much as 3 m below the floodplain and the permeable deposits often rest on gravels.

Brief profile descriptions of three of the main series are given below, and of Alun series in § 11.

Teme series
(Full description Palmer 1982, p.109)

0-25 cm — Ap
Dark greyish brown, stoneless silty clay loam.

25-60 cm — Bw1
Brown, stoneless silty clay loam; moderate medium subangular blocky structure.

60-100 cm — Bw2
Brown, stoneless silty clay loam; moderate medium prismatic or subangular blocky structure.

Clwyd series
(Full description Thompson 1982, p.51)

0-20 cm — Ap
Dark greyish brown, stoneless silty clay loam.

20-50 cm — Bw
Yellowish brown, stoneless silty clay loam; moderate fine subangular blocky structure.

50-75 cm — Bw(g)
Yellowish brown, slightly mottled, stoneless silty clay loam; moderate coarse angular blocky structure.

75-100 cm — Bg
Brownish grey, mottled, stoneless silty clay loam; moderate coarse prismatic structure.

Conway series
(Full description Thompson 1982, p.58)

0-20 cm — Apg
Dark greyish brown, stoneless silty clay loam.

20-80 cm — Bg1
Light brownish grey, mottled, stoneless silty clay loam; moderate coarse prismatic structure.

80-120 cm — Bg2
Light grey with many ochreous mottles, stoneless silty clay loam; moderate coarse prismatic structure.

Although the Teme series is generally dominant, on some floodplains other soils are more extensive, for example Clwyd series is the main soil along the Cleddau in south-west Dyfed and the Severn near Caersws. Near Caersws, and along the Tywi, the alluvium is underlain in places by gravel at relatively shallow depth. The Tywi valley is partly floored by fine silty typical brown alluvial soils belonging to Llandeilo series (Wright 1980) which rest on non-calcareous gravel. Nearby, close to the river, well drained but coarse loamy Tavy soils, originally mapped as Tywi series by Wright (1980) occur. There are small patches of ranker-like alluvial soils (Aled series, Wright 1981) in many valleys, especially their upper reaches. Fine loamy soils of the Wharfe series (Jones 1983) also occur as minor inclusions and in places where the alluvium is derived from Devonian rocks, reddish Lugwardine (Palmer 1982), Walford (Appendix 2, p.325) and Mathon soils (Palmer 1982) are included.

Key to component soil series

	Unmottled soils	1
	Mottled soils; fine silty	4
1.	Gravel within 80 cm	3
	Deeper soils	2
2.	Fine silty	TEME
	Coarse loamy	ALUN
	Fine loamy	Wharfe
3.	Fine silty over gravelly	Llandeilo
	Coarse loamy over gravelly	Tavy
	Gravel within 30 cm; sandy	Aled
4.	Subsoil faintly mottled above 60 cm or distinctly mottled between 40 and 80 cm	CLWYD
	Prominently mottled or greyish above 40 cm	CONWAY

Soil water regime

Although most of the soils are permeable and well drained or only occasionally waterlogged (Wetness Class I or II), there is a risk of flooding, particularly in winter. The frequency and duration of floods which depend on location and control measures, are major factors limiting land use. The associated Clwyd series (Wetness Class III) and Conway series (Wetness Class IV or V) are waterlogged for long periods so both benefit from attention to ditches and outfalls and in places they respond well to underdrainage. The deep silty soils of the association provide good reserves of water for grass and arable crops, but the shallower Llandeilo, Tavy and Aled series are more droughty.

Cultivation and cropping

The Teme association can produce valuable agricultural crops but permanent grassland prevails because it is least damaged by the floods. Natural fertility is enhanced periodically by the addition of fresh sediments from the flood waters. The good fertility stimulates soil faunal activity including that of earthworms, and this in turn increases the porosity of the soils. These productive meadows are much valued for fattening livestock and grass conservation. The growing season is usually about 7.5 to 9 months (Smith 1976) but the grazing period is often a month or two less to avoid poaching damage.

Arable crops are successful where flood risk is small (Thompson 1982, p.97). Teme soils are fairly easy to cultivate (Jones 1979) and require few rain-free days before ploughing. Their large silt content, however, renders them susceptible to capping. The permeable soils on level ground are attractive for slurry disposal but much of the land is too close to watercourses to be suitable (Lea 1979). The level ground provides for playing fields and temporary summer activities such as camping and caravanning.

§ 87. TRUSHAM ASSOCIATION
541n

The association covers 312 km² in England and Wales, though only 10 km² are in Wales. It occurs on basic igneous rocks, mostly in Devon and Cornwall (Findlay *et al.* 1984), but with small patches in the Peak District of Derbyshire, the Welsh Borderland and in Snowdonia. Most of the soils are well drained and fine loamy. On Snowdon, Trusham and Erisey series, typical brown earths overlie soft pumice-tuffs of the Bedded Pyroclastic series (Ball *et al.* 1969), and as the land is steep and rocky humic and brown rankers are also present. On the dolerite and andesite of the Breidden Hills near Welshpool there are Trusham and East Keswick soils on drift with rankers and bare rock on high ground (Thompson 1982). Brown podzolic soils, Malvern series, occupy the steepest ground in both locations.

The well drained soils (Wetness Class I) readily absorb surplus winter rainfall even on the steepest slopes. The shallowest soils, in Snowdonia and over scree on the Breidden Hills are droughty. Stone contents are very variable but the presence of surface boulders and exposed rock is the main limitation to agriculture. There are pastures on the flatter parts of the Breidden Hills, but much of the land is afforested or under natural vegetation containing the rare rock cinquefoil *(Potentilla rupestris)*, spiked speedwell *(Veronica spicata)* and red german catchfly *(Lychnis viscaria)*. In Snowdonia these soils support clover-rich pastures preferred by sheep (Hughes *et al.* 1975). Rock outcrops abound and have calcicolous plants such as the liverwort *Neckera crispa*, oak fern *(Thelypteris dryopteris)*, green spleenwort *(Asplenium viride)*, saxifrages *(Saxifraga* spp.*)* and early purple orchid *(Orchis mascula)*. Apart from small blocks of broad-leaved woodland above Llyn Gwynant, there is little forestry. West of the lake the association extends into Snowdonia National Nature Reserve.

§ 88. UNRIPENED GLEY SOILS
22

Formed under saltmarsh vegetation in marine alluvium of various textures, this association includes ripened as well as unripened soils. It occurs in estuaries and creeks in many places round the coast of England and Wales. Saltmarshes lie above the reach of daily tides but are covered by periodic spring tides. Pioneer plants bind the alluvium and, by checking the flow of water, increase sedimentation. Algae, especially seaweeds, are followed by flowering plants, such as glassworts *(Salicornia* spp.*)* and, later, by larger more efficient silt-trapping plants—cord-grass *(Spartina townsendii)*, herbaceous seablite *(Suaeda maritima)*, sea aster *(Aster tripolium)* and sea purslane *(Halimione portulacoides)*. Some marshes become completely colonized by *Spartina*. Since its appearance in Britain at about the turn of the century this tall vigorous perennial hybrid of *S. maritima* and *S. alterniflora* has formed almost pure stands over large areas.

A dendritic creek system channels much of the tidewater but when creeks overflow, the silt-laden waters are partially checked by the vegetation and coarse textured material is deposited, forming low banks or levées. Finer particles are carried further and settle later under relatively quiet conditions. The particle-size distribution of the constituent soils depends upon the history of marsh development but they are often finer towards the surface and coarser at depth, since progressively finer sediments accumulate as the marsh ages and its height increases.Variations on this simple pattern are common, partly through disturbance by wave action and burrowing animals; abrupt changes in texture can follow alterations in creek or river courses. Developing saltmarshes differ in maturity from one part to another and different plant communities can occur in distinct zones, approximately parallel to the primary source of sediment (Tansley 1939).

The highest and most mature parts of the firmer marshes have been periodically enclosed behind earthen embankments, derived from a perimeter ditch or borrow pit. Creeks are commonly filled in, drained into the landward ditch at the new sea wall, or left to provide field boundaries or barriers for livestock. In the widest estuaries, saltmarshes have formed over long periods with successive sections reclaimed as silting proceeded.

Ultimately many square kilometres have been enclosed and, as in the Humber and Ribble estuaries, successive innings have the same pattern of deposition.

Most common are saline raw gley soils with an unripened mineral horizon that starts within 20 cm of the surface. Particle-size class varies but there is always more than 8 per cent clay. There is often a distinct, humose or peaty topsoil. Some soils of the higher saltings resemble alluvial gley soils but, although ripened, remain highly saline until reclamation and drainage. Occasionally, non-saline, ripened alluvial gley soils are found.

Unripened gley soil
(Full description p.324)

0–20 cm — Ahg
Very dark grey, mottled, stoneless, humose silty clay.

20–40 cm — Cg
Greyish brown, mottled, stoneless silty clay.

40–100 cm — 2Cg
Greyish brown, mottled, stoneless silty clay loam, sandy silt loam or silt loam.

In Wales the association is most extensive in Carmarthen Bay but also occurs in the Dee, Dovey and several smaller estuaries. It includes most undrained and unprotected coastal marshland above mean high water mark. Land below this mark has not been included. Unripened gley soils occur at the mean high water mark under *Spartina* and *Salicornia* and consist of loamy (sandy silt loam to silty clay loam), soft and usually calcareous mud. This passes to a higher level of marsh, often by a small step, to vegetation commonly dominated by sea poa *(Puccinellia* sp.*)* with *Aster* along the creeks. The soils are similarly loamy, the coarser soils being found on creek levées. They are saline, partially ripened and often calcareous. Inland, and again often with a small step, there is a zone of *Festuca rubra*, and here the soils are fully ripened and often non-calcareous. The highest zone of the marsh is again to landward and under *Juncus maritimus*; here there is commonly the Tanvats series (§ 85). The pattern described in the Burry inlet is typical of most saltmarshes.

Land use
Saltmarshes are used for wildfowling and, on the higher, more stable ground, for grazing of sheep and ponies. They are valued as nature reserves and some are Sites of Special Scientific Interest. There is some conflict of interest because many areas could eventually be embanked, drained and cultivated. Reclamation has gone on for centuries and there is scope for more without undue disturbance of the conservation sites.

§ 89. WALLASEA 1 ASSOCIATION
813f

This is the main association on the marine alluvium of the coastal marshes of north Kent, Essex and Suffolk. It occurs in the Humber estuary, and there are small areas in north Norfolk, Devon, Wales and along the Hampshire coast. Sea walls, river embankments and other works protect the land from daily flooding as much is below high tide level. On the east coast the sea defences are designed to withstand tidal surges in the North Sea that may reach heights of 4–5 m O.D. Occasional breaches have occurred, sometimes with serious consequences. Drainage is by open ditches leading to tidal sluices or pumps. There is little microrelief partly a result of recent levelling.

The soil pattern is simple with non-calcareous clayey Wallasea pelo-alluvial gley soils (Sturdy 1976), and clayey Newchurch pelo-calcareous alluvial gley soils (Fordham and Green 1980) occupying almost all the land. There are inextensive inclusions of fine loamy Paglesham (Clayden and Hollis 1984), fine silty over clayey Pepperthorpe, fine silty Agney (Appendix 2, p.309) and clayey Dymchurch soils (Clayden and Hollis 1984). Humose-topped non-calcareous clayey soils of the Downholland series (Hodge *et al.* 1984) are frequent in Humberside, but rare elsewhere. Brief profile descriptions of Wallasea and Newchurch series and a key to the component soils are given below.

Wallasea series
(Full description p.325)

0–25 cm — Ap
Greyish brown, stoneless silty clay.

25–60 cm — Bg
Brownish grey, mottled, stoneless silty clay; moderate coarse blocky or prismatic structure.

60–100 cm — BCg
Brown with grey mottles, stoneless silty clay; weak coarse prismatic structure.

Newchurch series
(Full description p.320)

0–25 cm — Apg
Dark greyish brown, slightly mottled, stoneless silty clay; slightly calcareous.

25–60 cm — Bg
Greyish brown, mottled, stoneless silty clay; moderate coarse angular blocky or prismatic structure; calcareous.

60–100 cm — BCg
Brown, mottled, stoneless silty clay; moderate coarse prismatic structure; calcareous.

The association covers 30 km² in Wales at the mouth of the Vale of Clwyd and in the lower Usk valley. From Abergele and Prestatyn inland almost to Bodelwyddan, Wallasea and Newchurch soils, formerly described as Rhyl and Rhuddlan series by Ball (1960), are present in roughly equal amounts. Occasional Pepperthorpe and Stockwith soils (Reeve and Thomasson 1981, p.115) occur in fine silty over clayey alluvium. In south Wales the small area along the Usk has frequent Downholland soils.

Key to component soil series

Soils prominently mottled or greyish within 40cm		1
Subsoils faintly mottled above 60 cm or distinctly mottled between 40 and 80 cm; non-calcareous, clayey		Dymchurch
1. Subsoil calcareous above 40 cm		2
Subsoil non-calcareous above 40 cm		3
2. Clayey		NEWCHURCH
Fine silty		Agney
Silty over clayey		Stockwith
3. With clayey upper horizons		4
With fine loamy or fine silty upper horizons		5
4. With grey, distinct topsoil		WALLASEA
With dark, humose topsoil		Downholland
5. Fine loamy over clayey		Paglesham
Fine silty over clayey		Pepperthorpe

Soil water regime

High groundwater levels cause severe waterlogging (Wetness Class V). Some drainage improvement can be achieved by a close network of ditches as subsoils are relatively permeable but, since most land is in grass, water levels are kept high to contain stock. With pipe drainage the soils may be only seasonally waterlogged (Wetness Class IV). The soils have moderate available water capacity so with water from groundwater sources and small summer soil moisture deficits, grass does not suffer from drought.

Land use

Most of the land is in productive permanent grass, although infestation by rushes in hollows reduces the herbage value on the wettest land and grazing by cattle is restricted to the summer months when poaching risk is least. Grazing ceases towards the end of October as soils return to field capacity and the risk of poaching increases although the grass continues to grow for about a further six weeks. Slurry spreading is also confined to dry periods between late spring and early autumn; at other times trafficability is poor and permeability slow, leading to surface run-off and pollution of watercourses. In some places, for example near Rhuddlan where Newchurch soils are most extensive, better drainage and workability permit occasional winter cereal crops often as a break before reseeding grassland. Occasional liming is needed, but manganese deficiency can occur if the soils are over-limed. The soils contain large reserves of potassium and magnesium.

§90. WALTHAM ASSOCIATION
541q

This association is composed of soils in non-calcareous drift over Carboniferous Limestone at generally between 170 and 490 m O.D. It is found mainly in Cumbria but also in North Yorkshire, Northumberland, Lancashire and south-east Wales. The principal soils are the well drained Waltham series (§ 30), fine loamy typical brown earths, with limestone within 80 cm of the ground surface; the East Keswick series (§ 37), typical brown earths in deeper drift; and the Crwbin series (§ 30), brown rankers in shallow drift over limestone.

The association covers some 7 km² in Gwent, mid-Glamorgan and Powys. There is no bare rock but there are shallow soils of the Crwbin and Elmton series (Hartnup 1977, p.41) on knolls. The mild climate in lowland Gwent favours agriculture, with a moisture deficit of around 100 mm, and an average annual rainfall of about 1,000 mm. The land is used mainly for ley grassland but there is some arable and also permanent grassland. Near Merthyr Tydfil in mid-Glamorgan where the climate is cooler and wetter, most of the association is under permanent grass. The main soils are permeable and well drained (Wetness Class I), and they readily absorb excess winter rainwater. East Keswick soils have enough available water for most crops but the Waltham series, being shallower, can be slightly droughty. The association has been described more fully by Jarvis R.A. *et al.* (1984).

§ 91. WENALLT ASSOCIATION
721e

The Wenallt association consists of very acid wet peaty soils in reddish drift, mainly on moorland at 300 to 500 m O.D. It is mapped only in south Wales and south-west Shropshire over or adjacent to the Old Red Sandstone outcrop and is most extensive on the gentle dipslopes and footslopes (Fig. 56) of the range of hills extending from the Black Mountain near Ammanford to the Brecon Beacons.

Loamy cambic stagnohumic gley soils, belonging to the Wenallt series, predominate and the main ancillary soils are raw oligo-amorphous peat soils, Crowdy series (§ 28), and ironpan stagnopodzols, Cray series. Cambic stagnogley soils of the Fforest series (§ 43) with mineral topsoils occur on enclosed land. The soils are all developed in thick till or solifluction deposits and have impermeable subsoils often with a fragipan which holds up surface water. They are wet for most of the year (Wetness Classes V–VI) and the winter run-off is very rapid, a dense drainage pattern of rills and streams being characteristic of most slopes. The peaty topsoils are commonly 15 to 30 cm thick over a gleyed and bleached subsurface horizon depleted of iron. In the Wenallt series this overlies a dense gleyed horizon with orange and reddish mottles and prismatic structures, whereas in the Cray series there is a thin ironpan over an often ungleyed subsoil. Both series pass to little altered drift within a metre depth. The particle-size class of the soils varies with the source of the drift, but over much of the association the incorporation of coarse sandstones and conglomerates of the Upper Old Red Sandstones gives coarse textured soils, commonly sandy loam over sandy clay loam which are frequently bouldery (Wright 1981). Elsewhere the particle-size class is clay loam or silty clay loam.

Wenallt series
(Full description Wright 1981, p.66)

0–20 cm — Oh
Very dark brown, stoneless humified peat.

20–30 cm — Eag
Light brownish grey, slightly stony silty clay loam; moderate coarse prismatic structure.

30–45 cm — Bg1
Grey, mottled, moderately stony silty clay loam or sandy silt loam; moderate coarse prismatic structure.

45–66 cm — Bg2
Reddish brown, mottled, moderately stony silty clay loam or clay loam; moderate coarse prismatic structure.

66–100 cm — BCg
Reddish brown, mottled, moderately stony silty clay loam or clay loam; massive structure.

Crowdy series
(Full description p.311)

0–10 cm — Om
Black, semi-fibrous peat, moderate granular structure.

10–20 cm — Oh1
Dark brown, stoneless humified peat; massive structure.

20–100 cm — Oh2
Black, stoneless humified peat; massive structure.

Cray series
(Full description p.311)

0–20 cm — Oh
Black, humified peat.

20–30 cm — Eag
Pinkish grey, slightly stony clay loam; moderate fine angular blocky structure.

At 30 cm — Bf
Thin ironpan.

30–50 cm — Bs
Reddish brown, slightly stony clay loam; weak medium sub-angular blocky structure.

50–80 cm — Cu
Reddish brown, moderately stony clay loam; massive structure.

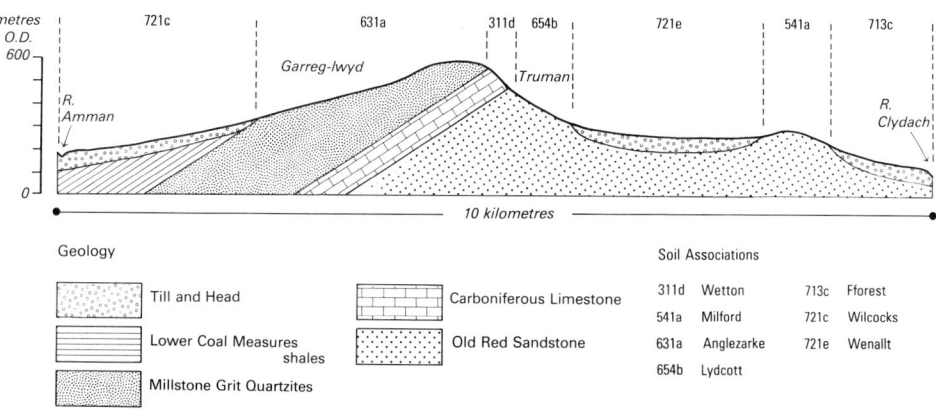

Figure 56. Soil associations, relief and geology on the Black Mountain, Dyfed

Although the association is most common on gentle moorland slopes, it is also mapped on steep-sided hills where drift is thick. The Cray series is a common associate, typically on better drained sites as on the edges of solifluction lobes. The Crowdy series is also common, especially above 400 m O.D. usually in depressions as basin peat but also as blanket peat. Upslope the association passes to the Lydcott association (§ 57) of stagnopodzols on red sandstone, the boundary being along a concave break of slope and spring line, with a vegetation change from *Molinia* to *Vaccinium* heath or *Nardus* grassland.

Key to component soil series

Mineral soils		1
Peat thicker than 40 cm		CROWDY
1.	With peaty topsoil; loamy	2
	With grey, distinct topsoil; fine silty	Fforest
2.	With thin ironpan	CRAY
	Without ironpan	WENALLT

The climate is cold and wet with annual rainfall about 2,000 mm and soil moisture deficits less than 40 mm. The semi-natural vegetation is *Molinia* grassland with some *Nardus* and moist heather moor of moderate or poor grazing value. Wet flushes and hollows are usually dominated by rushes and bog-mosses. Most of the land is used for common grazing but recreation and military training are also important activities and much is within the Brecon Beacons National Park. Wetness, boulders and poor nutrient status make grassland improvement costly. Conventional drainage techniques are prohibitively expensive on large areas of moorland, yet without removing excess water there can be little significant pasture improvement. Some farmland has been drained then rotavated shallowly rather than ploughed. Careful grazing management is required as the organic topsoil is soft and easily poached. Some of the land is used for forestry, the largest areas being Glasfynydd Forest owned by the Forestry Commission and private forests on Fan Fraith and Fan Gihirych. Sitka spruce is the preferred species and site preparation involves deep ploughing and opening deep drains to remove surface water and encourage root penetration as windthrow is a hazard.

§ 92. WETTON 2 ASSOCIATION
311d

This association covers 162 km² on Carboniferous Limestone mainly in the Yorkshire Dales, with small areas in south Wales around the northern rim of the coalfield. Here the land is uneven and rocky and lies at about 400–500 m O.D., broken by limestone crags and quarries. The climate is cold, wet and exposed. The dominant soils are the shallow Wetton series, humic rankers, locally with pockets of deeper, more or less stoneless, well drained silty Malham series (Hollis 1975). Locally, bare limestone covers more than half the ground. The soils are fully described by Jarvis, R.A. *et al.* (1984).

The thin humose surface horizons of Wetton soils retain water and on average are wet for less than half the year shedding winter rainwater laterally. However, the water moves only a short distance before it is intercepted by the numerous sink holes and enters the limestone. The pockets of Waltham soils readily absorb winter rainwater, so the combined effect is that there is little winter run-off from this land. The small available water capacity of the main soil makes it very susceptible to drought in summer dry spells, when plants begin to wilt. The land is of little agricultural value because of climate and shallow soil, and is used mainly for rough grazing, which is characterized by sheep's fescue *(Festuca ovina)*, common bent *(Agrostis capillaris)* and herbs of high grazing value, contrasting sharply with the boggy or heathy vegetation of the nearby Wilcocks and Anglezarke associations. Shallow soils, rocky ground and widespread exposure to wind, make this association unsuitable for forestry.

§ 93. WICK 1 ASSOCIATION
541r

The Wick association occurs widely throughout Northern England, the Midlands and Wales but is inextensive in Eastern and South West England. Deep well drained coarse loamy typical brown earths, Wick series, are intermixed with gleyic brown earths, Arrow series and typical brown sands, Newport series. Wick and Arrow series are sandy loam throughout but, whilst Wick soils are unmottled the Arrow series has dull colours and mottles, attributable to wetness below 40 cm. Both may be sandy at depth and may overlie gravel particularly in river terrace deposits. The soils are in glaciofluvial and river terrace drift of variable stoniness, and in local Head. Brief profile descriptions of the main series and a key to the component soils are given below.

Wick series
(Full description Hollis 1978, p.57)

0-30 cm — Ap
Dark brown, slightly stony sandy loam or sandy silt loam.

30-60 cm — Bw
Brown, slightly stony sandy loam or sandy silt loam; moderate medium subangular blocky structure.

60-80 cm — Bw
Yellowish brown, slightly or moderately stony loamy sand or sandy loam; weak medium angular blocky or single grain structure.

80-120 cm — 2BCu
Brownish yellow, slightly or moderately stony sand or loamy sand; weak coarse angular blocky or single grain structure.

Arrow series
(Full description Palmer 1982, p.72)

0-25 cm — Ap
Dark brown, stoneless or slightly stony sandy loam.

25-50 cm — Bw
Dark yellowish brown, slightly stony sandy loam; weak medium angular blocky structure.

50-100 cm — Bg
Brown, mottled, slightly stony sandy loam or loamy sand; weak coarse angular blocky structure.

Newport series
(Full description Jones 1975, p.38)

0-25 cm — Ap
Dark brown, slightly stony sandy loam or loamy sand.

25-55 cm — Bw
Brown, slightly stony loamy sand or sand; weak fine subangular blocky structure.

55-120 cm — Cu
Yellowish red or brownish yellow, slightly stony sand; single grain structure.

The association covers 355 km² in Wales from sea level to 275 m O.D. It occurs along the
Menai Strait, in the Alun, Dee and Usk valleys, the Vales of Glamorgan, Clwyd and
Conwy and in scattered patches elsewhere. In places there are pockets of fine loamy or
coarse loamy soils over gravel with hard quartzitic stones, Rheidol series, (§ 75) and Hall
series (Appendix 2, p.315), or more commonly, deep fine loamy brown earths, East
Keswick series (§ 36). There are also small patches of Nercwys (§ 67), Brickfield (§ 23),
Salwick (§ 80) or Clifton soils (§ 24). Some soils have been marled where calcareous red
clay is nearby. There are small pockets of Meline series (Clayden and Hollis 1984) west of
Conwy, but Newport and Ollerton soils (King 1977, p.65) are virtually absent. In south
Wales the soils occur on river terraces and hummocky glaciofluvial deposits in the valleys,
and on morainic drift at the northern edge of the Vale of Glamorgan, where they are
frequently related to kame and kettle terrain. The Wick and Hall series are the main soils
and the Newport series is rare.

Key to component soil series

	Unmottled soils	1
	Mottled soils	3
1.	Sandy	NEWPORT
	Loamy	2
2.	Deep soils, no gravel within 80 cm; coarse loamy	WICK
	Gravelly within 40 cm	Ellerbeck
	Gravelly between 40 and 80 cm; coarse loamy	Hall
3.	Prominently mottled or greyish above 40 cm; coarse loamy	Quorndon
	Subsoil faintly mottled above 60 cm or distinctly mottled between 40 and 80 cm	4
4.	Coarse loamy	ARROW
	Sandy	Ollerton
	Fine loamy	5
5.	Reddish	Salwick
	Brownish	Nercwys

Soil water regime

The major soils are permeable and well drained (Wetness Class I), and although the
Arrow series is waterlogged for short periods in winter where undrained (Wetness Classes
II or III), groundwater levels are easily controlled. The soils readily absorb winter
rainwater.

Cultivation and cropping

The main enterprise is dairying with supplementary arable crops which include barley,
wheat and some potatoes. Potential grass yields are large and as there is little risk of

poaching, grazing is feasible over much of the growing season. Early dressings of fertilizer are possible. Whilst growth can slow down in mid-season in drier parts of the country such as the Usk Valley and north-east Wales, the autumn flush provides valuable late grazing on these soils. Outwintering of stock is possible and slurry can be spread on suitable days throughout winter. Except where unusually stony the soils are easily cultivated and suitable for direct drilling. The land can be worked during dry periods in all seasons. Where large areas are being sown, autumn cultivation provides most insurance against damage to soil structure (Fig.57). Late harvesting of crops such as potatoes and sugar beet also minimizes structural damage, provided the season is not too wet. Deep-tine cultivation may be necessary to loosen compacted subsoil.

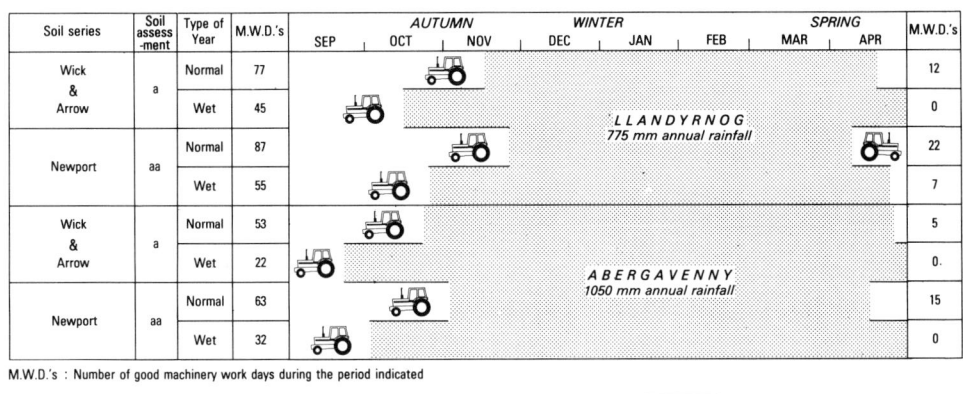

Figure 57. The effects of soil and climate on landwork, Wick 1 association

§ 94. WIGTON MOOR ASSOCIATION
831c

The fine and coarse loamy soils which dominate this association are developed in glaciofluvial and river terrace deposits associated with major river valleys and are affected by fluctuating groundwater. The association is most extensive in the Midlands and Northern England but small areas are mapped in Wales, Eastern and South West England. A full description of the association is given by Ragg et al. (1984).

The main soils belong to Wigton Moor series, fine loamy typical cambic gley soils, with permeable subsoil horizons affected by groundwater. The ancillary Quorndon series is coarse loamy but otherwise similar and the association also includes the Arrow (Palmer 1982) and Hopsford series (Whitfield and Beard 1977, p.50). The association covers approximately 4 km² north-west of Rossett in Clwyd on terrace deposits of the river Alun where Quorndon and Arrow soils are more common than elsewhere. Profile drainage is impeded by reddish till below about 1 m depth. The land has been mapped in detail as the Quorndon and the former Rossett series by Lea and Thompson (1978).

The coarse and fine loamy gley soils of this association are permeable but seasonally waterlogged (Wetness Class III). Underdrainage can improve the water regime of Quorndon soils to Wetness Class I or II depending on suitable outfalls but Wigton Moor soils remain seasonally waterlogged (Wetness Class III). Where undrained Arrow and Hopsford soils are waterlogged for short periods only in winter (Wetness Class II) but after appropriate measures they usually drain freely (Wetness Class I).

In Wales the land is used for grass and cereal production. Dairying is the main enterprise with barley grown for stock feed on the farm. Underdrainage is needed to combat surface wetness and groundwater. There is a serious risk of poaching on the main soils though this is less of a problem on Arrow and Hopsford soils.

Overall, Hopsford and Arrow series are more suitable for cultivation than Wigton Moor or Quorndon soils. Compaction and slaking occur under repeated cultivation and regular subsoiling is necessary for arable crops.

§ 95. WILCOCKS 1 ASSOCIATION
721c

The association contains strongly gleyed soils with peaty or humose topsoils. These belong to the loamy Wilcocks series and, less frequently, the fine loamy over clayey Kielder series, both cambic stagnohumic gley soils, and the coarse loamy Fordham series of typical humic gley soils. They are seasonally waterlogged, stony soils in greyish drift derived from Carboniferous and Lower Palaeozoic rocks, occurring on gently to moderately sloping land in the Pennines and adjacent upland districts and on the moorlands and in the valleys of upland Wales, between 150 and 600 m O.D. The Wilcocks series has an acid organic surface layer 10 to 40 cm thick, with underlying clay loam or sandy clay loam horizons which are grey and strongly mottled, although the mineral layer below the peat is normally stained with organic matter. The Kielder series is similar but becomes clayey at depth. Unlike these soils, the Fordham series is coarse loamy and gleying is not caused by a slowly permeable subsoil but by high groundwater levels. Stone content frequently increases with depth; stones are commonly sandstones and gritstones but include shale fragments, or occasionally igneous and other erratics. Brief profile descriptions of the main series and a key to the component soils are given below.

The Wilcocks 1 association is most extensive over the Carboniferous outcrop of south Wales (Plate 11) but is also widespread in central Wales. Near exposures of the Millstone Grit and Pennant sandstones, the Wilcocks series is frequently a very stony sandy clay loam. Over the Middle and Lower Coal Measures, shale fragments in the drift give more clayey soils, and the Roddlesworth (Clayden and Hollis 1984) or Kielder series are common. On moorland the Crowdy (§ 28), Freni (Clayden and Hollis 1984) and

Wilcocks series
(Full description Hollis 1975, p.75)

0–20 cm — Oh or Ah
Black, stoneless humified peat or humose clay loam.

20–50 cm — Bg
Light brownish grey, mottled, slightly stony clay loam or sandy clay loam; weak subangular blocky structure.

50–100 cm — BCg
Grey with many ochreous mottles, moderately stony clay loam; weak medium blocky or prismatic structure; high packing density.

Kielder series
(Full description p.317)

0–20 cm — Oh or Ah
Black, humified peat or humose clay loam.

20–45 cm — Eg
Dark grey, mottled, slightly stony clay loam or sandy clay loam; moderate coarse angular or subangular blocky structure.

45–100 cm — Bg
Grey, mottled, slightly stony clay; weak coarse prismatic structure.

Fordham series
(Full description p.314)

0–20 cm — Oh or Ah
Black, humified peat or slightly stony, humose sandy loam.

20–50 cm — Bg
Light brownish grey, mottled, moderately stony sandy loam; weak medium subangular blocky structure.

50–100 cm — Cg
Light brownish grey, mottled, moderately stony sandy loam or loamy sand; weak coarse prismatic or massive structure.

Fordham series (Appendix 2, p.314) are the main associates. The proportion of soils with mineral surface horizons increases on enclosed land, giving rise to the Brickfield series (§ 17). The Wilcocks series was formerly mapped as the Hirwaun series by Clayden and Evans (1974) in drift from Carboniferous rocks, and as the Ynys series by Wright (1980) in drift from Lower Palaeozoic rocks.

Key to component soil series

Peat thicker than 40 cm		Winter Hill or Crowdy
Mineral soils		1
1.	With dark, humose or peaty topsoil	2
	With grey, distinct topsoil	5
2.	Soil deeper than 80 cm	3
	Shale within 80 cm	4
3.	Fine loamy	WILCOCKS
	Fine loamy over clayey	KIELDER
	Coarse loamy	FORDHAM
	Clayey	Roddlesworth
4.	Fine loamy over clayey	Ipstones
	Clayey	Onecote
5.	Fine loamy	Brickfield
	Fine silty	Cegin

Soil water regime

The main soils are severely or permanently waterlogged near the surface (Wetness Class V or VI), the wetness being due to a combination of high rainfall, impermeable subsoil,

Plate 31. The northern slopes of Mynydd Preseli. The land showing erosion gullies carries wet soils of the Wilcocks 1 association. Drier Moretonhampstead and Manod associations occupy adjacent enclosed land with patches of Cegin association. Distant slopes are dominated by the Denbigh association.

gentle relief, and a flow of water from the nearby Crowdy association. Winter rainwater is not absorbed and runs off rapidly. This has produced gullying on the northern slopes of the Preseli Hills (Plate 31). The soils are not droughty because of their large available water capacity and the low moisture deficit in the districts where they occur.

Land use
The cold and wet climate has an overriding effect on land use. Annual rainfall generally ranges from 1,500 mm to, on the highest ground, over 2,000 mm. The relatively low temperatures, high rainfall and inherent wetness of the soils have resulted in semi-natural vegetation of *Molinia* and *Nardus* grassland or heather moor with cotton-grass and mosses on the wettest ground. Elsewhere the land is enclosed and the grazing improved but, when neglected, it quickly becomes infested with rushes. The natural trend towards acid mat formation can be arrested by using drainage, lime, fertilizer and mowing or

controlled grazing. Good permanent grassland can be maintained by careful management, but the soils are easily poached. The land is mainly used for sheep rearing, with seasonal grazing for cattle in the more favoured localities. *Nardus, Molinia* and rush-infested bent-fescue communities have moderate grazing value but heather moor generally gives poor grazing although it provides a valuable winter bite. On Gower these soils form much of the common land, which has important amenity value.

Being waterlogged for much of the time the soils are not suitable for regular cultivations. War-time attempts at reclamation demonstrated the futility of ploughing without suitable applications of lime and fertilizer. They also showed that when Wilcocks soils with peaty, rather than humose, surface horizons are ploughed, the moisture-retentive surface mat remains undecomposed after burial, increasing surface waterlogging and leading to rush-infestation.

Forestry

This is one of the common associations used for forestry because of its poor agricultural value. It forms large parts of forests in the south Wales coalfield, notably Coed y Rhaiadr and adjacent plantations. The main species is Sitka spruce with some Norway spruce. Deep ploughing is necessary to improve drainage and rooting. Wind damage is least where there is no thinning.

§ 96. WILCOCKS 2 ASSOCIATION
721d

This association of wet loamy soils with peaty tops is widespread over Palaeozoic rocks in upland Wales and Northern and South West England. It usually occupies drift-covered gentle slopes where natural drainage is impeded. The cold wet climate with low summer

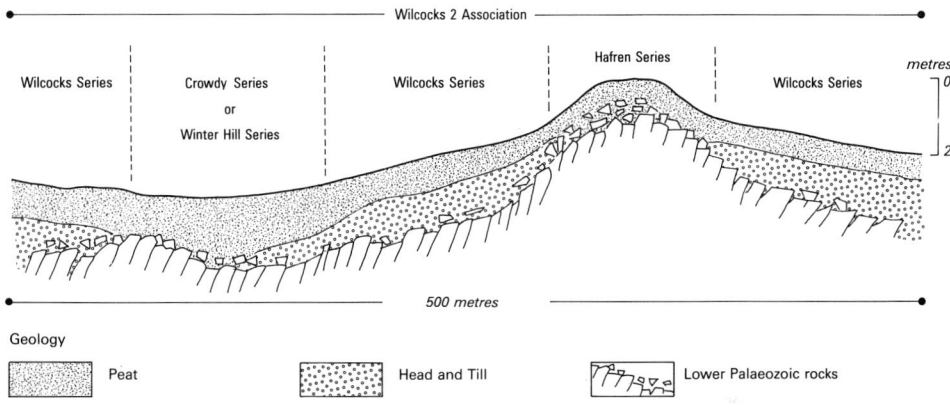

Figure 58. Soil series in Wilcocks 2 association in central Wales

temperatures have combined with slow subsoil permeability to produce the acid peaty surface horizons.

Stagnohumic gley soils of the Wilcocks series cover over half the association but on thick peat, amorphous Crowdy and fibrous Winter Hill soils are found. On steeper slopes and ridges there are stagnopodzols of the Hafren or Hiraethog series (§ 51)(Fig.58).

Brief profile descriptions of the Wilcocks, Crowdy and Winter Hill series are given below and of the Hafren series in § 51.

Wilcocks series
(Full description Hollis 1975, p.75)

0–20 cm — Oh or Ah
Black, stoneless humified peat or humose clay loam.

20–50 cm — Bg
Light brownish grey, mottled, slightly stony clay loam or sandy clay loam; weak subangular blocky structure.

50–100 cm — BCg
Grey with many ochreous mottles, moderately stony clay loam; weak medium blocky or prismatic structure; high packing density.

Crowdy series
(Full description p.311)

0–10 cm — Om
Black, semi-fibrous peat; moderate granular structure.

10–20 cm — Oh1
Dark brown, stoneless humified peat; massive structure.

20–100 cm — Oh2
Black, stoneless humified peat; massive structure.

Winter Hill series
(Full description Carroll *et al.* 1979, p.88)

0–10 cm — Om1
Black, semi-fibrous peat.

10–40 cm — Om2
Dark reddish brown, semi-fibrous or amorphous *Eriophorum-Sphagnum* peat; moderate coarse platy structure.

40–70 cm — Om3
Dark reddish brown, semi-fibrous *Eriophorum-Sphagnum* peat; weak coarse platy structure.

70–120 cm — Om4
Dark reddish grey, semi-fibrous *Eriophorum-Sphagnum* peat; massive structure.

Wilcocks series now includes soils formerly described as Ynys series by Ball (1960, 1963) (Clayden and Hollis 1984). Humic rankers belonging to the Skiddaw series (Appendix 2, p.323) and occasional loamy podzolic rankers occur near rock outcrops.

Key to component soil series

Shallow peaty soils, rock within 30 cm		Skiddaw
Deeper soils		1
1.	Peat thicker than 40 cm	4
	Peat thinner than 40 cm; fine loamy or fine silty mineral soils	2
2.	With thin ironpan or brightly coloured subsoil	3
	No podzolic features	WILCOCKS
3.	With brightly coloured subsoil; no ironpan	HAFREN
	With thin ironpan	Hiraethog
4.	Amorphous peat	CROWDY
	Fibrous peat with moss and cotton-grass remains	WINTER HILL

Soil water regime
The main soils lie wet for long periods or are permanently waterlogged (Wetness Classes V and VI). During winter excess rainwater moves rapidly from the saturated soil to streams.

Land use
Most of the land is permanent pasture or rough grazing (Plate 32) because the climate is cold and wet. The semi-natural vegetation is mainly *Molinia* bog on wetter sites though *Sphagnum* can be dominant on the peat. *Molinia* or *Nardus* grassland or heather moor

T.R.E. Thompson

Plate 32. The northern slopes of Moel Siabod. This almost treeless landscape is mainly Hafren and Wilcocks 2 associations with shallow Skiddaw soils in the far distance. In the valley are peat soils, and on the Glyders to the left mainly Moor Gate and Bangor soils.

occupy drier ground. Natural fertility is very low and these very acid soils require heavy liming if they are to be improved.

Reclamation also depends on improving soil drainage which in wet districts, is usually done with open ditches. In drier localities, pipe or tile drainage with permeable backfill improves the bearing strength of the organic surface sufficiently to support machinery for liming, fertilizing and reseeding, but poaching is still a risk. Rushes soon invade neglected pastures, but when reclaimed fields are well managed the increased transpiration of improved swards can significantly improve the strength, structure and stability of the surface.

Forests cover more than a tenth of the land (Plate 33) and Sitka spruce and Lodgepole pine yield moderately well but even with deep ploughing and draining there is a risk of shallow rooting so that trees on exposed sites are easily blown over. Phosphorus fertilizer is beneficial and timber extraction requires vehicles adapted to the soft ground.

Left undrained and unforested this land provides a wetland habitat of upland plant communities of some scenic but little other amenity value because the soft wet ground and often tussocky vegetation make walking difficult.

Plate 33. The Arenigs from the south. The seasonally waterlogged soils under rough grassland or forest in the middle ground belong to the Wilcocks 2 association whilst the rocky slopes of Arenig Fawr and Arenig Fach adjacent to Llyn Celyn carry Bangor association.

§ 97. WISBECH ASSOCIATION
812b

The Wisbech association, on stoneless marine alluvium, covers 874 km^2 in England and Wales, of which 546 km^2 fringes the Wash. The remainder occurs on the Lancashire and Cumbrian coasts and in the Mersey, Ribble and Dee estuaries and near the mouth of the river Winster where it discharges into Morecambe Bay. It consists mainly of coarse silty calcareous alluvial gley soils, Wisbech series (Seale 1975) and gleyic brown calcareous alluvial soils belonging to the Romney series (Reeve and Thomasson 1981). Other soils include fine silty Agney (Appendix 2, p.309 and Blacktoft (Reeve and Thomasson 1981, p.112), and the silty over clayey Stockwith series (Reeve 1981).

Below are brief descriptions of the two main soils.

Wisbech series
(Full description p.326)

0–30 cm — Ap
Dark greyish brown, slightly mottled, stoneless silt loam; calcareous.

30–50 cm — Bg
Greyish brown, mottled, stoneless silt loam or fine sandy silt loam; coarse blocky structure; calcareous.

50–100 cm — Cg
Grey, mottled, stoneless fine sandy silt loam; relic laminar structure; calcareous.

Romney series
(Full description Reeve and Thomasson 1981, p.111)

0–35 cm — Ap
Dark greyish brown, stoneless silt loam; slightly calcareous.

35–45 cm — Bw
Brown, stoneless silt loam; weak fine subangular blocky structure; calcareous.

45–120 cm — Bg
Brown, mottled, stoneless silt loam; weak fine angular blocky structure; calcareous.

The association is restricted to the Dee estuary where about half the soils are coarse silty, some containing much fine sand. They were formerly mapped as Dee series (Furness and King 1973). Romney soils are infrequent.

Key to component soil series

Soils prominently mottled or greyish above 40 cm		1
Subsoils faintly mottled above 60 cm or distinctly mottled between 40 and 80 cm		2
1.	Coarse silty	WISBECH
	Fine silty	Agney
	Silty over clayey	Stockwith
2.	Coarse silty	ROMNEY
	Fine silty	Blacktoft

Soil water regime

The soils are waterlogged for long periods in winter (Wetness Class IV), where there has been little improvement in land drainage and the soil mottling reflects the current water regime. Where there are ditches, the soils are only occasionally waterlogged (Wetness Class II).

Cultivation and Cropping

As most of the soils have not been effectively drained in the Dee estuary, they are under permanent grass but where drained are used mainly for winter cereals. Opportunities for working the land are summarized in Figure 59. The soils tend to erode easily when cultivated and seedbeds are often blown away, necessitating redrilling, or the seedlings are damaged by the abrasive sand particles, causing a reduction of yield. Where the soils have not long been reclaimed or have been in permanent grass for a long period phosphorus fertilizers are usually needed.

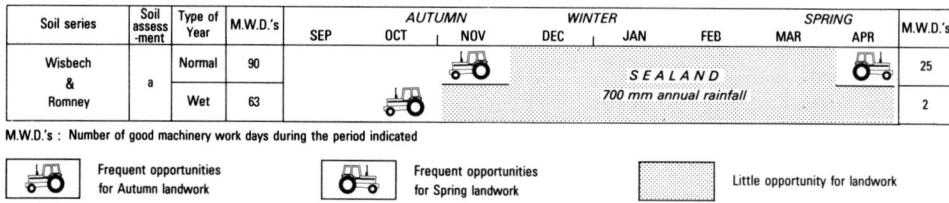

Figure 59. The effects of soil and climate on landwork, Wisbech association

§ 98. WITHNELL 1 ASSOCIATION
611d

The Withnell association consists mainly of loamy, typical brown podzolic soils over sandstone, often on steep slopes up to about 350 m O.D. It is most extensive in south Wales, but is also mapped in the Forest of Dean and north Staffordshire. The soils are predominantly coarse loamy (Withnell series) but fine loamy profiles (Loxhore series) also occur where there are thin beds of shale or the sandstone contains an appreciable proportion of silt and clay. Beneath the topsoil, which is often thin in uncultivated soils, there is a bright ochreous subsoil over sandstone or very stony Head usually within 50 to 80 cm depth; the deeper profiles generally occurring on lower slopes. There are some typical brown earths with duller, yellowish brown subsoils. Both the fine loamy Neath (§ 66) and the coarse loamy Rivington series occur, mainly on cultivated gentle slopes at low altitudes. The brown earths and brown podzolic soils are permeable and rarely wet (Wetness Class I) and readily absorb winter rain. Stagnogleyic brown earths (Nercwys series) are locally common in patches of thick, slowly permeable drift, while brown rankers (Newtondale series, Carroll and Bendelow 1981) occur sporadically around rock outcrops. The soils are usually stony with large stones widespread in topsoils. Brief descriptions of Withnell, Rivington and Loxhore series are given below and of Nercwys and Neath series in § 66.

Withnell series
(Full description Crompton 1966, p.79)

0–15 cm — Ah
Dark brown, slightly stony sandy loam or sandy silt loam.

15–50 cm — Bs
Strong brown, moderately stony sandy loam or sandy silt loam; moderate fine subangular blocky or granular structure.

At 50 cm — R or Cu
Hard sandstone or extremely stony sandy loam.

Rivington series
(Full description Reeve 1975, p.86)

0–20 cm — Ap
Dark greyish brown, slightly stony sandy loam or sandy silt loam.

20–50 cm — Bw
Yellowish brown, slightly or moderately stony sandy loam or sandy silt loam; weak medium subangular blocky structure.

At 50 cm — R or Cu
Hard or soft sandstone or extremely stony sandy loam.

Loxhore series
(Full description p.318)

0–5 cm — Ah
Dark brown, stoneless humose clay loam.

5–50 cm — Bs
Strong brown, slightly stony clay loam; fine granular or subangular blocky structure.

50–60 cm — Cu
Brownish yellow, very stony sandy loam; single grain structure.

At 60 cm — Cr
Shattered sandstone.

The association is extensive on the Pennant sandstone of the south Wales coalfield. Most of the land is steep (Plate 13), but above 150 m O.D. these soils are also found on gentle slopes. There is a small area on Silurian sandstone in the upper Conwy valley.

Key to component soil series

Subsoils faintly mottled above 60 cm or distinctly mottled between 40 and 80 cm; fine loamy	NERCWYS
Unmottled soils; rock within 80 cm	1
1. With bleached subsurface horizon and dark, humus-enriched subsoil; coarse loamy	Anglezarke
No bleached subsurface horizon or humus-enriched subsoil	2
2. Shallow soils, sandstone within 30 cm; loamy	Newtondale
Deeper soils	3
3. With brightly coloured subsoil	4
With brown subsoil	5
4. Coarse loamy	WITHNELL
Fine loamy	LOXHORE
5. Coarse loamy	RIVINGTON
Fine loamy	NEATH

The climate is wet, much of the land having more than 1500 mm of rainfall annually with a field capacity period of more than 250 days. A large part of the land is under acid, bent-fescue pasture which provides good grazing except where there is a dense bracken cover, but on high ground there is poorer, *Nardus* grassland. Most of the improved grassland is on slopes of less than 15 degrees but even here stones can hinder cultivation. There is little risk of poaching because the soils drain freely and topsoils retain little water. Nevertheless grazing sheep cause the soil to move on very steep slopes so forming terracettes. Where neither slope nor stoniness is limiting, these soils provide some of the most useful farmland in country where agriculture is generally constrained by topography, poor soil drainage and a wet climate. About a quarter of the land is afforested, Japanese larch being the preferred species. Old scrub and woodland survives on a few steep valley sides.

The industrial character of south Wales and the nearby presence of the Brecon Beacons National Park means that the land is little used for recreational activity, except where forest walks and drives are provided.

§ 99. WORCESTER ASSOCIATION
431

This association is dominated by reddish clayey soils developed in Permo-Triassic mudstone and clay shale. It is extensive throughout the Midlands and South West England on moderate to steep slopes. Small areas are also mapped in south Wales and Northern England. It includes non-calcareous Worcester soils, typical argillic pelosols, (Jones 1983 and Palmer 1982) and the calcareous Clayworth series, typical calcareous pelosols (Reeve and Thomasson 1981). Where there is a thin layer of fine loamy or fine silty drift over the mudstone, Whimple series (Whitfield and Beard 1980, p.52), stagnogleyic argillic brown earths, are included, mostly on slopes less than 8 degrees. Bands or lenses of greenish grey mudstone or clay shale and occasional harder siltstones (skerries) give soils of the Agardsley series (Jones 1983). On wetter, lower-lying land, profiles of the Brockhurst (Clayden and Hollis 1984) and Spetchley series (Palmer 1976, p.128) are included. The soils are fully described by Ragg *et al.* (1984). Most Worcester, Clayworth and Whimple soils are seasonally waterlogged (Wetness Class III). Winter run-off is rapid because subsoils are slowly permeable, though, where the soils are better drained winter rain is more readily absorbed. The land is mainly under permanent grass but being adjacent to Cardiff and Newport it is also partly built over.

4

LAND DRAINAGE

Land drainage design for agriculture has as its objective the removal of excess water from the soil or land surface (Plate 15). Its aims are to sustain or increase yields, improve management flexibility, reduce production costs and make farming more profitable. Since Roman times men have raised flood banks, cut dykes and formed channels to protect, reclaim or drain land for food production and since the mid-18th Century attempts have been made to improve subsurface water control by the installation of various types of underdrainage systems. Early stone drains were followed by baked clay horseshoe tiles and later still by round clay pipes. Plug drainage was an early primitive form of mole drainage in which a pole was laid in the trench bed and the clay packed over it. The pole was then pulled out end-wise leaving an unlined drainage channel in the clay. Brush and faggots were also used in an attempt to control groundwater levels and improve soil conditions (Trafford 1970).

Early drainage systems relied on close spacing and surface water interception as a means of controlling water levels but in this century, mainly in the last 30 years, the design of drainage systems has involved more attention to the hydrology of water movement in the soil.

The benefits of drainage improvements are well known. For example recent experimental work shows that waterlogging causes substantial reductions in the yields of winter wheat, peas and oilseed rape (Cannell *et al.* 1980, Belford *et al.* 1980, Cannell and Belford 1980).

§ 100. FACTORS AFFECTING FIELD DRAINAGE DESIGN

The design of a drainage scheme is influenced by many considerations, for example rainfall, cropping practice, soils and slope. It is convenient to consider these factors separately although in practice they interact with each other.

Rainfall mainly influences drainage design through the frequency of heavy storms. For example, a "design rate" of 20 mm per day may be used on the assumption that it will be exceeded only once in two years, as assessed from meteorological data (Smith and

Trafford 1976). The type of cropping also influences drainage design. High value horticultural crops justify higher design standards and it is wise to design for a rainfall which might only be exceeded 1 year in 10 years, whereas for cereals 1 year in 2.5 years is used, and for low value grassland a 1 in 1 year risk is acceptable (Bailey *et al.* 1980). Soil properties affect the rate at which water passes through the soil profile and determine the permeability class of the soils, which varies from very slow, less than 0.01 m/day, to very rapid, greater than 10 m/day. The factors affecting this variation are discussed in MAFF (1982a) and Thomasson (1975a). The slope of the land surface affects pipe size and layout and the type of drainage system appropriate for each particular section.

§ 101. VARIABLES IN DRAINAGE DESIGN

The main features which a designer can vary are depth, spacing, the diameter of the drains and the depth and type of secondary treatment such as moling or subsoiling. These aspects are now discussed to explain the classifications applied in Table 24 to the soils of Wales.

Drain depth
Drain depth controls the level of the water-table when drain flow has ceased. When drains are flowing, drain depth interacts with spacing to affect mid-spacing water-table height. Generally, greater depth allows a wider spacing to be used provided the soil is permeable above the drain.

When drain flow has ceased, the water-table depth represents the maximum which pipe drainage alone can achieve. It sets an upper limit to the soil moisture suction that drainage can exert at the surface. A deeper drain may achieve a drier soil by the time the drains stop flowing. Whether the increased drying is significant depends on soil properties, in particular on the relationship of water content to suction (§ 7).

Apart from the effect on the retained moisture content, the suction achieved by drainage also affects trafficability. Soil moisture suction influences the strength of the soil and its capacity to carry vehicles without rutting, or stock without poaching. Trafford (1975) has shown that traffic ruts on arable land and poaching on grassland can be significantly reduced if the water-table can be controlled so that it is kept below 50 cm depth.

Drain spacing
Drain spacing affects the rate at which the water-table level rises during rain and falls after it ceases. The closer the spacing the slower the rise and more rapid the fall. Spacing has no effect on the final, drained, moisture content at the surface or on the water-table position when drain flow has ceased. It affects the rate at which these states are reached.

Pipe size
Pipe size can affect both the way in which the pipe accepts water from the soil, taking into

account the degree of slotting and other relevant factors, and its capacity to transport water. If pipes are to act as effective drains they must not run under pressure so it is desirable that, under given conditions of drainage rate and pipe gradient, they are capable of transporting away all the water without running completely full. With the traditional 75 mm clay pipes used for laterals this is rarely a problem as such pipes usually have considerable reserve capacity, but where small-diameter corrugated plastic pipes are used, their capacity to transport peak flows should be carefully checked.

Filters

Soils requiring a filter usually have a small clay content and a large proportion of particles in the 50–150 μm range. Ideally the filter mesh should be open enough to allow the finest particles within this range to enter the pipe whilst retaining the larger ones so that the filter remains effective for the longest possible time. Several types of plastic drain, factory-wrapped with a synthetic or partially synthetic and organic material, are now available and may offer a possible solution. However, since filters cannot be replaced, they must not clog and should not be used where drains are likely to become blocked by ochre (Thorburn and Trafford 1976, Willardson and Walker 1979).

Permeable fill

Permeable fill, usually gravel, hard crushed stone or synthetic material, is commonly placed in the trench above the pipe to form a permanent connector for moling or subsoiling treatments. It ensures that concentrated flows of water have an easy entry to the pipe. These flows may come from water-bearing layers or old drains but more commonly from channels formed by moling or subsoiling. Although permeable fill is expensive, its use ensures that the pipe system remains effective for many years, enabling subsoiling and moling to be repeated when necessary.

Trenchless drainage

Trenchless pipe installation has become more popular recently because of cost advantages over conventional methods and smaller losses of yield when operating through growing crops. However, Fry (1983) reports that when working below the critical depth, defined as that depth below which the sideways extent of soil loosening is severely restricted, trenchless installation can, in certain soil conditions, cause smearing, compaction or soil shear near the pipe which restrict water entering the pipe. This effect is unimportant if permeable fill is used and where the dominant movement of water to the pipe is vertical through the disturbed trench. Soils mostly at risk are silty and clayey soils in marine alluvium, because they usually need deep pipes, are often weakly structured and have a natural macroporosity which may be destroyed during installation operations below the critical depth. Care must be exercised to reduce the chance of soil damage around the pipe. This may require alteration to equipment or modification of the installation technique.

§ 102. SECONDARY DRAINAGE TREATMENTS

Since many clay subsoils have hydraulic conductivities of 0.1 m/day or less, effective water movement is mainly confined to the surface horizon. In such soils pipe drains are largely ineffective unless the subsoil properties affecting water movement are modified by physically increasing the number and size of cracks and fissures, by subsoiling or by moling, to form unlined channels which convey water more easily through the soil to the pipe drainage system.

Moling

Mole drains are formed by a mole plough which consists of a steel shaft at the base of which is a steel foot or bullet of circular section, normally 75 mm in diameter, that trails a cylindrical expander of 100 mm diameter (Godwin *et al.* 1981). Mole drains are generally drawn at 45 to 60 cm depth and 2 to 3 m spacing. The operation should be carried out when the upper layers of the soil are dry enough to allow shattering and fissuring but the soil at moling depth is sufficiently plastic to form a stable circular drainage channel. The cracks created have a large influence on the hydraulic characteristics of the mole channel. They fill and empty at very low tensions and provide a direct route for water flow to the channels. This results in rapid removal of water from the upper soil layers (Leeds-Harrison *et al.* 1982).

Before moling, the effectiveness of the drainage system depends entirely on the natural porosity and hydraulic conductivity of the undisturbed soil. Because most soils suitable for moling are clayey and relatively impermeable the contribution of crack flow is large. The size of cracks generated by the mole plough increases with decreasing soil moisture content.

Many workers, for example Spoor *et al.* (1982), have attempted to quantify and define factors which influence the effective life of mole drainage. The clay content at moling depth should be not less than 30 or 35 per cent. The soil should be cohesive enough not to slump or slake but need not have a well developed structure. Bulk density should preferably be greater than 1.3 g/cm^3. Surface gradients should not be too steep, normally within 2 to 5 per cent (1 to 3 degrees). The channels are drawn over suitably designed pipe drain systems and permeable fill placed over the pipes to a depth at which it can be intercepted by the mole channel. The best months of the year for moling are May and June as the soil dries out and again in September and October provided there is a soil moisture deficit of about 50 mm. For long life, mole drainage systems must discharge freely and not be submerged for long periods. Outfalls must be kept clear.

Mole drainage is traditionally a treatment of arable land. Where used on grassland, especially in wetter climates, there has been a recent trend towards closer spacing of mole channels, and shallower working depth. If moling causes too uneven a surface, the ground needs to be levelled for forage harvesting.

In Table 24, column 2, soils are graded for moling as follows:

1. Highly suitable   Mole channels can be expected to survive for up to 10 years. Timing of moling, within the general limits (see above) is less critical than for lower grades.

2. Moderately suitable   Mole channels have a shorter period of active life; correct timing is needed.

3. Acceptable or   Frequent re-moling (1:3 years) is
slightly suitable   desirable; timeliness is critical.

4. Unsuitable

Subsoiling

Subsoiling aims at breaking up the subsoil to make it less dense and so improve air and water movement. The tools range from a small tine, working below the plough sole, to giant rippers used in civil engineering. The normal equipment is similar to a mole plough but is equipped with a tapered square share to heave the subsoil. Subsoiling is often done to eliminate plough pans or other impediments to downward movement of water, but more often the objective is the general improvement in the permeability of the subsoil with the aim of speeding horizontal movement of water to the drains.

Ideally subsoiling should be done when the soil is dry and brittle. In this condition the soil mass can be shattered to the full depth of subsoiling and abundant interconnecting fissures formed. Over subsequent years the soil gradually settles, closing the fissures, so the operation must then be repeated. Subsoiling is distinguished from moling by greater shattering and the absence of defined channels. In practice, especially in the west, these differences are less clear-cut. Many soils are subsoiled when they are too moist, forming a channel similar to a mole but of square cross section. Under moist soil conditions the mole plough should be preferred.

Depending on the soil type and its condition there is a maximum depth below which a satisfactory heave cannot be achieved. This limiting or critical depth can be increased by widening the leading edge of the subsoil tine, by loosening the surface layer of the soil ahead of the subsoiler and by carrying out the operation when the soil is friable, being neither too soft and wet nor too hard and dry. To subsoil properly the operations should be delayed until the soil is sufficiently dry, and if necessary the crop sequence should be adjusted to achieve this. Subsoiling in conjunction with pipe drainage is usually done at 1.5 m centres and 40 to 50 cm deep.

When subsoiling is used in a drainage context, the drains are rarely placed more than 30 m apart and may be as close as 10 m. A covering of permeable fill is commonly used as a connector and, where soil permeability diminishes with depth and regular subsoiling is envisaged, permeable fill over at least some drains is desirable.

The disturbance produced by the subsoiler can be improved considerably by attaching inclined blades or wings to the side of the foot so positioned to run in horizontal cracks. Although the winged subsoiler has a higher draught than one with a conventional foot, it is more efficient as the soil disturbance it creates is two to three times greater.

In Table 24, relevant soils are classified in terms of need for subsoiling as follows:

A. *Soils with drainage problems caused by slow permeability.* Soils with clayey subsoils frequently have coarsely developed structure which becomes less permeable with depth. In these soils fissures must be created mechanically for water to penetrate to the subsoil and to be conducted through it and permeable backfill to the drains. Mole drainage is usually very effective but occasionally a combination of moling and subsoiling is useful, for example where it is necessary to have shallow subsoiling to disrupt a pan just below plough depth.

B. *Soils with subsoil pans caused by farm machinery or stock.* These are intrinsically more permeable than in group A but compacted layers in the subsoil caused by machinery during field operations can build up gradually over several years as a result of smearing by implements and excessive wheel slip. Subsoiling is needed to disrupt pans induced by the system of cultivation.

Intensive grazing of weakly structured grassland soils in high rainfall districts often creates severe compaction within the top 30 cm of soil. This causes prolonged waterlogging with subsequent sward deterioration and increased susceptibility to pulling of grass by livestock.

C. *Soils with natural pans.* In some soils the natural downward movement of iron has produced a subsoil ironpan. Pans are also formed in coarse textured soils by the precipitation of iron/manganese at the upper limit of a water-table. These are commonly found in gravel soils overlying clay.

§ 103. WORKMANSHIP, MATERIALS AND MAINTENANCE

The success of a drainage design largely depends on satisfactory standards of workmanship, the use of good quality materials (MAFF 1983) and on maintenance. Recognized codes of practice should be followed and materials should conform to the appropriate British Standards where these exist.

Effective drainage on many farms relies on a ditch network which provides outfalls for pipes and conveys water to main channels and rivers. The depth of the ditch must be sufficient to allow discharge from outfall pipes at all times. On sloping ground ditches are used to intercept surface and subsurface water flows and, on low-lying sites where they are often supplemented by pumping and pipe drainage, they control groundwater in permeable soils to acceptable levels. All ditches should be regularly cleared and maintained.

Where blocking of pipes by ferruginous compounds (iron ochre) or by sand and silt is likely to occur, pipes should be laid with individual outfalls so that they can be cleared periodically by jetting or rodding. Where practicable steeper pipe gradients than usual can be used to reduce the rate of sedimentation.

§ 104. SOIL SERIES AND DRAINAGE DESIGN

The general drainage designs required for extensive soil series under their normal land use are summarized in Table 24. In the first column the soils are listed in alphabetical order. The second column indicates suitability for moling, graded as described above and similarly the third column shows the need for subsoiling. Suggested pipe depth and spacing is given in the fourth and fifth columns. The need for permeable backfill is indicated in the sixth column. Wetness Class for drained and undrained soils is shown in the seventh and eighth columns. The remarks in the ninth column indicate any special drainage features associated with the particular soil series and the final column shows the soil associations in which the series occurs over significant areas.

Table 24
Soil Series and Drainage Design

Soil series	Suitability for moling	Need for subsoiling	Pipe depth (cm)	Pipe spacing (m)	Permeable backfill	Wetness Class Drained land	Undrained land	Remarks	Soil associations
Adventurers'	4		100–120	15–20	Unnecessary	II & III	VI	Pipes laid at base of peat where possible. Risk of ochre accumulations	Adventurers' 1, Altcar 1
Altcar	4		100–120	15–20	Unnecessary	II & III	VI	Pipes laid at base of peat where possible. Risk of ochre accumulations	Adventurers', Altcar 1
Arrow	4	B	100	20–40	Unnecessary	I & II	III & IV	Complex topography may need individually designed systems	Arrow, East Keswick 1, Wick 1
Brickfield	3	A	75–90	10–30	Desirable	III & IV	IV & V	If mole drainage is used permeable backfill is essential	Brickfield 1, 2 & 3, Nercwys
Cegin	3	A	75–90	10–30	Desirable	III & IV	IV & V	If mole drainage is used permeable backfill is essential	Brickfield 1, Cegin
Clifton	3	A	75–90	15–20	Desirable	III	IV	Variation in texture makes moling unreliable	Clifton, Salop
Clwyd						II & III	III	Risk of flooding	Conway
Compton	2–3	B	75–90	20	Essential	IV	V	Normally too wet to subsoil Risk of flooding	Compton
Conway	3	B	100–120	20	Desirable	IV	V	Risk of flooding. Pipe depth at outfall depends on ditch levels	Conway, Fladbury 3, Teme
Crewe	2	A	75–90	20–25	Essential	IV	IV & V		Crewe, Salop
Crowdy	4		90–120	20	Unnecessary	IV	VI	Drainage worthwhile only on lower ground. Risk of ochre accumulations. Dewatering by temporary open trenches before pipe installation may be necessary	Bangor, Crowdy 1 & 2, Laployd, Revidge, Wenallt, Wilcocks 2
Denchworth	2	A	75–90	20–25	Essential	IV	IV & V		Denchworth

Disturbed soils 3	3	A	75–90	20	Essential			Very variable material. Ditches installed 1–2 years before pipes	Disturbed soils 3
Enborne	3	B	80–100	20–30	Desirable	IV	V	Pipe depth at outfall depends on ditch levels. Risk of flooding	Alun, Enborne, Fladbury 3
Fforest	3	A	75–90	15–20	Essential	III & IV	V	Complex topography and occurrence of springs may require individually designed systems	Fforest
Fladbury	2	B	75–90	20	Essential	IV	V	Risk of flooding. Often too wet to subsoil	Compton, Conway, Enborne, Fladbury 1 & 3, Midelney
Flint	3	B	75–90	15–40	Desirable	II & III	III & IV	Wide spacings can be used with permeable backfill	Flint, Salop
Foggathorpe	2	A	75–90	20	Essential	III & IV	IV & V		Foggathorpe 1 & 2
Gelligaer	4					III	IV	Pipe systems uneconomic. Ditching and subsoiling can be beneficial	Gelligaer
Hafren	4					III	IV	As for Gelligaer series	Crowdy 1, Hafren, Parc, Wilcocks 2
Hallsworth	2–3	A	75–90	20–25	Essential	IV	V		Brickfield 3, Hallsworth 1, Foggathorpe 1
Hexworthy	4	C				III	IV	Pipe systems uneconomic. Subsoiling necessary to break subsurface pan	Hexworthy, Parc
Hiraethog	4					III	IV	As for Hexworthy series	Hafren, Skiddaw
Kielder	3–4	C	75–90	15–20		V	V	Opportunity for moling or subsoiling limited. Risk of ochre accumulations	Wilcocks 1
Laployd	4		80–100	15–20	Desirable	III	VI	Drainage improvements limited by stones and boulders	Laployd
Lydcott	4					III	IV	Pipe systems uneconomic. Ditching and subsoiling can be beneficial	Lydcott

Table 24 (contd.)
Soil Series and Drainage Design

Soil series	Suitability for moling	Need for subsoiling	Pipe depth (cm)	Pipe spacing (m)	Permeable backfill	Wetness Class Drained land	Wetness Class Undrained land	Remarks	Soil associations
Nercwys	3	B	75–90	20	Desirable	III	III & IV		Brickfield 2, East Keswick 1, Neath, Nercwys, Withnell 1
Newchurch	2	A	80–100	20	Essential	III	IV	Pipe depth at outfall depends on ditch levels	Newchurch 2, Wallasea 1
Rockcliffe	4	B	80–100	20–25	Unnecessary	III	IV		Rockcliffe
Salop	3	A	75–90	20–30	Essential	III & IV	IV		Crewe, Flint, Salop
Salwick	4	B	75–90	20–30	Desirable	II	III		Clifton, Flint, Salwick
Sannan	3	B	75–90	20–30	Essential	II	III		Cegin, Denbigh 1
Tanvats	3	B	80–100	20–30	Desirable	III & IV	V	If mole drained permeable backfill is essential	Rockcliffe, Tanvats
Wallasea	2	A	80–100	20	Essential	III & IV	V	Pipe depth at outfall depends on ditch levels	Newchurch 2, Wallasea 1, Tanvats
Wenallt	3		75–90	15–20	Essential	IV & V	V & VI	Opportunity for subsoiling and moling very limited. Ochre accumulations locally	Fforest, Lydcott, Wenallt
Wilcocks	3		75–90	15–20	Essential	IV & V	V & VI	Opportunity for subsoiling and moling very limited. Ochre accumulations locally	Brickfield 1, Crowdy 1, Hafren, Parc, Wilcocks 1 & 2
Winter Hill	4		90–120	15–20		V	VI	Drainage economic only on lower ground. Ochre accumulations locally	Bangor, Crowdy 2, Revidge, Skiddaw, Wilcocks 2
Wisbech	4	B	80–100	25	Unnecessary	II	IV		Agney, Wisbech

5

AGRICULTURAL AND FORESTRY INTERPRETATIONS

Rural Wales is a land of livestock farmers who rely heavily on grass production to feed their stock. Most of the grass is consumed directly by grazing animals but in early summer sufficient must be preserved, either as hay or silage, to maintain stock through the winter months. Only in warmer, drier lowland districts is arable farming of importance, with barley and wheat the main crops. First early potatoes are of local importance in frost-free areas of south-west Dyfed, the Gower and Anglesey. Forestry is extensive in the uplands (§ 107).

Land suitability for crops or farming systems depends on climate and relief, as well as soil. In the assessments of suitability that follow, good management is assumed including, where appropriate, the use of suitable crop varieties, fertilizers, rotations, crop protection and drainage measures. Social and economic factors which often affect the choice of land use are mainly excluded. The aim is to offer a guide to the relative suitability of land for particular crops or systems, mainly by the use of tables in which land suitability is related to selected soil series. These ratings reflect average conditions and relate only to stated limits of other variables such as climate or slope. It is therefore essential that the explanatory sections be read before the tables are used. The actual suitability of a piece of land for any use depends on many variables including social and economic factors. Information given in this chapter is intended as an aid to the interpretation of land potential and as a guide when alternative land uses are being considered.

§ 105. GRASSLAND

Suitability for grassland
Grass needs water, warmth and nutrients to grow well. Soil conditions and climate regulate the supply of moisture while fertilizers, notably nitrogen, are applied to sustain the supply of nutrients. The suitability of land for grass depends not only on its capacity to grow the crop but also on its ability to support animals and machinery without damage to sward or soil. Interactions between climate and soil are complex and affect potential yield

and management of grassland in different ways. The circumstances ideal for growth are not always those that make management, grazing or cropping easy. The large potential yields that can be achieved in a moist climate may be offset by poor ground conditions. The grassland suitability scheme used here is based, with minor modifications, on those of Harrod (1979) and Harrod and Thomasson (1980). These assess potential yield, the risk of poaching (inversely related to trafficability) and the balance between them, to indicate relative fitness for intensive grass production.

Poaching risk

The factors affecting poaching by animals are essentially the same as those restricting trafficability by tractors and machinery. Soils are classed on a 1 to 5 scale of increasing restriction, referred to as poaching risk categories. Soil properties, including the amount of water retained in the topsoil at field capacity, the depth to slowly permeable horizons and the duration of waterlogging within shallow depth, form the basis of the classification. Adjustment is made for climatic differences between the dry lowlands and moister districts in the uplands and the west.

Yield categories

In the assessment of yield category, the volume of soil water available to grass is compared with maximum potential soil moisture deficit to derive four categories from a (highest) to d (lowest). These follow the droughtiness classes of Thomasson (1979), but are modified to allow for the length of the growing season, which is comparatively long in moist western districts and short in the uplands.

Grassland suitability classes

Poaching risk and yield categories form the basis of four grassland suitability classes, A to D (Table 25). On more difficult land, other soil properties and the terrain are also considered. Class A land has small poaching risk from grazing animals and allows safe access for machinery while having large yield potential due to substantial soil water

Table 25
Grassland Suitability Classes

		Increasing risk of poaching →				
		1	2	3	4	5
	d	Cy	Cyp	Cyp	Cyp	Cyp or D
Yield category increasing	*c*	B	B	Cyp	Cyp	Cyp or D
	b	A	B	B	Cp	Cp
	a	A	A	B	Cp	Cp

reserves or moist climate. Class B land has slight restrictions due either to poaching risk or yield. In Class C, major restrictions are identified by the following suffixes: p for poaching risk; y for restricted yield; yp for yield and poaching restriction; u for upland climate and terrain; g for steep slopes. Class D is unsuited to intensive grassland use by reason of climate, soil or terrain.

The suitability of extensive soil series in Wales is given in Table 26 for appropriate climatic zones, based on the range of meteorological field capacity days. Classification is given for sites where slope is not restricting, although many soil series and soil associations have areas steeper than 11 degrees (see comments). The ADAS grass growth class for nitrogen use (MAFF 1982b), based on soil available water and summer rainfall, is a guide to potential dry matter yield and nitrogen fertilizer use in lowland districts where droughtiness is often a limitation.

The moist climate of Wales favours vigorous grass growth in spring and early summer, declining in mid-summer as the soil dries out. It is at this time that soil conditions most strongly influence the pattern of grass growth. Late summer and autumn rainfall stimulates a flush of growth which is eventually curtailed by falling temperatures. There are variations within the Principality. Summer moisture deficits are greatest in the north-east while the shortest autumn flush is experienced in east Powys. In south-west Dyfed, grass continues to grow slowly for most of the winter.

Well suited land (Class A) is extensive in the lowlands of Wales wherever there are deep well drained brown soils. East Keswick series and most Denbigh soils are in this category. Manod soils are inherently well suited, but management is often restricted by the steep slopes upon which these soils are characteristically developed.

Coarse loamy, sandy and shallow soils in drier districts, which are unable to supply sufficient moisture in the summer, are included in Class B which also contains brown soils in cool districts with a short growing season (for example, Denbigh, Eardiston and Malvern series) and soils with restricted subsoil permeability (Nercwys, Flint and Salwick series) or a humose surface horizon which increases risk of poaching (Moor Gate and Parc series). Although capable of storing large amounts of moisture for the crop in summer, humose topsoils have low bearing strength and poach if grazed in wet conditions.

Land with serious limitations to intensive grassland use is placed in Class C which is subdivided on the kind of restriction. Much of Class C land lies wet because of slowly permeable subsoils or high groundwater. When compared with Class A land, growth begins later but continues into the dry part of summer particularly where there is movement of groundwater into the soil. Total grass yields can exceed those of drier land but wet ground conditions, which persist into late spring and hinder early fertilizer applications, return in autumn, effectively shortening the grazing season. These soils are classed Cp and include the Cegin, Brickfield, Clifton, Fforest, Foggathorpe, Hallsworth and Salop series which have slowly permeable subsoils and Conway, Compton, Newchurch and Wallasea soils which are affected mainly by groundwater.

Application of the ADAS grass growth classification for nitrogen use suggests that

Table 26
Land Suitability for Grass Production

Soil series	Field capacity range (days)	Poaching risk and yield category	Suitability class	Growth class[1]	Soil associations	Comment
Brickfield	175–200 >200	4a 5a or 5b	Cp Cp	Average–Good Good–very good	Brickfield 1, 2 and 3, Cegin, Nercwys	175–200 F.C. days range mainly in west Anglesey and east Clwyd; poaching risk limits optimum use of nitrogen; category 5b in cool districts[2]
Bromyard	175–200	2a	A	Good	Bromyard, Eardiston 1	
Cegin	175–200 >200	4a 5a or 5b	Cp Cp	Good Very good	Cegin, Brickfield 1	175–200 F.C. days range mainly in east Powys; poaching risk limits optimum use of nitrogen; category 5b in cool districts[2]
Clifton	150–200 200–250	4a 5a	Cp Cp	Average–good Good	Clifton, Salop	150–200 F.C. days range mainly in east Clwyd; poaching risk limits optimum use of nitrogen
Compton	150–175	5a	Cp	Good	Compton, Hollington	Possible flood risk; poaching risk limits optimum use of nitrogen
Conway	150–200 >200	3a 4a	B Cp	Good Very Good	Conway, Fladbury 3, Teme	150–200 F.C. days range mainly in Vale of Clwyd and east Powys; possible flood risk; poaching risk limits optimum use of nitrogen
Denbigh	175–225 >225	2b 2a or 2b[3]	B A or B[3]	Average–Good Good	Denbigh 1, Cegin, Manod, Neath	175–225 F.C. days range mainly on coast and in Severn valley

Series	F.C. days	Category	Class	Quality	Associations	Notes
Eardiston	175–225	2b	B	Average–Good	Eardiston 1 and 2, Bromsgrove, Bromyard, Goldstone	175–225 F.C. days range mainly in east Gwent
	>225	2a or 2b[3]	A or B[3]	Good		
East Keswick	175–250	2a	A	Good	East Keswick 1 and 3, Brickfield 2, Escrick 2	
Escrick	175–200	2a	A	Good	Escrick 1 and 2	
Fforest	>250	5a	Cp	Very good	Fforest	Poaching risk limits optimum use of nitrogen
Flint	175–225	3a	B	Average–good	Flint, Salop	
Foggathorpe	>175	5a	Cp	Good	Foggathorpe 1 and 2	Poaching risk limits optimum use of nitrogen
Hallsworth	>200	5a	Cp	Very good	Hallsworth 1, Brickfield 3, Cegin, Foggathorpe 1	Poaching risk limits optimum use of nitrogen
Malham	200–250	2a	A	Good	Malham 2, Crwbin	Droughtiness in shallow soils and in coastal districts
Malvern	200–275	2a	A	Good	Malvern	Gradient restricts use in parts, droughtiness in shallow soils; category 2b in cool districts[2]
	>275	2b	B	Good		
Manod	175–225	2b	B	Average–good	Manod, Denbigh 1	175–225 F.C. days range mainly in coastal districts. Steep slopes common
	>225	2a or 2b[3]	A or B[3]	Good		
Milford	175–225	2b	B	Average–good	Milford, Eardiston 2	175–225 F.C. days range mainly in south-west Dyfed and east Gwent. Gradient restricts use in parts
	>225	2a or 2b[3]	A or B[3]	Good		
Moor Gate	225–300	3a	B	Good	Moor Gate	Rockiness and gradient severely limit use in some parts
Moretonhampstead	>200	2a	A	Good	Moretonhampstead, Malvern, Moor Gate	Rockiness and gradient severely limit use in some parts

Table 26 (contd).
Land Suitability for Grass Production

Soil series	Field capacity range (days)	Poaching risk and yield category	Suitability class	Growth class[1]	Soil associations	Comment
Neath	225–300	2a	A	Good	Neath, Withnell 1	Gradient restricts use in parts
Nercwys	200–250	3a	B	Good	Nercwys, Brickfield 2, East Keswick 1, Neath, Withnell 1	
Newchurch	150–200	5a	Cp	Good–very good	Newchurch 2, Wallasea 1	Poaching risk limits optimum use of nitrogen
Newport	150–200	1c	B	Fair	Newport 1, Wick 1	150–200 F.C. days range mainly in Clwyd
	200–250	1b	A	Average		200–250 F.C. days range mainly in Gwynedd and Dyfed
Oglethorpe	200–250	2a	A	Good	Oglethorpe	Category 2b in cool districts[2]
	250–300	2b	B			
Parc	>250	3a or 3b	B	Very good	Parc	Category 3b in cool districts[2]
Restored opencast coal workings	225–300	5b or 5c	Cp or D	Good	Disturbed soils 3	
	>300	5c	Cu or D			
Rheidol	175–200	2b	B	Good	Rheidol, Rowton	
	200–300	2a	A	Good		
Rivington	225–250	2a or 2b	A or B	Good	Rivington 2, Eardiston 2 Withnell 1	Gradient restricts use in parts; category 2b, suitability B in cool districts[2]

Salop	150–175	4b	Cp	Average	Salop, Crewe, Flint }	150–175 F.C. days range mainly in east Clwyd; poaching risk limits optimum use of nitrogen
	175–225	5a	Cp	Average-good		
Salwick	175–250	3a	B	Average-good	Salwick, Clifton, Flint	Droughtiness in shallow soils
Ston Easton	200–250	2b	B	Good	Ston Easton	Possible flood risk
Teme	175–300	2a	A	Good-very good	Teme	
Wallasea	150–200	5a	Cp	Very good	Wallasea 1, Newchurch 2, Tanvats	Poaching risk limits optimum use of nitrogen
Wick	150–175	2b	B	Average	Wick 1, Escrick 2, Newport 1, Salwick }	150–175 F.C. days range mainly in east Clwyd
	175–300	2a	A	Good		
Wisbech	150–175	2a	A	Good	Wisbech, Agney	
Withnell	>250	2a or 2b	Cg	Good	Withnell 1	Gradient restricts use on much of this land; category 2b in cool districts[2]

[1] ADAS grass growth class for nitrogen use—not applied to cool districts[2]
[2] Cool districts have a short growing season of less than 225 days, roughly equivalent in Wales to accumulated temperatures of less than 1250 day degrees above 0 degrees C between January and June (see Fig. 8)
[3] Category 2b, suitability B in cool districts[2]

most soils in suitability Class A should achieve yields in excess of 11 tonnes of dry matter per hectare at optimum nitrogen application rates of over 400 kg per hectare. Yields for Class B soils are likely to be 1 tonne per hectare less, and the optimum application rate is also lower at 370 kg per hectare. Class Cp soils pose problems in interpretation of ADAS grass growth classes because although large yields are possible there are difficulties with early fertilizer applications on wet ground. The effectiveness of nitrogen use on wet land is commonly limited by the risk of denitrification or loss in solution. Utilization is again difficult in autumn. The ADAS classes have not been applied to soil series which occur mainly on land with a short growing season.

Upland grazings
The vegetation of upland grazings, which are widespread in Wales, varies with soil. Brown podzolic soils on steep slopes commonly carry bent-fescue *(Agrostis-Festuca)* grasslands, often infested by bracken *(Pteridium aquilinum)*, or gorse *(Ulex* spp.), which is commoner where the soils are shallow. Grazings on wet stagnogley and stagnohumic gley soils, can be very rushy, while those with the wettest peaty topsoils frequently carry dense tussocks of purple moor-grass *(Molinia caerulea)*. Stagnopodzols usually have vegetation dominated by mat-grass *(Nardus stricta)* which is replaced by heather moor where grazing is light. Heather *(Calluna vulgaris)* and bilberry *(Vaccinium myrtillus)* moors are also found on lightly-grazed brown podzolic soils. The peat soils of the hills carry blanket bog dominated by cotton-grasses *(Eriophorum* spp.) or *Molinia*. Associated plants include *Sphagnum* mosses, cross-leaved heath *(Erica tetralix)* and deer-grass *(Trichophorum cespitosum)*. Slightly drier peat land may carry heather moor, while lowland bogs like those at Borth and Tregaron have distinctive raised moss floras dominated by *Sphagnum*, sedges and heathers. Locally in the mountains the vegetation is of special interest. Base-rich crags have assemblages of alpine species which flourish free from grazing. The flora is richest where the rocky ledges are flushed by seepage water. Exposed summits are bare, or sparsely vegetated with woolly hair-moss *(Rhacomitrium lanuginosum)*, the alpine club-moss *(Lycopodium alpinum)* or viviparous fescue *(Festuca vivipara)*.

   Upland grazings are increasingly being improved by a combination of methods appropriate to soil type, topography and climate. Stagnogley soils and, where slopes permit, brown podzolic soils can be ploughed and reseeded. Stagnopodzols can be improved by fertilizing, flail mowing, burning, shallow cultivation and seeding with suitable grasses and clovers (§ 51). Careful management is required to prevent regrowth of native species.

Relative grazing value
Methods of improving hill grazings are discussed for relevant soil associations in Chapter 3, but improvement is not always economic. Assessments of grazing value of unimproved land are made by the method of Bibby *et al.* (1982), who describe 84 plant communities based on the classification of Scottish plant associations (Birse and Robertson 1976, Birse 1980, Robertson 1984). The communities are identified by the proportions of different

species, but also by the presence of "character" species (Robertson 1984). The method relies on assigning grazing values to individual species ranging from -1 (for poisonous species) to 8 (for the most palatable and nutritious grasses such as *Lolium perenne*). The relative grazing value of plant communities is calculated from quadrat samples–the grazing value of each species is multiplied by its percentage cover and the sum of the products is divided by 100. From a number of quadrats an average grazing value has been determined for each plant community. The relative grazing value is described as good, moderate or poor corresponding to the high, moderate or low grazing divisions described by Bibby *et al.* (1982).

Moorlands dominated by heather and most bog communities have poor grazing value. *Molinia* and *Nardus* grasslands have moderate or poor grazing value according to their composition. Grasslands dominated by bent and fescues are generally rated as good.

§ 106. ARABLE CROPS

Land suitability for arable crops

Suitability is assessed for *sustained* production in a rational cropping system (McRae and Burnham 1981, FAO 1976). In a good year and with good management, the range of environmental conditions under which temperate crops can be grown is wide. However, the chances of failure to establish, mature or harvest a particular crop must be considered when land suitability is assessed. For land to be judged suitable for a crop there must be reasonable confidence that it can be grown regularly and make a predictable contribution to the farm economy. Inevitably some land is well suited to many crops and other land poorly suited to all. The aim in this text is the objective assessment of the range of cropping possibilities, without emphasizing the ideal crop for any piece of land. The suitability of the more important series in Wales for common arable crops is given in Table 29 employing the following suitability classes:

Well suited	Potential production is high and sustainable from year to year. There is adequate opportunity to establish the crop in average years at or near the optimum sowing time and harvesting is rarely restricted by poor ground conditions. Even in wet years (up to a frequency of about one in four) working conditions are acceptable and do not prevent crop establishment. There are sufficient soil water reserves to meet the average requirements of the crop.
Moderately suited	Potential production is usually moderate or high but can be lower in years when soil water is insufficient to sustain full growth or when crop establishment is unsatisfactory owing to untimely sowing or poor soil structure. In favourable years, which may be wet years on droughty land or dry years in a generally wet climate, outstanding yields can be achieved, but production is not predictable. For root crops, harvesting can be difficult and soil structure may be damaged with consequent penalties for the following crop.
Marginally suited	Potential production varies from year to year. There are considerable risks, high costs, or difficulties in maintaining continuity of output, which are due to the interaction of climate with soil properties, disease or pest problems. In some years there may be a failure to establish the intended crop. Often marginal suitability does not imply high risks in producing the particular crop but rather reflects difficulties in fitting it into a continuous farming system.

Unsuited   Criteria of unsuitability vary with individual crops but are chiefly climate, gradient
                    and, for root crops, stoniness. Near the selected climatic limits for a crop there will be
                    years with favourable weather which allows efficient production and other years which
                    are too wet or cool. The climatic limits chosen, however, are expected to exclude
                    normal production of the crop at least one year in four; a higher degree of risk was
                    judged unacceptable. In practice, the chosen limits are mainly beyond the current
                    geographic limits of the crops.

The class limits to define the level of crop suitability, for individual soils in various
climates, were reached after considering the distribution and the performance of farm
crops. Available field experimental data were taken into account. Most field experiments
however were not designed for this purpose and improved management techniques have
recently extended the limits for some crops. The gradings allocated in Table 29 are
consistent with the climatic and soil information available, but are necessarily
provisional. They may require revision in the light of future research.

Winter cereals
Most autumn-sown cereals are now grown in predominantly arable farming systems.
Land is considered to be well suited for winter wheat or winter barley if it can support
more or less continuous arable cropping, giving yields over a period of years similar to or

Table 27
Allocation of Suitability Classes[1] for Winter Cereals[2]

Machinery work days after 1 Sept.		Available water (A.P.)[3]–Moisture deficit (M.D.)[3] in mm				
		Over 40	>20 to 40	>0 to 20	0 to –19	–20 & lower
Over 80	Increasingly	Well	Well	Well	Moderate	Marginal
>50 to 80	restricted	Well	Well	Moderate	Marginal	
>20 to 50	workability	Moderate	Moderate	Moderate	Marginal	
20 & less		Marginal	Marginal	Marginal		
					Increasing droughtiness	

[1] The full names for these classes are *well suited*; *moderately suited*; *marginally suited*; and *unsuited*.
[2] Land is classified as *unsuited* for winter cereals where the accumulated temperature is less than 1125 day°C above 0°C (January to June), *or* the maximum P.S.M.D. is less than 75 mm, *or* where slope is more than 11°.
[3] Crop-adjusted value.

above the national average. Soil working and trafficking conditions must be adequate for
control of weeds and establishment of the crop over a large area of land during September
and October. The number of autumn machinery work days (M.W.D.) is an indication of
this (Table 27). On large farms a tendency to reduce the number of tillage operations, or to
use direct drilling, often indicates a risk of few machinery work days in a wet autumn.
However, it is desirable that there be an adequate period for landwork in an average
autumn so any necessary ploughing or other ameliorative treatment, such as subsoiling or
moling, can be undertaken between harvesting and drilling. Ground conditions for crop
protection and fertilizer application in winter or spring are less critical because damage if

any is confined to wheel rutting. The size of soil water reserves (A.P.) in relation to transpiration demands (M.D.) during the crop growth period strongly affects the suitability class.

The criteria for unsuitability are based on climate and slope rather than on soil properties. A maximum potential soil moisture deficit (P.S.M.D.) of 75 mm is close to the wet limit of cereal production, as there are few dry days for combining in July and August. A slope of greater than 11 degrees (1 in 5) is too steep for efficient use of combine harvesters. Accumulated temperature is not limiting for cereals in England and Wales as most land unsuited to cereals is so designated because the climate is too wet rather than too cool.

There are slight differences in the pattern of water uptake between winter wheat and winter barley, because of differences in the onset of ripening. In most respects however the two crops have the same requirements so they are classed together. The choice between winter wheat and winter barley depends mainly on the incidence of take-all, on the need to spread the harvesting and tillage work-load, and, in some farming systems, on the need to cultivate early for the following crop, rather than on soil or climatic differences.

Spring barley
Suitability criteria are particularly difficult to allocate for this crop. The yield potential is generally lower than for winter cereals, and land well suited to spring barley is usually well suited to other, more profitable crops. In Eastern and South East England spring barley is often sown as a last resort, but the crop is grown in Wales because of its tolerance of adverse conditions and its common use on the farm. In drier districts it is a useful crop after late harvested roots, when there is insufficient opportunity to plant winter cereals. In cooler districts it can be grown where the winter wheat harvest lasts into September. The emphasis is on good working conditions in spring to allow early establishment. Other criteria are similar to those for winter cereals (Table 28).

Table 28
Allocation of Suitability Classes[1] for Spring Barley[2]

Machinery work days 1 Jan–30 April		Available water (A.P.)[3]–Moisture deficit (M.D.)[3] in mm				
		Over 40	>20 to 40	>0 to 20	0 to –19	–20 & lower
Over 30	Increasingly	Well	Well	Moderate	Moderate	Marginal
>20 to 30	restricted	Well	Well	Moderate	Marginal	Marginal
>10 to 20	workability	Moderate	Moderate	Marginal	Marginal	
10 & less		Marginal	Marginal	Marginal		
					Increasing droughtiness →	

[1] The full names for these classes are *well suited; moderately suited; marginally suited;* and *unsuited.*
[2] Land is classified as *unsuited* for spring barley where the accummulated temperature is less than 1025 day °C above 0°C (January to June), *or* the maximum P.S.M.D. is less than 75 mm, *or* where slope is more than 11°.
[3] Crop-adjusted value.

Table 29
Suitability of selected Soil Series for sustained Cereal Cropping

Soil series	Field capacity range[1] (days)	Winter cereals[2]	Spring barley	Soil associations	Comment
Bromyard	175–225	Well	Marginal	Bromyard, Eardiston 1	
Cegin	175–225	Marginal	Marginal	Cegin	Only small proportion of Cegin series is suitable for arable cropping because of climate
Clifton	175–200	Moderate	Marginal	Clifton, Salop	
Denbigh	175–250	Moderate	Moderate	Denbigh 1, Cegin, Neath	Only small proportion of Denbigh series is suitable for arable cropping because of climate and gradient
Eardiston	175–225	Moderate	Moderate	Eardiston 1, Bromsgrove, Bromyard	Gradient restricts use locally
East Keswick	175–250	Moderate and well	Moderate	East Keswick 1 and 3, Brickfield 2, Escrick 2	
Escrick	175–200	Well	Moderate	Escrick 1 and 2	
Flint	175–225	Moderate	Marginal	Flint, Salop	
Malham	200–225	Moderate	Marginal	Malham 2	
Manod	200–250	Moderate	Moderate	Manod, Denbigh 1	Only small proportion of Manod series is suitable for arable cropping because of gradient and climate
Milford	175–200 200–250	Well Moderate	Moderate Marginal }	Milford	Gradient restricts use locally
Moretonhampstead	200–250	Moderate	Marginal	Moretonhampstead	This range (200–250) is only in south-west Dyfed

Neath	200–250	Moderate	Marginal	Neath	Only small proportion of Neath series is suitable for arable cropping because of gradient and climate
Newport	150–225	Well	Moderate	Newport 1, Bridgnorth, Wick 1	Droughtiness can depress yields in dry districts
Salop	150–200	Moderate	Marginal	Salop, Flint	
Ston Easton	200–225	Well	Marginal	Ston Easton	
Wick	150–200 / 200–225	Well / Well	Moderate / Marginal }	Wick 1, Escrick 2, Newport 1, Salwick	
Wisbech	150–200	Well	Moderate	Wisbech, Agney	Caps easily, wind erosion possible.

[1] Field Capacity ranges for which suitability gradings are given represent the main occurrences, but not necessarily the full climatic range, of each series.

[2] Oilseed rape is usually well suited on soils that are well or moderately suited to winter cereals, except in districts with a potential maximum soil moisture deficit of more than 100 mm.

Factors affecting suitability for cereal crops

Table 29 lists the soils capable of supporting sustained cereal production and indicates their suitability and the climatic range over which cereals may be grown. In Wales land suited to sustained cereal cropping is concentrated in parts of Gwent, East Powys and Clwyd. The prime limitation is climatic wetness and suitable land has an average maximum P.S.M.D. of at least 75 mm. Continuous arable cropping is rare and most cereals are grown within predominantly grassland systems. Livestock farms grow barley as a supplement to bought-in concentrates. These crops are mainly on ploughed-in grass fields which will return to grass following harvest so that structural problems are seldom encountered. Quality grain is not essential for home consumption and with small acreages involved a successful crop under such conditions does not necessarily indicate suitability for sustained arable cropping. In areas of Wales climatically unsuited to sustained production, the well drained soils listed in Table 29 are those on which single crops could best be concentrated. The Welsh climate favours autumn cultivations when soils, particularly subsoils, are drier than in spring. Sufficient machinery work days must be available for cultivating and sowing a large acreage and perhaps for remedial work such as subsoiling. Despite its desirability for weed control, continued spring working even on well drained soils such as Denbigh series often leads to subsoil compaction, suppressed yields and the eventual need for a remedial grass crop.

First early potatoes

First early potatoes are those capable of providing an economic return when lifted before the end of June. Profit depends on earliness, and good husbandry is vital. A warm early climate and freedom from frosts in spring is essential and this restricts production to western or southern coastal areas.

Land is cultivated during winter and is planted from late January to March. Most crops are planted between early February and early March. Good tilth helps accurate placement and ridging. Early potatoes are shallow rooting and are unable to exploit soil moisture much below 60 cm depth. Cultivation and planting in wet soil is common and causes subsoil compaction and further restricts effective rooting. Irrigation is widely used to give early bulking and improves lifting in dry seasons by reducing cloddiness.

Well suited land has a climate and soil that allow successful early production and mechanized harvesting over several successive years. Moderately suited land has climatic or soil limitations which impose restrictions on the earliness or continuity of production without too high a risk of crop failure. Early warmth in spring is needed, and areas with an accumulated temperature (January to June) of less than 1,450 day degrees above 0 degrees Centigrade or a maximum potential soil moisture deficit of less than 75 mm cannot successfully sustain extensive long cropping. Current production is mostly limited to land near coasts with more than 200 continuous frost-free days and only those areas with more than 250 days and 1,550 day degrees Centigrade are capable of consistent very early cropping without recourse to specialist techniques. A recent innovation is the use of plastic film which helps conserve soil moisture and radiant energy.

Appreciable variations in frost risk and warmth related to aspect, slope and exposure occur within a district and these should be considered when assessing local suitability. Slopes greater than 7 degrees are normally too steep for mechanized harvesting and have an erosion risk, but, where the aspect is favourable, larger than average yields and earlier lifting dates offset the disadvantages of hand picking. Such favoured slopes can sometimes give much larger returns than land suited to mechanical harvesting.

In wet districts where the field capacity period exceeds 200 days, only well drained soils with coarse loamy, sandy or calcareous loamy topsoils are considered to be well suited to long-term continuous production. Other soils are moderately suited or unsuited because of adverse ground conditions during the winter. However, much of the crop in Wales is grown in rotation with grass to maintain soil structure so production is also successful on well drained fine loamy soils. Although stones can be dealt with by windrowing machines very stony soils are unsuited to mechanized production because of potato damage and wear to machinery; hand-lifting can alleviate both problems.

Within Wales, first early potatoes are grown in south-west Dyfed, the south coast of the Gower, parts of Anglesey and to a limited extent on the tip of the Lleyn. In Dyfed the main localities are around Marloes and St David's. Almost all this land has at least 200 field capacity days and certain favoured slopes near Pembroke have a frost-free period of more than 250 days. Land below about 60 m O.D. has a January to June median accumulated temperature exceeding 1,550 day degrees Centigrade and the remainder of south-west Dyfed has at least 1,450 day degrees Centigrade. Figures for the Gower are similar but in north Wales, no land receives 1,550 day degrees Centigrade and only land below 90 m O.D. on the Lleyn and 60 to 70 m O.D. on Anglesey has more than 1,450 day degrees Centigrade. All soils used for early potatoes are well drained, many of them with rock at shallow depth. They include the Milford, Moretonhampstead, Manod, Neath, Denbigh, Malham and East Keswick series. Most sites have a southerly or easterly aspect and slopes are up to 7 degrees. Only on the Moretonhampstead series and some Manod soils is continuous cropping practised. On nearly all other soils early potatoes are grown in rotation with grass. On the Gower early potatoes are grown on well drained soils in the East Keswick and Malham associations. In Anglesey they are grown along the Menai Straits and in the north-east of the island mainly on East Keswick association.

§ 107. FORESTRY

The mild moist oceanic climate makes much of Wales well suited to forestry, although exposure on coasts and mountains is a limitation. Frosts in early autumn and late spring are comparatively rare in the west where sensitive species like Douglas fir can be grown. Young spruces and Corsican pines are also at risk though their selection for planting is as much influenced by rainfall and temperatures as frost. Sitka spruce thrives where annual rainfall exceeds 1000 mm as in most of Wales. Green (1964) suggests that incidence of drought is a more important limitation on some species than total rainfall. In Table 31 a mean maximum potential soil moisture deficit (PSMD) of 100 mm is used to divide climatically moist areas (<100 mm PSMD) suitable for example for Sitka spruce, from dry generally lowland areas (>100 mm PSMD) in which species like Corsican pine are more likely to thrive. The success of Norway spruce is also possibly linked more with drought risk than total rainfall. Stress from drought causes vertical cracks in the bark which provide focal points for attack by pests such as the great spruce bark beetle.

Exposure to strong winds affects choice of species. Near coasts, salt-laden winds limit foliage growth of sensitive species, and cold northerly and easterly winds on the exposed mountains restrict the range of suitable trees. Leyland cypress, Douglas fir, the larches, oaks and ash all risk damage by strong winds, Lodgepole pine and Sitka spruce withstand exposure better although windthrow is likely where rooting is restricted.

Soil and site strongly influence species selection (Pyatt 1977). Three soil limitations are particularly widespread; wetness, acidity and shallowness. Wetness is mostly surface waterlogging caused by slowly permeable fine textured subsoils or on the higher uplands by a silty compact horizon or an ironpan and seriously reduces tree stability. Better rooting can be achieved by drainage and deep ripping. Trees intolerant of wet ground include the larches, Douglas fir, beech, ash and sweet chestnut. Sitka spruce yields well in damp sites, whilst Lodgepole pine tolerates a wide range of soils from well drained sands to wet peats. Soil acidity and nutrient deficiency occur in peaty and podzolic soils in both the lowlands and the uplands though with applications of phosphorus fertilizers at planting and as necessary thereafter, growth is usually satisfactory. In contrast, trees on shallow rendzinas and rankers over limestone, and brown calcareous earths, suffer check on growth from lime-induced chlorosis before normal rotation age is reached. Rock at shallow depth restricts root growth and tree stability in large parts of Snowdonia dominated by humic rankers, and locally elsewhere.

Site limitations which influence management include very steep and rocky slopes, where timber extraction may require the use of overhead cables rather than ground vehicles.

Soil Suitability for Forestry

In Wales, there are large areas of brown and brown podzolic soils capable of growing large yields from a wide range of tree species, both in the drier lowlands and the moist uplands.

The suitability of the soils for forestry is summarized in Tables 30 and 31 below. The major soil series in Wales are allocated to a Suitability Group using the alphabetical list in Table 30. The Suitability Groups are lettered nationally; some groups do not occur in the Principality so they are not included.

Soil characteristics and site limitations of each suitability group are summarized in column two of Table 31. In this column Forestry Commission Terrain Classes affecting management (Rowan 1977) are given where their range is not too great. Three elements are considered in allocating these classes: ground condition, ground roughness and slope. For example, Terrain Class 454 indicates poor ground condition on very rough land that is steeply sloping. Ground condition represented by the first of the three digits ranges from class 1 (very good) to class 5 (very poor) the latter for example occurring on humic stagnogley and peat soils in the uplands. Specialized equipment and low ground pressure vehicles may be required to avoid 'bogging' in classes 4 and 5. Ground roughness, given by the second digit, limits forestry work on bouldery and rocky ground particularly in Snowdonia, and locally elsewhere. The scale ranges from 1 (very even) to 5 (very rough). Slopes, represented by the third digit, range from class 1 (level) to class 5 (very steep), land in classes 4 and 5 requires specialized vehicles for timber extraction. Where a range of conditions often applies it is shown by two figures in brackets. The likely windthrow hazard (Booth 1977), based on regional wind values, elevation, exposure and rootable soil depth is shown in column 2.

The soil codes in the third column are those used for soil and site classification by the Forestry Commission and developed from those of Pyatt (1982).

The species indicated in column four are those best suited to the site. Only commonly used species are included and in some cases no species is well matched to a suitability group, especially for wet groups in the uplands so the species listed are those least limited.

In the last column special measures and management problems relating to the suitability groups are noted.

Table 30
Forestry Suitability Groups for selected Soil Series showing the
Soil Associations in which the Soil Series occur

Soil series	Suitability group	Soil association	Soil series	Suitability group	Soil association
Adventurers'	JJ	Adventurers' 1	Clifton	X	Clifton
		Altcar 1			Salop
Agney	AA	Agney	Compton	Z	Compton
Altcar	JJ	Adventurers' 1			Hollington
		Altcar 1	Conway	Z	Conway
Alun	L	Alun			Teme
		Teme			Fladbury 3
Anglezarke	R	Anglezarke	Cray	U	Wenallt
Arrow	CC	Arrow	Crewe	X	Crewe
		Wick 1			Salop
		East Keswick 1	Crowdy	HH	Crowdy 1
Bangor	A	Bangor			Crowdy 2
Barton	H & I	Barton			Revidge
		Denbigh 1			Bangor
Belmont	U	Gelligaer			Wilcocks 2
Blackwood	CC	Blackwood			Wenallt
		Everingham	Denbigh	H	Denbigh 1
		Newport 1			Manod
Brickfield	V & X	Brickfield 1			Cegin
		Brickfield 2	Denchworth	X	Denchworth
		Brickfield 3	Eardiston	H & I	Eardiston 1
		Nercwys			Eardiston 2
		Cegin			Bromyard
Bromsgrove	H & I	Eardiston 1	East Keswick	H & I	East Keswick 1
Bromyard	I	Bromyard			East Keswick 3
		Eardiston 1			Brickfield 2
Burcombe	U	Lydcott	Enborne	Z	Enborne
Cegin	V & X	Cegin			Alun
		Brickfield 1			

Table 30 (contd.)
Forestry Suitability Groups for selected Soil Series showing the Soil Associations in which the Soil Series occur

Soil series	Suitability group	Soil association	Soil series	Suitability group	Soil association
Escrick	I	Escrick 1	Parc	Q	Parc
		Escrick 2	Pinder	X	Pinder
Everingham	CC	Everingham	Powys	B	Denbigh 1
Fforest	V & X	Fforest			Manod
Fladbury	Z	Fladbury 1	Revidge	A	Revidge
		Fladbury 3			Anglezarke
		Enborne	Rheidol	H	Rheidol
		Conway			Rowton
		Compton	Rivington	H & I	Rivington 2
Flint	P	Flint			Eardiston 2
		Salop			Withnell 1
Foggathorpe	W	Foggathorpe 1	Rockcliffe	Z	Rockcliffe
Gelligaer	U	Gelligaer			Tanvats
Goldstone	R	Goldstone	Rowton	I	Rowton
Hafren	U	Hafren	Salop	X	Salop
		Parc			Flint
		Crowdy 1	Salwick	P	Salwick
		Wilcocks 2			Flint
Hallsworth	W	Hallsworth 1			Clifton
		Hallsworth 2	Sandwich	E	Sandwich
		Foggathorpe 1	Skiddaw	A	Skiddaw
		Brickfield 3	Stanway	X	Stanway
Hexworthy	U	Hexworthy			Curtisden
		Moor Gate	Ston Easton	N	Ston Easton
Hiraethog	U	Skiddaw	Tanvats	Z	Tanvats
		Hafren			Rockcliffe
Hollington	Z	Hollington	Teme	L	Teme
		Lugwardine	Wallasea	Z	Newchurch 2
Laployd	EE	Laployd			Tanvats
Lugwardine	L	Lugwardine			Wallasea 1
Lydcott	U	Lydcott	Waltham	N	Waltham
Malham	N	Malham 2			Crwbin
Malvern	H	Malvern	Wenallt	Y	Wenallt
Manod	H	Manod			Fforest
		Denbigh 1			Lydcott
Middleton	P	Middleton	Whitcott	H	Milford
		Bromyard	Wick	I	Wick 1
Midelney	Z	Midelney			Newport 1
Milford	H & I	Milford			Escrick 2
Moorgate	Q	Moor Gate			Salwick
Moretonhampstead	H & I	Moretonhampstead	Wilcocks	Y	Parc
		Malvern			Hafren
		Moor Gate			Crowdy 1
Neath	H & I	Neath			Brickfield 1
		Withnell 1			Wilcocks 1
Nercwys	V	Nercwys			Wilcocks 2
		Neath	Wilderhope	N	East Keswick 3
		East Keswick 1	Windrush	AA	Midelney
		Withnell 1	Winter Hill	HH	Revidge
		Brickfield 2			Skiddaw
Newbiggin	H & I	Oglethorpe			Bangor
		Milford			Wilcocks 2
Newchurch	AA	Newchurch 2			Crowdy 2
		Wallasea 1	Wisbech	AA	Wisbech
Newnham	H & I	Newnham	Withnell	H	Withnell 1
		Oglethorpe			Rivington 2
Newport	J & K	Wick 1	Worcester	X	Worcester
		Newport 1	Restored Opencast	FF	Disturbed soils 3
Oglethorpe	H & I	Oglethorpe			

Table 31
Land Suitability for Forestry

Suit-ability group	Soil characteristics and site limitations	Forestry Commission Soil code	Best suited species	Notes on forest management
A	Shallow very acid peaty-topped soils of the wet uplands. Steep, irreg-ular slopes, fragmented rock outcrops. Wind-throw hazard Class 4–6. Terrain Class 454	13p and 13r	Lodgepole pine	Terrain and rockiness make drainage and cultivation difficult and restrict harvesting methods. Eric-aceous vegetation locally may lead to checked growth. Phosphorus is essential at planting, and is beneficial with potassium after 5–15 years. This group is not suited to forestry on high exposed sites
B	Shallow loamy soils with some bare siliceous rock and steep slopes. Windthrow hazard Class 5–6. Terrain Class 1(2–4)(2–4)	13b and 13r	Lodgepole pine	These are small areas of limited productivity associated with deeper soils more suited to forestry
E	Deep, permeable, usually calcareous sandy soils of coastal dunes. Irregular often unstable terrain. Exposed to salt-laden winds. Windthrow hazard Class 2. Terrain Class 145	15d	Corsican pine, Lodgepole pine (except on shell sand)	Where active, the dunes need to be stabilised before planting
H	Well drained loamy moderately acid soils of the uplands. Usually stony. Slopes often steep. Windthrow hazard Class 1–2. Terrain Class 21(2–4)	1 and 1d	Sitka spruce, Douglas fir (except where exposed), Larches, Noble fir, Grand fir and Western Hemlock with Sessile oak and Beech where sheltered	Direct planting, except where shallow cultivation needed to suppress weeds. Brambles and bracken. Ericaceous vegetation checks growth locally. Specialized harvesting machinery needed on steep slopes
I	Well drained loamy soils of the dry lowlands. Some steep slopes. Windthrow hazard Class 1–2. Terrain Class 21(2–4)	1b	Most common species except Sitka spruce, Japanese and Hybrid larch and poplars	Some weed control required. Specialized harvesting machinery needed on steep slopes
J	Well drained sandy moderately acid soils in moist areas. Windthrow hazard	1	Scots pine, Lodgepole pine	

Table 31 (contd.)
Land Suitability for Forestry

Suit-ability group	Soil characteristics and site limitations	Forestry Commission Soil code	Best suited species	Notes on forest management
J (contd.)	Class 1–2. Terrain Class 11(2–3)			
K	Well drained sandy moderately acid soils of the dry lowlands. Windthrow hazard Class 1–2. Terrain Class 11(1–3)	1	Corsican pine, Lodgepole pine	
L	Well drained soils in alluvium. Risk of flood and frost damage. Windthrow hazard Class 1–3. Terrain Class 211	1/v	Norway spruce (where frost damage not serious), Oak, Poplars, Alders	Drainage required locally
N	Well drained loamy lowland soils over limestone. Some clayey subsoils. Windthrow hazard Class 1–2. Terrain Class 211	12t	Most common species except Corsican, Maritime and Radiata pine and Poplars	Weed control required
P	Deep loamy soils with slight seasonal waterlogging in dry areas. Windthrow hazard Class 2–3. Terrain Class 311	1bg and part of 7b	European larch, Douglas fir, Norway spruce, English oak, Sweet chestnut	Weed control required
Q	Well drained loamy acid soils of the moist uplands with humose surface horizon over siliceous rock. Often bouldery. Some steep slopes. Windthrow hazard Class 2–3. Terrain Class 332	1u	Lodgepole pine, Sitka spruce, Norway spruce	Shallow cultivation for suppressing bracken and in places heather. Phosphorus needed at planting. Specialized harvesting equipment needed where bouldery or steep
R	Well drained very acid, stony coarse loamy or sandy upland soils over sandstone. Locally bouldery or rocky. Windthrow risk where shallow on rock. Terrain Class 13(2–4)	3	Lodgepole pine, Sitka spruce, Douglas fir, Japanese and Hybrid larch, Western hemlock and Sessile oak on sheltered sites	Heather growth should be suppressed before and after planting. Phosphorus required. Specialized harvesting machinery needed where rocky or steep
U	Loamy very acid permeable upland soils over rock, with seasonally waterlogged peaty	4,4z and 4b	Sitka spruce to withstand exposure. Lodgepole pine	Cultivation essential to improve aeration to suppress vegetation, and to break ironpan where present.

Table 31 (contd.)
Land Suitability for Forestry

Suit-ability group	Soil characteristics and site limitations	Forestry Commission Soil code	Best suited species	Notes on forest management
U (contd.)	surface horizon and bleached subsurface horizon. Thin ironpans common locally. Wind-throw hazard Class 2–4. Terrain Class 32(2–4)		to withstand competition from ericaceous vegetation	Phosphorus essential for Sitka spruce and beneficial for Lodgepole pine
V	Slowly permeable, seasonally waterlogged loamy soils in moist areas. Windthrow hazard Class 4–5. Terrain Class 421	7l	Sitka spruce usually first choice. Norway spruce also suited particularly for frosty sites	Double mouldboard ploughing down slope at 4 m spacing with deep cross drains. Tall grass growth to be suppressed. Phosphorus top dressing beneficial after tree crop established
W	Slowly permeable, seasonally waterlogged clayey soils in moist areas. Windthrow hazard Class 4-5. Terrain Class 421	7 and part of 7b	Sitka spruce usually first choice. Norway spruce also suited particularly for frosty sites	Double mouldboard ploughing down slope at 4 m spacing with deep cross drains. Tall grass growth to be suppressed. Phosphorus top dressing beneficial after tree crop established
X	Slowly permeable, seasonally waterlogged soils of the dry lowlands. Frost risk. Windthrow hazard Class 3–4. Terrain Class 311	7	Norway and Pedunculate oak, Western red cedar	Deep drains and cultivation to increase rooting depth
Y	Slowly permeable loamy upland soils with waterlogged peaty surface horizons. Windthrow hazard Class 5–6. Terrain Class 52(1–2)	6	Sitka spruce, Lodgepole pine	Deep double mouldboard ploughing downslope with cross-drains. Phosphorus beneficial, particularly where vegetation heathery. Timber extraction requires vehicles adapted to soft ground
Z	Loamy and clayey soils in alluvium affected by groundwater and liable to flood. Frost risk common inland. Windthrow hazard Class 3–4. Terrain Class (3–4)11	5/v	Poplars, Alders, Norway spruce, Sitka spruce	Deep drains to lower water table. Strong growth of tall grasses lead to protracted weeding if establishment is slowed by frosting. Soils in marine alluvium near the coast not generally suitable for forestry
AA	Deep seasonally waterlogged calcareous silty and clayey soils in alluvium. Flood	5k/v	Poplars, Alders	Calcareous soils and need to drain by pumping generally makes forestry uneconomical

Table 31 (contd.)
Land Suitability for Forestry

Suitability group	Soil characteristics and site limitations	Forestry Commission Soil code	Best suited species	Notes on forest management
AA (contd.)	risk locally. Windthrow hazard Class 3–4. Terrain Class 311			
CC	Deep loamy and sandy lowland soils affected by groundwater. Windthrow hazard Class 3–4. Terrain Class 311	5	Norway and Omorika spruce, Pedunculate oak, Alders	Deep drains required
EE	Permeable loamy upland soils with waterlogged peaty surface horizons affected by groundwater. Windthrow hazard Class 4–5. Bouldery locally. Terrain Class 54(1–2)	5h	Sitka spruce, Lodgepole pine Norway spruce in more sheltered sites	Deep drains and ploughing. Phosphorus beneficial. Timber extraction requires machinery adapted to soft and bouldery ground
FF	Upland restored after opencast coal working. Seasonally waterlogged, compact, loamy or clayey-Usually very stony with thin topsoils. Erosion risk on steep land. Windthrow hazard Class 5. Terrain Class (2–4) (2–4)(2–4)	2m	Lodgepole pine, Hybrid and Japanese larch	
GG	Lowland restored after opencast working. Stony, coarse textured. Windthrow hazard Class 1–2. Terrain Class 32(1–3)	2s		
HH	Thick mainly amorphous and semi-fibrous raw peat soils. Perennially wet; hagged and eroded in places. Windthrow hazard Class 5–6. Terrain Class 531	9,10 and 11	Lodgepole pine or Sitka spruce	Peat ploughs for ridge planting and cross drains. Phosphorus and potassium at planting and after 5–15 years. Nitrogen fertilizer and herbicides may be needed to establish Sitka spruce on heathery sites
JJ	Thick, relatively base-rich peat of the dry lowlands. Windthrow hazard Class 2–4. Terrain Class 511	8	Sitka spruce (or Lodgepole pine in frosty sites), Alders and Poplars, Birch	Peat ploughs for ridge planting and cross drains. Phosphorus after 5–15 years Strong growth of grasses may cause trouble on frosty sites

REFERENCES

AITCHISON, J.W. (1979). The agricultural landscape of Wales. Part 1 The structure of agricultural holdings 1964–74. *Cambria* **6**, 32–53.

ALLISON, J.W. and HARTNUP, R. (1981). *Soils in North Yorkshire VI: Sheet SE39 (Northallerton)*. Soil Surv. Rec. No. 68.

ARCHER, A.A. (1968). *Geology of the South Wales Coalfield. The Upper Carboniferous and later formation of the Gwendraeth valley and adjoining areas.* Mem. geol. Surv. U.K.

ARGENT, C.J. and FURNESS, R.R. (1979). Corrosion in soils. In: *Soil Survey applications* (ed. M.G. Jarvis and D. Mackney). Soil Surv. Tech. Monogr. No. 13, 135–147.

ARNOLD, P.W. and CLOSE, B.M. (1961). Release of non-exchangeable potassium from some British soils cropped in the glasshouse. *J. agric. Sci.* **57**, 295–304.

AVERY, B.W. (1980). *Soil classification for England and Wales. (Higher categories)*. Soil Surv. Tech. Monogr. No. 14.

AVERY, B.W. and BASCOMB, C.L. (Ed.)(1982). *Soil Survey laboratory methods*. Soil Surv. Tech. Monogr. No. 6.

AVERY, B.W. and BULLOCK, P. (1977). *Mineralogy of clayey soils in relation to soil classification.* Soil Surv. Tech. Monogr. No. 10.

BAILEY, A.D., DENNIS, C.W., HARRIS, G.L. and HORNER, M.W. (1980). *Pipe size design for field drainage.* Land Drainage Service Research and Development Report No. 5. MAFF, London.

BALL, D.F. (1960). *The soils and land use of the district around Rhyl and Denbigh.* Mem. Soil Surv. Gt Br.

BALL, D.F. (1963). *The soils and land use of the district around Bangor and Beaumaris.* Mem. Soil Surv. Gt Br.

BALL, D.F. and GOODIER, R. (1968). Large sorted stone-stripes in the Rhinog Mountains, North Wales. *Geogr. Annlr* **50**, 54–9.

BALL, D.F. and GOODIER, R. (1970). Morphology and distribution of features resulting from frost-action in Snowdonia. *Fld Stud.* **3**, 193–217.

BALL, D.F., MEW, G. and MACPHEE, W.S.G. (1969). Soils of Snowdon. *Fld Stud.* **3**, 69–107.

BASCOMB, C.L. (1961). A calcimeter for routine use on soil samples. *Chemy Ind.*, 1826–7.

BASCOMB, C.L. and BULLOCK, P. (1982). Sample preparation and stone content. In: *Soil Survey laboratory methods.* (ed. B.W. Avery and C.L. Bascomb). Soil Surv. Tech. Monogr. No.6.

BEARD, G.R. (1984). *Soils in Warwickshire V: Sheet SP27/37 (Coventry).* Soil Surv. Rec. No. 81.

BELFORD, R.K., CANNELL, R.Q., THOMSON, R.J. and DENNIS, C.W. (1980). Effects of waterlogging at different stages of development on the growth and yield of peas (*Pisum sativum* L.) *J. Sci. Fd Agric.* **31**, 857–69.

BENDELOW, V.C. and HARTNUP, R. (1980). *Climatic classification of England and Wales.* Soil Surv. Tech. Monogr. No. 15.

BIBBY, J.S., DOUGLAS, H.A., THOMASSON, A.J. and ROBERTSON, J.S. (1982) *Land capability classification for agriculture.* Macaulay Institute for Soil Research, Aberdeen.
BIRSE, E.L. (1971). *Assessment of climatic conditions in Scotland. 3 The bioclimatic sub-regions.* Soil Survey of Scotland.
BIRSE, E.L. (1980). *Plant communities of Scotland.* Bull. Soil Surv. Scot. No. 4.
BIRSE, E.L. and ROBERTSON, J.S. (1976). *Plant communities and soils of the lowland and southern upland regions of Scotland.* Macaulay Institute for Soil Research, Aberdeen.
BISCOE, P.V. and GALLAGHER, J.N. (1978). A physiological analysis of cereal yield. I. Production of dry matter. *Agric. Prog.* **53**, 34–50.
BOOTH, T.C. (1977). *Windthrow hazard classification.* Forestry Commission Research Information Note 22/77.
BOWEN, D.Q. (1966). Dating Pleistocene events in south-west Wales. *Nature, Lond.* **211**, 475–6.
BOWEN, D.Q. (1974). The Quaternary of Wales. In: *The Upper Palaeozoic and Post-Palaeozoic rocks of Wales.* (ed. T.R. Owen), 373–426. University of Wales Press, London.
BOWEN, D.Q. (1977). The land of Wales. In: *Wales. A new study.* (ed. D. Thomas). David and Charles, Newton Abbot.
BOWEN, D.Q. (1980). *The Llanelli landscape: the geology and geomorphology of the country around Llanelli.* Lanelli Borough Council, Llanelli.
BRADLEY, R.I. (1976). *Soils in Dyfed III: Sheet SN13 (Eglwyswrw).* Soil Surv. Rec. No. 38.
BRADLEY, R.I. (1980). *Soils in Dyfed V: Sheet SN24 (Llechryd).* Soil Surv. Rec. No. 63.
BREWER, R. (1964). *Fabric and mineral analysis of soils.* John Wiley. New York.
BROWN, E.H. (1960). *The relief and drainage of Wales.* University of Wales Press, Cardiff.
BURTON, R.G.O. (1981). *Soils in Cambridgeshire II: Sheet TF00E/10W (Barnack).* Soil Surv. Rec. No. 69.
CANNELL, R.Q. and BELFORD, R.K. (1980). Effects of waterlogging at different stages of development on the growth and yield of winter oilseed rape (*Brassica napus* L.). *J. Sci. Fd Agric.* **31**, 963–65.
CANNELL, R.Q., BELFORD, R.K., GALES, K., DENNIS, C.W. and PREW, R.D. (1980). Effects of waterlogging at different stages of development on the growth and yield of winter wheat. *J. Sci. Fd Agric.* **31**, 117–32.
CARROLL, D.M. and BENDELOW, V.C. (1981). *Soils of the North York Moors.* Soil Surv. Spec. Surv. No. 13.
CARROLL, D.M., HARTNUP, R. and JARVIS, R.A. (1979). *Soils of South and West Yorkshire.* Bull. Soil Surv. Gt Br.
CHALLINOR, J. (1930). The hill-top surface of north Cardiganshire. *Geography* **15**, 651–6.
CHAMBERS, F.M. (1983). Three radiocarbon-dated pollen diagrams from upland peats north-west of Merthyr Tydfil, South Wales. *J.Ecol.* **71**, 475–87.
CHURCH, B.M., BOYD, D.A., EVANS, J.A. and SADLER, J.I. (1968). A type of farming map based on agricultural census data. *Outl. Agric.* **5**, 191–6.
CHURCH, B.M. and LEWIS, D.A. (1977). Fertilizer use on farm crops in England and Wales: Information from the survey of fertilizer practice, 1942–1976. *Outl. Agric.* **9**, 186–93.
CLAYDEN, B. (1971). *Soils of the Exeter district.* Mem. Soil Surv. Gt Br.
CLAYDEN, B. and EVANS, G.D. (1974). *Soils in Dyfed I: Sheet SN41 (Llangendeirne).* Soil Surv. Rec. No. 20.
CLAYDEN, B. and HOLLIS, J.M. (1984). *Criteria for differentiating soil series.* Soil Surv. Tech. Monogr. No. 17.
COLBORNE, G.J.N. (1981). *Soils in Gloucestershire III: Sheet SO61 (Cinderford).* Soil Surv. Rec. No. 73.
COPE, D.W. (1973). *Soils in Gloucestershire I: Sheet SO82 (Norton).* Soil Surv. Rec. No. 13.

CORBETT, W.M. and TATLER, W. (1970). *Soils in Norfolk: Sheet TM 49 (Beccles North).* Soil Surv. Rec. No. 1.

COURTNEY, F.M. and FINDLAY, D.C. (1978). *Soils in Gloucestershire II: Sheet SP12 (Stow-on-the-Wold).* Soil Surv. Rec. No. 52.

CRAMPTON, C.B. (1961). An interpretation of the micro-mineralogy of certain Glamorgan soils: the influence of ice and wind. *J. Soil Sci.* **12**, 158–71.

CRAMPTON, C.B. (1966). Certain effects of glacial events in the Vale of Glamorgan, South Wales. *J. Glaciol.* **6**, 261–6.

CRAMPTON, C.B. (1972). *Soils of the Vale of Glamorgan.* Mem. Soil Surv. Gt Br.

CRAMPTON, C.B. and TAYLOR, J.A. (1967). Solifluction terraces in South Wales. *Biul. peryglacjalny* **16**, 15–36.

CROMPTON, A. and MATTHEWS, B. (1970). *Soils of the Leeds district.* Mem. Soil Surv. Gt Br.

CROMPTON, E. (1966). *The soils of the Preston district of Lancashire.* Mem. Soil Surv. Gt Br.

DAY, P., DEADMAN, A.J., GREENWOOD, B.D. and GREENWOOD, E.F. (1982). A floristic appraisal of marl pits in parts of north-western England and northern Wales. *Watsonia* **14**, 153–65.

FAO (1976). *A framework for land evaluation.* Soils Bulletin No. 32. FAO, Rome.

FINDLAY, D.C. (1965). *The soils of the Mendip district of Somerset.* Mem. Soil Surv. Gt Br.

FINDLAY, D.C. (1976). *Soils of the southern Cotswolds and surrounding country.* Mem. Soil Surv. Gt Br.

FINDLAY, D.C., COLBORNE, G.J.N., COPE, D.W., HARROD, T.R., HOGAN, D.V. and STAINES, S.J. (1984). *Soils and their use in South West England.* Bull. Soil Surv. Gt Br.

FITZPATRICK, E.A. (1956). An indurated soil horizon formed by Permafrost. *J. Soil Sci.* **7**, 248–54.

FORDHAM, S.J. and GREEN, R.D. (1976). *Soils in Kent III: Sheet TQ86 (Rainham).* Soil Surv. Rec. No. 37.

FORDHAM, S.J. and GREEN, R.D. (1980). *Soils of Kent.* Bull. Soil Surv. Gt Br.

FORESTRY COMMISSION. (1982). *Sixty-first annual report and accounts of the Forestry Commission for the year ended 31 March 1981.* H.M.S.O., London.

FRY, R.K. (1983). Trenchless drainage tines—soil disturbance, field performance and practical implications. *The Agricultural Engineer.* **38**, 9–12.

FURNESS, R.R. (1971). *Soils in Cheshire I: Sheet SJ65 (Crewe West).* Soil Surv. Rec. No. 5.

FURNESS, R.R. and KING, S.J. (1973). *Soils in Cheshire II: Sheet SJ37 (Ellesmere Port).* Soil Surv. Rec. No. 17.

FURNESS, R.R. and KING, S.J. (1978). *Soils in North Yorkshire IV: Sheet SE63/73 (Selby).* Soil Surv. Rec. No. 56.

GEORGE, H. (1978). *Soils in Northumberland I: Sheet NZ07 (Stamfordham).* Soil Surv. Rec. No. 53.

GEORGE, H. and JARVIS, M.G. (1979). Land for camping and caravan sites, picnic sites and footpaths. In: *Soil survey applications.* (ed. M.G. Jarvis and D. Mackney). Soil Surv. Tech. Monogr. No. 13, 166–83.

GEORGE, T.N. (1970). *South Wales.* British Regional Geology. H.M.S.O.

GEORGE, T.N. (1974). The Cenozoic evolution of Wales. In: *The Upper Palaeozoic and post-Palaeozoic rocks of Wales.* (ed. T.R. Owen). University of Wales Press, Cardiff.

GODWIN, R.J., SPOOR, G. and LEEDS-HARRISON, P. (1981). An experimental investigation into the force mechanics and resulting soil disturbance of mole ploughs. *J. agric. Engng Res.* **26**, 477–97.

GREEN, F.H.W. (1964). The climate of Scotland. In: *The vegetation of Scotland.* (ed. J.H. Burnett). Oliver and Boyd Ltd, Edinburgh.

GREEN, J.O. (1982). *A sample survey of grassland in England and Wales 1970*-1972. Grassland Research Institute, Maidenhead.

HALL, B.R. and FOLLAND, C.J. (1967). *Soils of the south-west Lancashire coastal plain.* Mem. Soil Surv. Gt Br.

HALL, D.G.M., REEVE, M.J., THOMASSON, A.J. and WRIGHT, V.F. (1977). *Water retention, porosity and density of field soils.* Soil Surv. Tech. Monogr. No. 9.

HARROD, T.R. (1979). Soil suitability for grassland. In: *Soil survey applications* (ed. M.G. Jarvis and D. Mackney). Soil Surv. Tech. Monogr. No. 13, 51–70.

HARROD, T.R. (1981). *Soils in Devon V: Sheet SS61 (Chulmleigh).* Soil Surv. Rec. No. 70.

HARROD, T.R., HOGAN, D.V. and STAINES, S.J. (1976). *Soils in Devon II: Sheet SX65 (Ivybridge).* Soil Surv. Rec. No. 39.

HARROD, T.R. and THOMASSON, A.J. (1980). *Grassland suitability. Map of England and Wales (1:1,000,000).* Ordnance Survey, Southampton.

HARTNUP, R. (1977). *Soils in South Yorkshire I: Sheet SK59 (Maltby).* Soil Surv. Rec. No. 42.

HAZELDEN, J. (n.d.). *Soils in Oxfordshire II: Sheet SP60 (Tiddington).* Soil Surv. Rec. (in press).

HAZELDEN, J. and JARVIS, M.G. (1979). Age and significance of alluvium in the Windrush valley, Oxfordshire. *Nature, Lond.* **282**, 291–2.

HEAVEN, F.W. (1978). *Soils in Lincolnshire III: Sheet TF28 (Donington on Bain).* Soil Surv. Rec. No. 55.

HENDERSON, H.J.R. (1977). Present farming patterns. In: *Wales, a new study.* (ed. D. Thomas). David and Charles, Newton Abbot.

HIGGINS, L.S. (1933). An investigation into the problem of sand dune areas on the south Wales coast. *Archaeologia Cambrensis* **88**, 26–67.

HODGE, C.A.H., BURTON, R.G.O., CORBETT, W.M., EVANS, R. and SEALE, R.S. (1984). *Soils and their use in Eastern England.* Bull. Soil Surv. Gt Br.

HODGSON J.M. (1972). *Soils of the Ludlow district.* Mem. Soil Surv. Gt Br.

HODGSON, J.M. (Ed.)(1976). *Soil survey field handbook.* Soil Surv. Tech. Monogr. No. 5.

HOGAN. D.V. (1977). *Soils in Devon III: Sheet SX47 (Tavistock).* Soil Surv. Rec. No. 44.

HOGAN, D.V. (1981). *Soils in Devon VI: Sheet SS63 (Brayford).* Soil Surv. Rec. No. 71.

HOLLIS, J.M. (1975). *Soils in Staffordshire I: Sheet SK05 (Onecote).* Soil Surv. Rec. No. 29.

HOLLIS, J.M., JONES, R.J.A. and PALMER, R.C. (1977). The effects of organic matter and particle size on the water-retention properties of some soils in the west Midlands of England. *Geoderma* **17**, 225-38.

HOLLIS, J.M. (1978) *Soils in Salop I: Sheet SO79E/89W (Claverley).* Soil Surv. Rec. No. 49.

HOLLIS, J.M. (n.d.). *Soils in Staffordshire IV: Sheet SK00/10 (Lichfield).* Soil Surv. Rec. (in press).

HOLLIS, J.M. and HODGSON, J.M. (1974). *Soils in Worcestershire I: Sheet SO87 (Kidderminster).* Soil Surv. Rec. No. 18.

HUGHES, D.O. and ROBERTS, E. (1958). *Soil map of Pwllheli* (Scale 1:63,360). Ordnance Survey, Chessington.

HUGHES, R.E., DALE, J., MOUNTFORD, M.D. and WILLIAMS, I.E. (1975). Studies in sheep populations and environment in the mountains of north-west Wales. II. Contemporary distribution of sheep populations and environment. *J. appl. Ecol.* **12**, 165–78.

HYDE, H.A. (1977). *Welsh timber trees native and introduced.* 4th Edition revised by S.G. Harrison. The National Museum of Wales, Cardiff.

INSTITUTE OF GEOLOGICAL SCIENCES. (1977). *Geological survey ten mile map: South sheet (Quaternary).* Ordnance Survey, Southampton.

INSTITUTE OF GEOLOGICAL SCIENCES. (1979). *Geological map of the United Kingdom South. 3rd Edition Solid.* Ordnance Survey, Southampton.

JARVIS, R.A. (1973). *Soils in Yorkshire II: Sheet SE60 (Armthorpe).* Soil Surv. Rec. No. 12.

JARVIS, R.A. (1977). *Soils of the Hexham district.* Mem. Soil Surv. Gt Br.

JARVIS, R.A., BENDELOW, V.C., BRADLEY, R.I., CARROLL, D.M., FURNESS, R.R., KILGOUR, I.N.L. and KING, S.J. (1984). *Soils and their use in Northern England.* Bull. Soil Surv. Gt Br.

JONES, R.J.A. (1975). *Soils in Staffordshire II: Sheet SJ82 (Eccleshall).* Soil Surv. Rec. No. 31.

JONES, R.J.A. (1979). Soils of the Western Midlands grouped according to ease of cultivation. In: *Soil survey applications* (ed. M.G. Jarvis and D. Mackney). Soil Surv. Tech. Monogr. No. 13, 24–42.

JONES, R.J.A. (1983). *Soils in Staffordshire III: Sheets SK 02/12 (Needwood Forest).* Soil Surv. Rec. No. 80.

KILGOUR, I.N.L. (1979). *Soils in Cumbria II: Sheet NY36/37 (Longtown).* Soil Surv. Rec. No. 59.

KING, S.J. (1977). *Soils in Cheshire III: Sheet SJ45E/55W (Burwardsley).* Soil Surv. Rec. No. 43.

LEA, J.W. (1975). *Soils in Powys I: Sheet SO09 (Caersws).* Soil Surv. Rec. No. 28.

LEA, J.W. (1979). Slurry acceptance. In: *Soil survey applications.* (ed. M.G. Jarvis and D. Mackney). Soil Surv. Tech. Monogr. 13, 83–100.

LEA, J.W. and THOMPSON, T.R.E. (1978). *Soils in Clwyd I: Sheet SJ35 (Wrexham North).* Soil Surv. Rec. No. 48.

LEEDS-HARRISON, P., SPOOR, G. and GODWIN, R.J. (1982). Water flow to mole drains. *J. agric. Engng Res.* **27**, 81–91.

LEWIS, C.A. (Ed.)(1970). *The glaciations of Wales and adjoining regions.* Longman, London.

LINNARD, W. (1982). *Welsh woods and forests: history and utilization.* National Museum of Wales, Cardiff.

LUTHIN, J.N. (1957). *Drainage of agricultural lands.* American Society of Agronomy. Madison, Wisconsin.

MACKNEY, D. and BURNHAM, C.P. (1966). *The soils of the Church Stretton district of Shropshire.* Mem. Soil Surv. Gt Br.

MAFF (1981). Agricultural returns—England and Wales, regions and counties—final results of June 1980 census. MAFF, Guildford.

MAFF (1982a). *The design of field drainage pipe systems.* ADAS Land and Water Service, Reference Book 345. MAFF, London.

MAFF (1982b). *Nitrogen for grassland.* Grassland practice No. 2. Booklet 2042. MAFF, Alnwick.

MAFF (1983). *Technical note on workmanship and materials for field drainage schemes.* ADAS. MAFF, London.

McRAE, S.G. and BURNHAM, C.P. (1981). *Land evaluation.* Clarendon Press, Oxford.

MOORE, P.D. (1977). Vegetational history. In: Studies in Welsh Quaternary. *Cambria* **4**, 73–83.

PALMER, R.C. (1972). *Soils in Herefordshire III: Sheet SO34 (Staunton-on-Wye).* Soil Surv. Rec. No. 11.

PALMER, R.C. (1976). *Soils in Herefordshire IV: Sheet SO74 (Malvern).* Soil Surv. Rec. No. 36.

PALMER, R.C. (1982). *Soils in Hereford and Worcester I: Sheets SO85/95 (Worcester).* Soil Surv. Rec. No. 76.

PALMER, R.C. and JARVIS, M.G. (1979). Land for winter playing fields, golf course fairways and parks. In: *Soil survey applications* (ed. M.G. Jarvis and D. Mackney). Soil Surv. Tech. Monogr. No. 13, 152–65.

PEACOCK, J.M. (1975a). Temperature and leaf growth in *Lolium perenne.* I. The thermal microclimate: its measurement and relation to crop growth. *J. appl. Ecol.* **12**, 99–114.

PEACOCK, J.M. (1975b). Temperature and leaf growth in *Lolium perenne.* II. The site of temperature perception. *J. appl. Ecol.* **12**, 115–23.

PEACOCK, J.M. (1976). Temperature and leaf growth in four grass species. *J. appl. Ecol.* **13**, 225–32.

PEARSALL, W.H. (1950). *Mountains and moorlands.* Collins, London.

PENMAN, H.L. (1948). Natural evaporation from open water, bare soil and grass. *Proc. Roy. Soc. Series A* **193**, 120–45.

PYATT, D.G. (1977). *Guide to site types in forests of north and mid Wales.* Forest Rec., Lond. No. 69.

PYATT, D.G. (1982). *Soil classification.* Forestry Commission Research Information Note 68/82/SSN.

RAGG, J.M., BEARD, G.R., GEORGE, H., HEAVEN, F.W., HOLLIS, J.M., JONES, R.J.A., PALMER, R.C., REEVE, M.J., ROBSON, J.D. and WHITFIELD, W.A.D. (1984). *Soils and their use in Midland and Western England.* Bull. Soil Surv. Gt Br.

RANWELL, D. (1959). Newborough Warren, Anglesey. I. The dune system and dune slack habitat. *J. Ecol.* **47**, 571–601.

RATCLIFFE, D. (1977). *A nature conservation review: The selection of biological sites of national importance to nature conservation in Britain.* University Press, Cambridge.

REEVE, M.J. (1975). *Soils in Derbyshire II: Sheet SK32E/42W (Melbourne).* Soil Surv. Rec. No. 27.

REEVE, M.J. (1978). *Soils in Northamptonshire I: Sheet SP66 (Long Buckby).* Soil Surv. Rec. No. 54.

REEVE, M.J. and THOMASSON, A.J. (1981). *Soils in Nottinghamshire IV: Sheet SK78N/79S (Gringley on the Hill).* Soil Surv. Rec. No. 72.

REEVE, M.J., SMITH. P.D. and THOMASSON, A.J. (1973). The effect of density on water retention properties of field soils. *J. Soil Sci.* **24**, 355-67.

ROBERTS, E. (1958). *The county of Anglesey. Soils and agriculture.* Mem. Soil Surv. Gt Br.

ROBERTSON, J.S. (1984). *A key to the common plant communities of Scotland.* Macaulay Institute for Soil Research, Aberdeen.

ROBSON, J.D. and THOMASSON, A.J. (1977). *Soil water regimes: A study of seasonal waterlogging in English lowland soils.* Soil Surv. Tech. Monogr. No. 11.

ROWAN, A.A. (1977). Terrain classification. *Forest Rec., Lond.* No 114.

RUDEFORTH, C.C. (1967). Upland soils from Lower Palaeozoic sedimentary rocks in mid-Wales. *Rep. Welsh Soils Discussion Group No. 8.* 42–51.

RUDEFORTH, C.C. (1970). *Soils of north Cardiganshire.* Mem. Soil Surv. Gt Br.

RUDEFORTH, C.C. (1974). *Soils in Dyfed II: Sheets SM90/91 (Pembroke/Haverfordwest).* Soil Surv. Rec. No. 24.

RUDEFORTH, C.C. and THOMASSON, A.J. (1970). *Hydrological properties of soils in the River Dee catchment.* Soil Surv. Spec. Surv. No. 4.

SEALE, R.S. (1975). *Soils of the Ely district.* Mem. Soil Surv. Gt Br.

SMITH, B. and GEORGE, T.N. (1961). *North Wales.* British Regional Geology. H.M.S.O.

SMITH, C.V. (1977). Work days from weather data. *The Agricultural Engineer.* **32**, 95-97.

SMITH, L.P. (1967). *Potential Transpiration.* Tech. Bull. Minist. Agric. Fish Fd, Lond. No. 16.

SMITH, L.P. (1976). *The agricultural climate of England and Wales. Areal averages 1941-70.* Tech. Bull. Minist. Agric. Fish Fd, Lond. No. 35.

SMITH, L.P. and TRAFFORD, B.D. (1976). *Climate and drainage.* Tech. Bull. Minist. Agric. Fish Fd, Lond. No. 34.

SPOOR, G., LEEDS-HARRISON, P.B., and GODWIN, R.J. (1982). Some fundamental aspects of the formation, stability and failure of mole drainage channels. *J. Soil Sci.* **33**, 411-25.

STAINES, S.J. (1976). *Soils in Cornwall I: Sheet SX18 (Camelford).* Soil Surv. Rec. No. 34.

STAPLEDON, R.G. (Ed.)(1936). *A survey of the agricultural and waste lands of Wales.* Faber and Faber, London.

STRAHAN, A., CANTRILL, T.C., DIXON, E.E.L. and THOMAS, H.H. (1907). *The geology of the South Wales coal-field. Part VII. The country around Ammanford.* Mem. geol. Surv. U.K.

STURDY, R.G. (1976). *Soils in Essex II: Sheet TQ99 (Burnham-on-Crouch)*. Soil Surv. Rec. No. 40.

TANSLEY, A.G. (1939). *The British Isles and their vegetation*. University Press, Cambridge.

TAYLOR, J.A. (Ed.)(1967). *Weather and agriculture*. Pergamon Press, Oxford.

TAYLOR, J.A. (1973). Chronometers and chronicles, a study of palaeo-environments in west central Wales. *Progress in Geography* **5**, 247–334.

THOMASSON, A.J. (Ed.)(1975a). *Soils and field drainage*. Soil Surv. Tech. Monogr. No. 7.

THOMASSON, A.J. (1975b). Gley morphology and soil water regimes in some soils in south-central England—A discussion. *Geoderma* **13**, 373–5.

THOMASSON, A.J. (1978). Towards an objective classification of soil structure. *J. Soil Sci.* **29**, 38-46.

THOMASSON, A.J. (1979). Assessment of soil droughtiness. In: *Soil survey applications* (ed. M.G. Jarvis and D. Mackney). Soil Surv. Tech. Monogr. 13, 43–50.

THOMASSON, A.J. (1982). Soil and climatic aspects of workability and trafficability. *Proceedings of the 9th Conference of the International Soil Tillage Research Organization, Osijek, Yugoslavia, 1982*. 551–7.

THOMPSON, T.R.E. (1978). *Soils in Clwyd II: Sheet SJ17 (Holywell)*. Soil Surv. Rec. No. 50.

THOMPSON, T.R.E. (1982). *Soils in Powys II: Sheet SJ21 (Arddleen)*. Soil Surv. Rec. No. 75.

THORBURN, A.A. and TRAFFORD, B.D. (1976). *Iron ochre in drains–A summary of present knowledge*. ADAS Land Drainage Service, Technical Bulletin No. 76.1. MAFF, London.

TRAFFORD, B.D. (1970). Field drainage. *Jl R. agric. Soc.* **131**, 129–52.

TRAFFORD, B.D. (1975). Drainage design. In: *Soils and field drainage*. (ed. A.J. Thomasson). Soil Surv. Tech. Monogr. No. 7, 5–17.

VEIHMEYER, F.J. and HENDRICKSON, A.H. (1931). The moisture equivalent as a measure of the field capacity of soils. *Soil Sci.* **32**, 181–93.

WATSON, E. (1971). Remains of pingos in Wales and the Isle of Man. *Geol. J.* **7**, 381–92.

WEBSTER, R. and BECKETT, P.H.T. (1972). Matric suctions to which soils in South Central England drain. *J. Agric. Sci., Camb.* **78**, 379-87.

WELSH OFFICE. (1983a). *Digest of Welsh statistics. No. 29*. Government Statistical Service, Cardiff.

WELSH OFFICE. (1983b). *Welsh agricultural statistics No. 4 1982*. Government Statistical Service, Cardiff.

WHITFIELD, W.A.D. and BEARD, G.R. (1977). *Soils in Warwickshire III: Sheet SP47/48 (Rugby West/Wolvey)*. Soil Surv. Rec. No. 45.

WHITFIELD, W.A.D. and BEARD, G.R. (1980). *Soils in Warwickshire IV: Sheet SP29/39 (Nuneaton)*. Soil Surv. Rec. No. 66.

WHITFIELD, W.A.D. (n.d.). *Soils in Warwickshire VI: Sheet SP25/35 (Stratford-upon-Avon)*. Soil Surv. Rec. (in press).

WILLARDSON, L.S. and WALKER, R.E. (1979). Synthetic drain envelope—soil interactions. *Journal of the Irrigation and Drainage Division*. **105**, 369–73.

WRIGHT, P.S. (1980). *Soils in Dyfed IV: Sheet SN62 (Llandeilo)*. Soil Surv. Rec. No. 61.

WRIGHT, P.S. (1981). *Soils in Dyfed VI: Sheet SN72 (Llangadog)*. Soil Surv. Rec. No. 74.

APPENDIX 1
TERMS USED IN SOIL PROFILE DESCRIPTIONS

Representative soil profile descriptions in Appendix 2 are for soil series mentioned in the main body of the book for which no published description is available elsewhere. Although many soil series common in the region are included, the list is not comprehensive. To save space, previously published descriptions of many extensive and well-known soil series are referenced in the main text.

Most of the terms used in the profile descriptions (pp 309-327) are fully described by Hodgson (1976); these are summarized below. The horizon notation, used here and in the summary profiles in Chapter 3, is that of Avery (1980). The analytical methods used are more fully described by Avery and Bascomb (1982). The basis of the brief soil series definitions at the head of each profile description is outlined in Chapter 2 (§ 6) and more fully described by Clayden and Hollis (1984).

TERMINOLOGY USED IN FIELD DESCRIPTIONS
Soil profiles are described in freshly dug profile pits, where component horizons are identified and then described from the surface downwards using standard terminology.

Site description
The site is described by its locality, map reference, relief (including aspect, elevation, angle and shape of slope), vegetation and land use.
The following general slope classes are used to describe the ground within 100 m of a profile.

Level	0–1°	*Moderately steeply sloping*	12–15°
Gently sloping	2–3°	*Steeply sloping*	16–25°
Moderately sloping	4–7°	*Very steeply sloping*	26–35°
Strongly sloping	8–11°	*Precipitous*	>35°

Depth and clarity of horizons
The depths of horizon boundaries are measured from the surface of the soil excluding fresh litter (L layer), and where they are not horizontal the range of depth is recorded. Horizon boundaries are described as *smooth* if at the same depth across the face, *wavy* if there are broad shallow pockets, *irregular* if pockets are deeper than their width and *broken* if a horizon is discontinuous.

The clarity of a boundary is *sharp* if the transition zone is less than 0.5 cm thick, *abrupt* if between 0.5 and 2.5 cm, *clear* if between 2.5 and 6 cm, *gradual* if between 6 and 13 cm, or *diffuse* if the transition is more than 13 cm thick.

Colour
Colours are referred to Munsell Soil Color Charts. Moist soil fragments are compared with colour chips to determine standard names and notations; in some instances the colour of air-dry soil or of crushed soil rubbed between finger and thumb is diagnostically important. In mottled horizons the chief colours are described in terms of abundance, size, contrast and distinctness of boundary.

Particle-size class
This refers to the size distribution of mineral soil particles smaller than 2 mm (the fine earth); the proportions of clay (<2 μm), silt (2–60 μm) and sand (60–2000 μm) determine the class (Fig.60). A soil sample can be assigned to a class either by analysis in the laboratory (Avery and Bascomb 1982) or in the field by working moist soil between finger and thumb.

The sand, loamy sand, sandy silt loam and sandy loam classes shown in this diagram can be divided into fine, medium and coarse subclasses according to the size of the sand fraction.

Fine–more than two-thirds of the sand fraction (60 μm–2 mm) is between 60–200 μm.

Coarse–more than one-third of the sand fraction is larger than 600 μm.

Medium–less than two-thirds of the sand fraction is between 60–200 μm and less than one-third of the sand fraction is larger than 600 μm.

In analyses before 1971 silt was classed as the 2–50 μm size fraction but these determinations have been converted to the now standard 2–60 μm silt fraction:

$$\%(2\text{–}60\ \mu m) = \%(2\text{–}50\ \mu m) + \%(50\text{–}100) \times 0.26.$$

This formula has been used for several representative profile descriptions in Appendix 2, as indicated by an asterisk (*).

The limits of the particle-size classes are shown in Fig.60.

For classification purposes (Chapter 2, § 6), particle-size classes are combined into particle-size groups and subgroups, as follows:

Group	Subgroup	Class
Sandy	Coarse sandy	Coarse sand, loamy coarse sand
	Medium sandy	Medium sand, loamy medium sand
	Fine sandy	Fine sand, loamy fine sand
Loamy	Coarse loamy	Sandy loam, Sandy silt loam
	Fine loamy	Sandy clay loam, clay loam
	Coarse silty	Silt loam
	Fine silty	Silty clay loam
Clayey		Sandy clay, silty clay, clay

Soil materials with less than 15 per cent of the mineral fine earth coarser than 100 μm are grouped with the silt loams as coarse silty for some classification purposes.

Organic matter content

With increasing organic matter content, soil horizons are termed *humose* or *organic (peaty)* on the basis of organic carbon percentage related to the clay percentage of the mineral fraction (Fig.61).

Stoniness

Classes of stone abundance (per cent by volume) are:

Stoneless (or rare stones)	<1	*Moderately stony (many)*	16–35
Very slightly stony (few)	1–5	*Very stony (abundant)*	36–70
Slightly stony (common)	6–15	*Extremely stony*	>70

Size, shape and kind of stone are also described. Size classes by diameter (cm) are:

Very small	0.2–0.6	*Large*	6–20
Small	0.6–2	*Very large*	20–60
Medium	2–6	*Boulders*	>60

The term *skeletal* refers to mineral materials containing more than 35 per cent stones by volume, and those with less than 70 per cent stones are further categorized as *sandy-skeletal*, *loamy-skeletal* or *clayey-skeletal*.

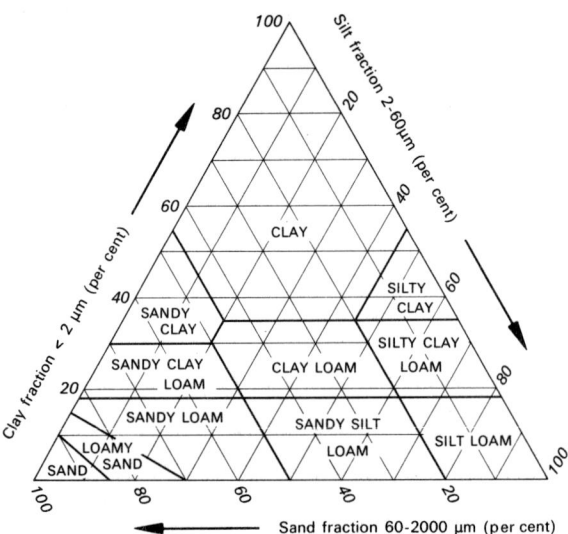

Figure 60. Particle size (textural) classes

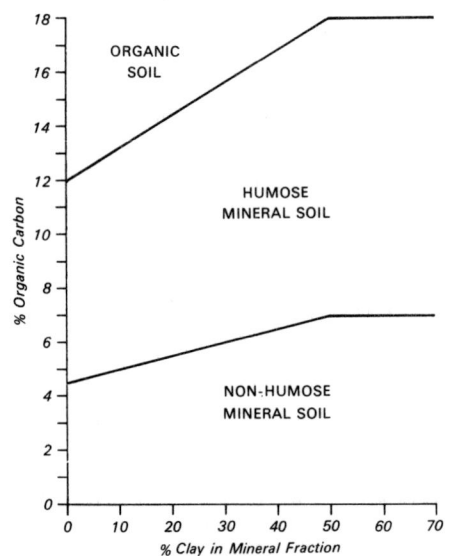

Figure 61. Limits of organic and humose soil material

Structure

Primary mineral particles are commonly arranged to form aggregates (*peds*) that are separated from each other by voids or planes of weakness.

Soil structure is defined by the degree of ped development and by ped size and shape.

Degree of ped development

Structureless:	no observable aggregation (*apedal*); *massive* if coherent and *single grain* if non-coherent.
Weak(ly):	poorly formed peds that are barely observable in place; when disturbed, the soil breaks into much unaggregated material.
Moderate(ly):	well formed moderately durable peds, that are evident but not distinct in undisturbed soil; little unaggregated material when the soil is disturbed.
Strong(ly):	durable peds that are quite distinct in undisplaced soil.

Shape

Platy:	peds with predominantly horizontal cleavage and short vertical axis.
Prismatic:	peds with vertical axis longer than the horizontal; vertical faces are well defined, vertices usually angular.
Blocky:	dimensions of peds are of the same order and surfaces are casts of the moulds formed by the faces of adjacent peds; subdivided into *angular blocky* and *subangular blocky*.
Granular:	spheroids or polyhedrons having plane or curved surfaces with slight or no accommodation to the faces of surrounding peds.

Peds are divided into size classes from fine to very coarse.

Consistence

This refers to soil cohesion and adhesion and is described in terms of strength, characteristics of failure, cementation, maximum stickiness and maximum plasticity. Strength and characteristics of failure vary widely with soil-water state.

Strength	This is the resistance to crushing of a 3 cm cube of soil when pressed either between extended thumb and forefinger, between the hands or under foot. In some cases both soil strength and ped strength can be recorded. Strength classes range from *loose*, through *weak* and *firm* (increasing force required for failure between thumb and forefinger) to *strong* and *rigid* (fails between hands and under foot).
Failure	Characteristics of failure are assessed only on very moist or wet soil. Either the compressibility of a 3 cm cube is tested (*brittle* to *deformable*) or the flow of soil material through the fingers on squeezing in the hand is observed (*slightly fluid* to *very fluid*).
Cementation	This is caused by substances such as calcium carbonate, silica or iron and aluminium compounds. Soil that slakes when an air-dry 3 cm cube is placed in water for one hour is *uncemented*; a surviving cube is assigned to one of four cementation classes (*very weakly* to *very strongly cemented*)

Voids and porosity

Estimates of total volume of voids are unreliable in the field but estimates of the volume of pores greater than 60 μm diameter can be made from particle-size class and estimated packing density (bulk density + 0.009 (% clay) g cm^{-3}). Packing density is more easily estimated in the field than bulk density.

Packing density		Field properties
	g cm^{-3}	
Low	<1.40	Loose when moist if single grain; peds if present are fine or medium and easily displaced; weak ped and/or soil strength when moist.
Medium	1.40–1.75	Neither strong nor loose consistence; peds not easily displaced, but may be well formed; weakly developed fine or medium peds, or strongly developed coarse peds with many macropores and moderately firm ped strength.
High	>1.75	Compact if single grain; peds are coarse (angular or prismatic) and structure is normally weakly developed; soils with strongly developed structure have very firm or strong ped strength and few macropores.

ANALYTICAL DATA

All determinations except pH are expressed on an oven-dry basis.

Particle-size analysis

This is carried out on air-dry soil passing a 2 mm sieve. Organic matter is removed by treatment with hydrogen peroxide and samples are dispersed by overnight shaking with alkaline sodium hexametaphosphate. Clay (<2 μm) and silt (2–60 μm) are determined by the pipette method and the sand fractions (>60 μm) by British Standard sieves. Fine clay (<0.2 μm) is measured by centrifuging the suspension left after total clay content has been determined. The percentages of all size fractions are expressed on the basis of oven-dry (105°C) weight of peroxidized soil.

Particle-size analyses are used to verify field estimates, to confirm classification of profiles, and to quantify or elucidate lithological discontinuities, parent material weathering and clay translocation. For profiles where the latter is suspected it is usual to determine fine clay to total clay ratios because fine clay is the more mobile fraction. Thus in argillic soils the ratio in the illuvial (Bt) horizon should be significantly larger than in the horizon above.

Calcium carbonate equivalent

This is measured by a calcimeter (Bascomb 1961) upon treatment of the fine earth with acid.

	CaCO$_3$ equivalent %
Non-calcareous	<0.5
Very slightly calcareous	0.5–1
Slightly calcareous	1–5
Calcareous	5–10
Very calcareous	>10

The calcimeter is also used to determine the amount of carbonate in the clay fraction.

Soil reaction

This is determined in a 1:2.5 suspension of soil in water and in 0.01M CaCl$_2$ solution, using a glass electrode assembly.

	pH (in water)
Strongly acid	<4.5
Moderately acid	4.5–5.5
Slightly acid	5.6–6.5
Neutral	6.6–7.5
Alkaline	>7.5

Organic carbon

This is determined by a wet oxidation method. For well-humified, cultivated soils, organic matter is about 1.7 times the organic carbon content.

	Organic matter %
Very low	<1.5
Low	1.5–2.9
Moderate	3.0–4.4
High	4.5–6.0
Very high	>6.0

Pyrophosphate and dithionite extractions

Soil shaken overnight with cold 0.1M potassium pyrophosphate is centrifuged and carbon, iron and aluminium are determined in the extract. Iron is also determined in a subsequent sodium dithionite extract of the residue (Residual dithionite ext. Fe%) (Avery and Bascomb 1982). Alkaline pyrophosphate is thought to extract chiefly complexes of iron and aluminium with organic matter, namely those that can move down-profile during podzolization. Sodium dithionite extraction gives the total amounts of poorly ordered and crystalline hydrous iron oxides (the so-called 'free' iron oxides). Pyrophosphate and dithionite extractable iron values together give a good indication of the amounts and types of iron present in the profile, and can be used to distinguish horizons of both iron and/or aluminium depletion (Ea, Eg) or accumulation (Bs, Bfe). On some samples sodium dithionite has been used to extract 'free' iron from the whole of the fine earth fraction (Dithionite ext. Fe %).

MICROMORPHOLOGICAL DESCRIPTION

Thin sections are prepared according to the method of Bascomb and Bullock (1982). Undisturbed soil samples taken in 7.5 × 6.5 × 4 cm Kubiena tins are dried, impregnated with a polyester resin and left to harden. A slice from each sample is fixed to a microscope slide and ground to 20–30 μm thickness with a diamond impregnated wheel. Examination under the optical microscope yields information which is used in conjunction with analytical data to supplement field descriptions. It also aids classification of the soil profiles and helps confirm horizon nomenclature. The terminology used in description of thin sections is largely that of Brewer (1964).

Clay coats and intrapedal concentrations

Clay coats (argillans) are concentrations of clay around voids, mineral grains or peds. Microscopically, they have strong optical continuity, strongly preferred orientation, a sharp boundary with adjacent material and often a laminated appearance.

Intrapedal concentrations are similar microscopically to argillans but are unrelated to voids or other free surfaces. They are believed to be clay coats which have been disrupted and integrated into the matrix by soil fauna, shrinking and swelling and other pedoturbation processes, or are totally infilled voids.

When assessing the amount of clay translocation and illuviation for classification purposes both clay coats and intrapedal concentrations are counted.

Abundance	(%)	Size	(μm thickness)
Rare	<0.2	*Very fine*	<25
Few	0.2–2	*Fine*	25–150
Common	2–4	*Medium*	150–500
Many	4–10	*Thick*	>500
Very many	10–20		
Abundant	>20		

PHYSICAL DATA
(after Hall *et al.* (1977))

Sampling and laboratory methods

Undisturbed triplicate core samples are taken from selected horizons in tinned metal sleeves of 222 cm³ volume and 7.6 cm diameter using a special coring tool (Hodgson 1976). The water retained by the samples is measured at suctions of 0.05 bar (approximately field capacity), 0.1 and 0.4 bar on sand and kaolin tension tables and at 2 and 15 bar (wilting point for most plants) in pressure membrane cells. Each replicate is dispersed and sieved to determine stone content. Particle density is determined for all surface horizons, but its range for most subsoils is sufficiently small (±0.1 g cm⁻³) to accept a general value of 2.65 g cm⁻³ (Reeve *et al.* 1973).

Mean values for the water content at each suction are expressed in terms of mass (%) of oven-dry soil or, more usually, as a volumetric water content (%) on the basis of total soil volume including stones. Bulk density is similarly expressed as the ratio of mass of oven-dried soil to initial sample volume.

Bulk density and packing density

Bulk density is an important field property influencing available water, air capacity and retained water; its effects vary with particle-size class and between surface and subsurface horizons.

Soil organic matter also significantly affects bulk density so that most soils with a large organic carbon content have low values. Under permanent grass, the bulk density of surface horizons is normally medium to low and remains relatively stable. On arable land, topsoil bulk density is easily altered by cultivations so it varies during the year.

The bulk density of mineral subsoils with little or no organic matter is determined principally by the development of structure, as affected by factors such as soil parent material, particle-size distribution, faunal activity, seasonal wetting and drying and compaction by cultivations. Sandy soils often have bulk densities higher than clayey soils despite having a larger volume of coarse pores. Bulk density, therefore, cannot be used as a simple diagnostic property when comparing different soils but packing density fulfils this need.

Packing density, L_d, is a modification of fine earth bulk density which takes into account the amount of unavailable water held in fine pores associated with clay particles. It is defined by the expression: $L_d = D_{bf} +$ 0.009C, where D_{bf} is the bulk density of fine earth and C is the gravimetric clay percentage. The factors influencing packing density are similar to those which affect bulk density, although air capacity is more closely related to packing density than to bulk density. The terms used to classify both packing and bulk density are given in Table 32.

Table 32   Class Limits for Soil Physical Data

Class	Retained water % vol	Available water % vol	Air capacity and porosity >60 μm % vol
Very small	0– 9.9	0– 4.9	0– 4.9 (very slightly porous)
Small	10–19.9	5– 9.9	5– 9.9 (slightly porous)
Moderately small	20–29.9	10–14.9	10–14.9 (moderately porous)
Moderately large	30–39.9	15–19.9	15–19.9 (very porous)
Large	40–49.9	20–24.9	20–24.9 (extremely porous)
Very large	>50	>25	>25

Class	Bulk density g cm^{-3}	Packing density g cm^{-3}
Very low	<0.20	
Low	0.20–0.80	<1.40
Medium	0.80–1.30	1.40–1.75
High	1.30–1.80	>1.75
Very high	>1.80	

Total pore space

The total pore space, T, whether water or air filled, is calculated from the formula:

$$T = \left(1 - \frac{D_b}{D_p}\right) 100 \qquad \%$$

where D_b is the bulk density and D_p the particle density.

This represents the volume of the sample not occupied by solids. It varies from 40 per cent in very dense material to 70 per cent in loose topsoil with a large organic matter content, and is inversely related to dry bulk density.

Air capacity

The air capacity, C_a, is the volume of air-filled pores (as a percentage of soil volume) at 0.05 bar suction and is approximately equivalent to the volume of pores >60 μm diameter. It indicates the relative aeration of soils when excess moisture has drained away under the influence of gravity and is calculated from:

$$C_a = T\Theta_v(0.05) \qquad \% \text{ vol.}$$

where T is total pore space and $\Theta_v(0.05)$ is retained water capacity.

Air capacity decreases with an increase in both clay content and packing density and varies from almost nothing in dense clayey subsoils to 30 per cent or more in loose sand. Classes of air capacity and their relationship to porosity are shown in Table 32. In general, the more porous the soil, the better its permeability. Slightly porous horizons with high packing density are relatively impervious and impede downward or horizontal redistribution of water. Very slightly porous horizons (air capacity less than 5 per cent) have horizontal hydraulic conductivities less than 10 cm/day on average, and are considered impermeable for soil classification purposes. Thomasson (1978) gives the range and significance of differences in air capacities between the main particle-size groups.

Retained water

Retained water is the volumetric water content of a soil at 0.05 bar suction and is approximately equivalent to the volume of pores <60 μm diameter. It is a consistently measurable approximation of the minimum water content of most soils in lowland Britain during the winter, when evapotranspiration is negligible (Webster and Beckett 1972) and, as such, corresponds to the moisture content at field capacity. It is used to assess the relative poaching risk and workability class of soils. For example, soils with large or very large retained water which exceeds the plastic limit are very susceptible to poaching whereas those with moderately small or small amounts of retained water have a low poaching risk. Classes of retained water are given in Table 32.

In surface horizons, the main factors influencing retained water capacity are clay and organic carbon content (Hollis *et al.* 1977). In subsurface horizons retained water is governed by clay content and bulk density.

Available water

The available water, A_v, of a soil is calculated from:

$$A_v = \Theta_v(0.05) - \Theta_v(15) \qquad \text{\% vol.}$$

where $\Theta_v(0.05)$ = retained water capacity (at 0.05 bar suction)
and $\Theta_v(15)$ = water retained at 15 bar suction

It is the amount of water in a soil available for plant growth after excess moisture has drained away under the influence of gravity. Classes of available water are given in Table 32. A value for easily available water can be calculated similarly, using the water retained at 2 bar instead of 15 bar suction. Available and easily available water can be summed to various depths according to crop (Hall *et al* 1977) to assess the relative droughtiness of soils.

Many factors affect available water and there are complex interactions between those of greatest importance, namely organic carbon content, bulk density and the mineral particle-size fraction 20–100 μm. In surface horizons the amount of available water has a strong positive correlation with organic carbon content. In subsurface horizons, it generally increases with decreasing density and, for sandy, coarse loamy and silty soils, increases with increasing content of fine mineral particles (20–100 μm).

APPENDIX 2
SELECTED SOIL PROFILE DESCRIPTIONS

AGNEY SERIES
Profile no.: TF 45/1584
Definition: Calcareous alluvial gley soils. Fine silty; marine alluvium.
Elevation: 1 m O.D.   Slope and aspect: Level.
Land use: Cereals.
Horizons:
  0-28 cm Ap   Dark greyish brown (10 YR 4/2) stoneless silty clay loam with few very fine mottles; slightly moist; moderately developed coarse subangular blocky; medium packing density; very firm soil strength; many very fine fibrous roots; slightly calcareous; abrupt smooth boundary.

  28-55 cm Bg   Brown (10 YR 5/3) stoneless silty clay loam with common medium strong brown (7.5 YR 5/6) mottles; slightly moist; moderately developed coarse angular blocky with dark grey (10 YR 4/1) faces; medium packing density; very firm ped strength; many very fine fibrous roots; slightly calcareous; few organic coats; clear smooth boundary.

  55-100 cm BCg   Very dark greyish brown (10 YR 3/2) stoneless silty clay loam with common fine strong brown (7.5 YR 5/6) mottles; moist; moderately developed medium prismatic with dark grey (10 YR 4/1) faces; medium packing density; very firm ped strength; many very fine fibrous roots; calcareous; few organic coats.

Horizon: Depth (cm)	Ap 0-28	Bg 28-55	BCg 55-100
Sand 600 µm-2 mm %	0	0	0
200-600 µm %	1	0	0
60-200 µm %	3	5	6
Silt 2-60 µm %	66	67	65
Clay <2 µm %	30	28	29
CaCO₃ equivalent %	2	4	7
Organic carbon %	3.7	0.9	1.1
pH in water (1:2.5)	7.3	7.6	7.9
pH in 0.01M CaCl₂ (1:2.5)	7.0	7.2	7.5

BANGOR SERIES
Profile no.: NY 21/2329
Definition: Humic ranker. Loamy or peaty, lithoskeletal acid crystalline rock.
Elevation: 777 m O.D.   Slope and aspect: 3° ESE, convex.
Land use and vegetation: Montane rough grazing with Festuca sp., heather, Sphagnum sp. and Galium sp.
Horizons:
  0-25 cm Om   Black (5 YR 2/1) slightly stony semi-fibrous peat; large subrounded, rhyolite; wet; many fine fibrous roots; gradual wavy boundary.

  25-33 cm Oh/Cu   Black (5 YR 2/1) extremely stony humose coarse sandy loam; medium subrounded, rhyolite; wet; sharp smooth boundary.

  At 33 cm R   Grey (7.5 YR 6/0) very hard massive rhyolite.

Horizon: Depth (cm)	Om 0-25	Oh/Cu 25-33
Sand 600 µm-2 mm %		29
200-600 µm %		16
60-200 µm %		5
Silt 2-60 µm %		31
Clay <2 µm %		9
Organic carbon %	31	
pH in water (1:2.5)	4.3	4.9
pH in 0.01M CaCl₂ (1:2.5)	3.4	4.1

BASCHURCH SERIES
Profile no.: SJ 50/6927
Definition: Typical brown podzolic soils. Loamy-gravelly, sandstones, siltstones, mudstones or slates.
Elevation: 105 m O.D.   Slope and aspect: 1° S, straight.
Land use and vegetation: Cereals - winter wheat.
Horizons:
  0-29 cm Ap   Very dark greyish brown (10 YR 3/2) moderately stony sandy silt loam; medium rounded and tabular, micaceous siltstone, sandstone and greywacke; moist; weakly developed medium subangular blocky; medium packing density; very weak soil strength; many very fine fibrous roots; non-calcareous; abrupt smooth boundary.

  29-55 cm Bs   Dark yellowish brown (10 YR 4/4) very stony sandy loam; moist; medium rounded, micaceous siltstone, sandstone and greywacke; weakly developed fine subangular blocky; medium packing density; very weak soil strength; common very fine fibrous roots; non-calcareous; clear wavy boundary.

  55-80 cm Cu   Dark yellowish brown (10 YR 4/4) very stony sandy loam; medium rounded and tabular, micaceous siltstone, sandstone and greywacke; dry; weakly developed fine granular; medium packing density; loose soil strength; few very fine fibrous roots; non-calcareous.

Horizon: Depth (cm)	Ap 0-29	Bs 29-55	Cu 55-80
Sand 600 µm-2 mm %	13	13	21
200-600 µm %	17	25	26
60-200 µm %	19	17	19
Silt 2-60 µm %	36	33	25
Clay <2 µm %	15	12	9
CaCO₃ equivalent %	0	0	0
Organic carbon %	2.1	0.7	
pH in water (1:2.5)	6.5	6.9	6.9
pH in 0.01M CaCl₂ (1:2.5)	6.2	6.4	6.4
Pyrophosphate extract			
Fe %	0.1	0.7	0.6
Al %	0.0	0.1	0.0
C %	0.8	0.2	0.1
Residual dithionite ext.			
Fe %	1.1	1.2	1.0

BEACON SERIES
Profile no.: SN 71/2196
Definition: Cambic stagnohumic gley soils. Reddish loamy over lithoskeletal sandstone.
Elevation: 365 m O.D.   Slope and aspect: 10° NE, convex.
Land use and vegetation: Nardus grassland.
Horizons:
  0-12 cm Oh   Black (5 YR 2/1) stoneless humified peat; wet; moderately developed fine granular; abrupt smooth boundary.

  12-21 cm Eag   Light brownish grey (10 YR 6/2) very slightly stony clay loam; small rounded, quartz; moist; weakly developed, adherent coarse prismatic with brown to dark brown (7.5 YR 4/2) faces; low packing density; moderately weak soil strength; common fine fibrous roots; common organic coats; abrupt smooth boundary.

  21-32 cm Bg   Pinkish grey (5 YR 6/2) moderately stony clay loam with many medium reddish brown (2.5 YR 5/4) mottles; small rounded and tabular, sandstone; moist; moderately developed coarse prismatic; medium packing density; very weak soil and ped strength; common fine fibrous roots; non-calcareous; clear smooth boundary.

  32-50 cm BCu   Reddish brown (2.5 YR 5/4) moderately stony coarse sandy silt loam; small subangular and tabular, sandstone; moist; moderately developed fine subangular blocky with reddish brown (5 YR 5/3) faces; medium packing density; moderately weak soil and ped strength; common fine fibrous roots; non-calcareous; clear smooth boundary.

  50-70 cm Cu   Reddish brown (2.5 YR 4/4) very stony coarse sandy silt loam; small to large angular and tabular, sandstone; wet; massive; high packing density; few fine fibrous roots; non-calcareous; clear irregular boundary.

  At 70 cm R   Reddish brown (2.5 YR 4/4) hard shattered sandstone.

Horizon: Depth (cm)	Oh 0-12	Eag 12-21	Bg 21-32	BCu 32-50	Cu 50-70
Sand 600 µm-2 mm %		1	2	16	14
200-600 µm %		7	7	9	8
60-200 µm %		26	20	14	19
Silt 2-60 µm %		48	53	46	48
Clay <2 µm %		18	18	15	11
Organic carbon %	32	2.1			
pH in water (1:2.5)	3.9	4.1	4.4	4.6	4.8
pH in 0.01M CaCl₂ (1:2.5)	3.3	3.5	3.7	4.0	4.1
Pyrophosphate ext.					
Fe %			0.2	0.3	0.0
Al %			0.1	0.1	0.1
C %			0.4	0.3	0.1
Residual dithionite ext.					
Fe %			0.4	2.4	2.3
Bulk density g cm⁻³	0.30	1.15		1.45	1.75
Available water capacity					
% vol. <2 bar		13		7	6
<15 bar		22		16	13
Air capacity % vol.	7	10		7	5
Retained water capacity					
% vol.	76	46		39	29
Illuvial clay %		0	0	<0.2	

BECKFOOT SERIES

Profile no.: NY 05/9808
Definition: Typical sand rankers. Sandy; stoneless drift
Elevation: 7 m O.D.   Slope and aspect: 5° N, convex
Land use and vegetation: Lowland heath with Ling, Purple Heather and
    Common Gorse.
Horizons:

0-12 cm Ah   Dark reddish brown (5 YR 2/2) sand; dry; single grain;
low packing density; moderately weak soil strength; abundant very fine
fibrous roots; sharp smooth boundary.

12-30 cm BCu   Dark brown (7.5 YR 3/2) stoneless sand; dry; single
grain; low packing density; moderately weak soil strength; common fine
fibrous roots; gradual smooth boundary.

30-68 cm Cu   Dark yellowish brown (10 YR 3/4) stoneless sand; dry;
single grain; low packing density; moderately weak soil strength; common
fine fibrous roots; sharp smooth boundary.

68-88 cm bABh   Very dark greyish brown (10 YR 3/2) very slightly
stony sand; moderately rounded and tabular, igneous; dry; single grain;
medium packing density; moderately firm soil strength; very weakly
cemented; few fine fibrous roots; clear smooth boundary.

88-101 cm bBh   Dark yellowish brown (7.5 YR 3/2) stoneless sand; slightly
moist; single grain; low packing density; very weak soil strength; few
fine fibrous roots; clear smooth boundary.

101-120 cm bCu   Dark yellowish brown (10 YR 4/8) stoneless sand;
slightly moist; single grain; low packing density; very weak soil
strength; few fine fibrous roots.

Horizon: Depth (cm)	Ah 0-12	BCu 12-30	Cu 30-68	bABh 68-88	bBh 88-101	bCu 101-120
Sand 600 µm-2 mm %	4	4	5	8	6	5
200-600 µm %	58	67	64	64	58	63
60-200 µm %	32	27	28	21	32	31
Silt 2-60 µm %	2	2	2	4	1	1
Clay <2 µm %	4	0	1	3	3	0
Organic carbon %	2.8	0.7	0.7	1.1	1.5	
pH in water (1:2.5)	4.8	5.1	5.3	5.2	5.0	5.3
pH in 0.01M CaCl$_2$ (1:2.5)	3.7	4.2	4.6	4.2	4.3	4.6
Pyrophosphate ext.						
Fe %	0.2	0.1	0.1	0.2	0.3	0.1
Al %	0.1	0.1	0.1	0.1	0.1	0.1
C %	0.9	0.3	0.2	0.4	0.5	0.2
Residual dithionite ext.						
Fe %	0.5	0.6	0.4	0.4	0.4	0.4

CHERUBEER SERIES

Profile no.: SS 61/1711
Definition: Stagnogleyic brown earths. Fine loamy over clayey; drift
    with siliceous stones.
Elevation: 153 m O.D.   Slope and aspect: 3° ENE.
Land use: Ley grassland.
Horizons:

0-26 cm Ap   Dark brown (10 YR 3/3) slightly stony clay loam; medium
subangular, micaceous sandstone, and small, Carboniferous limestone;
slightly moist; weakly developed fine subangular blocky; medium packing
density; moderately strong soil strength; moderately firm ped strength;
many very fine fibrous roots; non-calcareous; abrupt smooth boundary.

26-49 cm Bw   Yellowish brown (10 YR 5/6) moderately stony clay loam
with very many coarse dark brown (10 YR 3/3) mottles as inclusions of Ap
around worm burrows; medium subangular, micaceous sandstone; slightly
moist; weakly developed fine subangular blocky; medium packing density;
very firm soil strength; common very fine fibrous roots; non-calcareous;
gradual wavy boundary.

49-76 cm 2Bw(g)   Light yellowish brown (10 YR 6/4) moderately stony
silty clay loam with common fine yellowish red (5 YR 5/8) mottles and
common coarse inclusions of brown to dark brown (10 YR 4/3) along worm
burrows; medium subangular, micaceous sandstone; moist; weakly developed
medium prismatic with yellowish brown (10 YR 5/4) faces; medium packing
density; very firm ped strength; common very fine fibrous roots; non-
calcareous; few ferri-manganiferous coats; clear smooth boundary.

76-110 cm 2BCg1   Dark yellowish brown (10 YR 4/4) moderately stony
clay with many medium light grey (10 YR 7/1) mottles; large to very small
subangular, micaceous sandstone; moist; massive; high packing density;
moderately strong soil strength; few very fine fibrous roots; non-
calcareous; common ferri-manganiferous coats; clear smooth boundary.

110-120 cm 2BCg2   Strong brown (7.5 YR 5/8) moderately stony clay
with many medium brownish yellow (10 YR 6/6) and common fine light grey
(10 YR 7/1) mottles; medium subangular, micaceous sandstone; moist;
massive; high packing density; moderately strong soil strength; few fine
roots confined to occasional worm burrows; non-calcareous; few ferri-
manganiferous coats.

Horizon: Depth (cm)	Ap 0-26	Bw 26-49	2Bw(g) 49-76	2BCg1 76-110	2BCg2 110-120
Sand 600 µm-2 mm %	4	4	3	7	4
200-600 µm %	3	4	3	5	5
60-200 µm %	18	13	9	6	7
Silt 2-60 µm %	52	52	52	38	35
Clay <2 µm %	23	27	33	44	49
<0.2 µm %	5	7	10	11	12
Organic carbon %	2.4	0.5			
pH in water (1:2.5)	6.0	6.5	6.5	6.6	6.7
pH in 0.01M CaCl$_2$ (1:2.5)	5.2	5.6	5.7	6.0	5.9

BISHAMPTON SERIES

Profile no.: SO 95/8315
Definition: Stagnogleyic argillic brown earths. Fine loamy; drift with
    siliceous stones.
Elevation: 34 m O.D.   Slope and aspect: Level.
Land use: Permanent grassland.
Horizons:

0-13 cm Ap   Dark brown (7.5 YR 3/2) stoneless humose sandy clay loam;
moist; strongly developed fine subangular blocky; low packing density;
moderately weak soil and ped strength; abundant very fine fibrous roots;
non-calcareous; abrupt smooth boundary.

13-38 cm Eb1   Brown to dark brown (7.5 YR 4/3) very slightly stony
sandy clay loam; small rounded, quartzite; slightly moist; moderately
developed medium subangular blocky; medium packing density; moderately
firm soil strength; moderately weak ped strength; many very fine fibrous
roots; non-calcareous; clear smooth boundary.

38-59 cm Eb2   Brown to dark brown (10 YR 4/3) slightly stony sandy
clay loam; medium rounded, quartzite, and subrounded sandstone and flint;
slightly moist; moderately developed medium subangular blocky; medium
packing density; moderately firm soil strength; moderately weak ped
strength; many very fine fibrous roots; non-calcareous; abrupt smooth
boundary.

59-85 cm Bw(g)   Brown to dark brown (7.5 YR 4/3) slightly stony sandy
clay loam with common extremely fine strong brown (7.5 YR 5/6) and
yellowish brown (10 YR 5/4) mottles; medium rounded, quartzite; moist;
moderately developed medium angular blocky; medium packing density;
moderately firm soil and ped strength; non-calcareous; common clay coats;
abrupt smooth boundary.

85-112 cm Bt(g)   Brown to dark brown (7.5 YR 4/4) moderately stony
sandy clay with very many coarse yellowish red (5 YR 4/6) mottles; medium
rounded, quartzite; moist; weakly developed medium prismatic with brown
(7.5 YR 5/3) faces; high packing density; moderately firm soil and ped
strength; few very fine fibrous roots; non-calcareous; common irregular
soft ferri-manganiferous concentrations; many clay coats around stones
and few filling pores.

Horizon: Depth (cm)	Ap 0-13	Eb1 13-38	Eb2 38-59	Bw(g) 59-85	Bt(g) 85-112
Sand 600 µm-2 mm %	3	3	3	5	3
200-600 µm %	41	36	35	34	28
60-200 µm %	15	15	14	14	8
Silt 2-60 µm %	22	24	24	24	24
Clay <2 µm %	19	22	24	23	37
<0.2 µm %	12	14	14	14	19
CaCO$_3$ equivalent %					0
Organic carbon %	5.5	1.8			
pH in water (1:2.5)	5.1	5.3	5.5	5.9	6.8
pH in 0.01M CaCl$_2$ (1:2.5)	4.5	4.5	4.8	5.2	6.2
Bulk density g cm^{-3}	1.05	1.30	1.50	1.55	1.55
Available water capacity					
% vol. <2 bar	16	9	9	7	4
<15 bar	22	13	11	12	4
Air capacity % vol.	17	25	19	16	8
Retained water capacity					
% vol.	41	27	25	25	34
Illuvial clay %		<0.2	1-3	>2	>2

CLAVERLEY SERIES

Profile no.: SO 89/1521
Definition: Typical stagnogley soils. Reddish coarse loamy; drift with
    siliceous stones.
Elevation: 77 m O.D.   Slope and aspect: Level.
Land use and vegetation: Permanent grassland with abundant Rye Grass,
    Creeping Soft Grass, Clover.
Horizons:

0-18 cm Ah   Dark brown (7.5 YR 3/2) very slightly stony sandy loam;
small and very small rounded and subrounded, quartz and quartzite
(Bunter); moist; moderately developed coarse, medium and fine subangular
blocky; low packing density; moderately weak soil and ped strength; many
fine and very fine fibrous roots, few medium woody roots; clear smooth
boundary.

18-53 cm Eg   Light reddish brown to light brown (6.5 YR 6/4) slightly
stony sandy loam with common medium and coarse light grey (10 YR 7/1),
strong brown (7.5 YR 5/8) and yellowish red (5 YR 4/6) mottles; medium
and small rounded and subrounded quartz and quartzite, some rounded and
subrounded, igneous and sandstone; moist; weakly developed coarse and
medium subangular blocky; medium packing density; moderately firm soil
strength; very weak ped strength; common fine and very fine fibrous
roots, few fine, medium and coarse woody roots; few indistinct soft
ferri-manganiferous concentrations; abrupt smooth boundary.

53-84 cm Btg1   Yellowish red (4 YR 4/6) very slightly stony sandy clay loam
with common medium and fine yellowish red (5 YR 4/6) and coarse light
grey (10 YR 7/1) mottles; moist; strongly developed very coarse
prismatic; high packing density; moderately strong ped strength; common
fine and very fine fibrous roots, few medium and coarse woody roots
mainly down ped faces; many continuous prominent light grey (10 YR 7/1)
sandy and clayey coats on ped faces; gradual smooth boundary.

84-110 cm Btg2   Reddish brown (2.5 YR 4/4) very slightly stony clay
loam with few medium mottles of slightly duller colour within peds;
medium and small rounded and subrounded, quartz and quartzite; moist;
weakly developed coarse and very coarse subangular blocky; occasional
large structure faces going to depth; high packing density; moderately
strong ped strength; few fine and very fine fibrous roots mainly down ped
faces, few coarse woody roots; many prominent light grey (10 YR 7/1)
sandy and clayey coats on the main structure faces and also lining pores
and root channels.

Horizon: Depth (cm)	Ah 0-18	Eg 18-53	Btg1 53-84	Btg2 84-110
Sand 600 µm-2 mm %	4	3	2	2
200-600 µm %	23	20	12	12
60-200 µm %	32	34	30	27
Silt 2-60 µm %	27	28	34	36
Clay <2 µm %	14	15	22	23
<0.2 µm %	6	6	8	8
Organic carbon %	3.3	0.6		
pH in water (1:2.5)	6.9	6.9	5.8	5.9
pH in 0.01M CaCl$_2$ (1:2.5)	6.6	6.4	5.2	5.2
Bulk density g cm-3	1.05	1.55	1.70	1.80
Available water capacity				
% vol. <2 bar	18	9	6	5
<15 bar	27	15	11	10
Air capacity % vol.	14	15	6	2
Retained water capacity				
% vol.	44	27	29	30
Illuvial clay %		0	>2	>2

CLAYWORTH SERIES
Profile no.: SP 35/0055
Definition: Typical calcareous pelosols. Reddish clayey passing to clay or mudstone.
Elevation: 57 m O.D.   Slope and aspect: 2° NW, straight.
Land use: Ley grassland.
Horizons:

0-25 cm Ap Dark brown (7.5 YR 3/2) stoneless clay; slightly moist; moderately developed medium angular blocky; medium packing density; moderately firm soil and ped strength; many fine fibrous roots along ped faces; very slightly calcareous; abrupt smooth boundary.

25-43 cm Bt1 Reddish brown (5 YR 4/3) stoneless silty clay; slightly moist; strongly developed medium prismatic; high packing density; moderately firm soil and ped strength; common fine fibrous roots; calcareous; common organic coats; gradual smooth boundary.

43-69 cm Bt2 Reddish brown (5 YR 4/3) stoneless silty clay loam; slightly moist; strongly developed coarse prismatic; high packing density; moderately strong ped strength; common very fine fibrous roots; calcareous; few rounded ferri-manganiferous nodules; common clay coats; clear smooth boundary.

69-119 cm Cr Reddish brown to yellowish red (5 YR 4/5) moderately stony silty clay loam; medium angular, micaceous mudstone; slightly moist; massive; high packing density; moderately strong ped strength; very calcareous.

Horizon: Depth (cm)	Ap 0-25	Bt1 25-43	Bt2 43-69	Cr 69-119
Sand 600 μm-2 mm %	1	0	0	0
200-600 μm %	11	1	0	1
60-200 μm %	10	2	4	4
Silt 2-60 μm %	43	52	68	68
Clay <2 μm %	35	45	28	27
CaCO₃ equivalent %	<1	8	8	12
Organic carbon %	2.5	0.6		
pH in water (1:2.5)	7.1	7.7	7.8	7.8
pH in 0.01M CaCl₂ (1:2.5)	6.9	7.2	7.3	7.1
Bulk density g cm⁻³	1.45	1.55	1.60	
Available water capacity				
% vol. <15 bar	13	10	11	
Air capacity % vol.	7	6	6	
Retained water capacity				
% vol.	40	34	33	
Illuvial clay %			>3	1-3

CLIFTON SERIES
Profile no.: SK 00/0162
Definition: Typical stagnogley soils. Reddish fine loamy; drift with siliceous stones.
Elevation: 130 m O.D.   Slope and aspect: 1° NNE.
Land use: Ley grassland.
Horizons:

0-23 cm Ap Dark greyish brown (10 YR 4/2) slightly stony sandy clay loam; large rounded, quartzite; moist; strongly developed coarse subangular blocky; high packing density; moderately firm soil and ped strength; many very fine fibrous roots; non-calcareous; sharp wavy boundary.

23-37 cm Eg Light grey to grey (10 YR 6/1) slightly stony sandy loam with many fine yellowish brown (10 YR 5/6) mottles; large rounded, quartzite; moist; weakly developed, adherent medium subangular blocky; high packing density; moderately firm soil and ped strength; common very fine fibrous roots; non-calcareous; few irregular ferri-manganiferous nodules; abrupt wavy boundary.

37-86 cm Btg Reddish brown (5 YR 4/4) slightly stony clay loam with common fine light grey to grey (10 YR 6/1) and strong brown (7.5 YR 5/6) mottles; large rounded, quartzite; moist; strongly developed very coarse prismatic with greyish brown (10 YR 5/2) faces; high packing density; very firm soil strength; few very fine fibrous roots; non-calcareous; many clay coats; gradual smooth boundary.

86-107 cm BCtg Reddish brown (2.5 YR 4/4) slightly stony clay loam with common medium grey (N 5/0) mottles; medium rounded, quartzite; moist; massive; high packing density; moderately strong ped strength; common clay coats.

Horizon: Depth (cm)	Ap 0-23	Eg 23-37	Btg 37-86	BCtg 86-107
Sand 600 μm-2 mm %	2	2	2	3
200-600 μm %	27	29	19	22
60-200 μm %	28	30	22	21
Silt 2-60 μm %	25	22	27	26
Clay <2 μm %	18	17	30	28
<0.2 μm %	8	8	13	12
CaCO₃ equivalent %	0	0	0	0
Organic carbon %	1.3	0.5		
pH in water (1:2.5)	6.5	7.0	7.2	7.6
pH in 0.01M CaCl₂ (1:2.5)	5.8	6.3	6.5	6.8
Bulk density g cm⁻³	1.70	1.75	1.75	1.75
Available water capacity				
% vol. <2 bar	7	6	5	5
<15 bar	12	11	9	10
Air capacity % vol.	5	5	1	3
Retained water capacity				
% vol.	31	29	32	31
Illuvial clay %		<2		1-3

CRAY SERIES
Profile no.: SN 72/9989
Definition: Ironpan stagnopodzols. Reddish loamy; drift with siliceous stones.
Elevation: 396 m O.D.   Slope and aspect: 2° SE, straight.
Land use and vegetation: Molinia grassland; open moorland.
Horizons:

0-17 cm Oh Black (5 YR 2/1) stoneless humified peat; moist; massive; many fine fibrous roots; abrupt smooth boundary.

17-24 cm Eag Pinkish grey (7.5 YR 6/2) slightly stony clay loam; small subangular, red sandstone; moist; moderately developed fine angular blocky with dark brown (7.5 YR 3/2) faces; medium packing density; moderately weak soil and ped strength; many fine fibrous roots; sharp wavy boundary.

At 24 cm Bf Dark reddish brown (5 YR 3/2) continuous ironpan 2 mm thick.

24-27 cm Bg Grey (5 YR 5/1) slightly stony sandy silt loam with common fine yellowish red (5 YR 5/6) mottles; small subangular, red sandstone; moist; weakly developed, adherent medium subangular blocky; medium packing density; moderately weak soil strength; sharp irregular boundary.

27-50 cm Bw(g) Reddish brown (2.5 YR 4/4) slightly stony coarse sandy silt loam with common fine yellowish red (5 YR 4/8) mottles; medium subangular, red sandstone; moist; weakly developed, adherent fine subangular blocky; medium packing density; moderately weak soil strength; sharp smooth boundary.

50-100 cm BCu Reddish brown (2.5 YR 4/4) moderately stony sandy silt loam; small subangular, red sandstone; moist; massive; high packing density; very firm soil strength.

Horizon: Depth (cm)	Oh 0-17	Eag 17-24	Bf At 24	Bw(g) 27-50
Sand 600 μm-2 mm %		2	32	14
200-600 μm %		6	13	9
60-200 μm %		12	8	10
Silt 2-60 μm %		61	35	51
Clay <2 μm %		19	12	16
Organic carbon %	35	3.4		
pH in water (1:2.5)	4.9	4.1	4.4	4.7
pH in 0.01M CaCl₂ (1:2.5)	3.5	3.5	3.8	4.0
Pyrophosphate ext.				
Fe %		0.1	1.0	0.4
Al %		0.2	0.2	0.2
C %		0.7	1.2	0.4
Residual dithionite ext.				
Fe %		0.3	9.9	2.1

CROWDY SERIES
Profile no.: SN 50/5584
Definition: Raw oligo-amorphous peat soils. Humified peat.
Elevation: 150 m O.D.   Slope and aspect: Level.
Land use and vegetation: Rough grazing with dominant rushes and Purple Moor Grass.
Horizons:

0-10 cm Oh1 Black (10 YR 2/1) very slightly stony humified peat; angular and tabular stones are artefacts; wet; moderately developed fine granular; very weak soil strength; abundant fine fibrous roots; clear smooth boundary

10-20 cm Oh2 Dark brown (7.5 YR 3/2) slightly stony humified peat; massive; very weak soil strength; many fine fibrous roots; clear smooth boundary.

20-70 cm Oh3 Black (10 YR 2/1) stoneless humified peat; wet; massive; clear smooth boundary.

70-100 cm Oh4 Very dark greyish brown (10 YR 3/2) stoneless humified peat; wet; massive.

Horizon: Depth (cm)	Oh2 10-20	Oh3 20-45	Oh3 45-70	Oh4 70-100
Organic carbon %	37	39	39	56
pH in water (1:2.5)	3.9	3.5	3.4	4.1
pH in 0.01M CaCl₂ (1:2.5) (undried sample)	3.4	3.2	3.2	3.4
Rubbed fibre % on ash free dry matter basis	9.0	2.5	3.0	9.3
Pyrophosphate ext.				
C %	12.7	13.7	10.1	12.8

CRIBBIN SERIES
Profile no.: SD 96/5254
Definition: Brown rankers. Loamy, lithoskeletal limestone.
Elevation: 396 m O.D.   Slope and aspect: 4° SSW, straight.
Land use and vegetation: Permanent grassland with Bent sp. and
  Fescue sp.
Horizons:
0-24 cm Ah   Brown to dark brown (10 YR 4/3) stoneless clay loam;
moist; weakly developed fine subangular blocky with dark greyish brown
(10 YR 4/2) faces; medium packing density; moderately weak soil strength;
many very fine fleshy roots; non-calcareous; sharp wavy boundary.

At 24 cm   R   Light grey to grey (10 YR 6/1) very hard shattered
limestone.

Horizon: Depth (cm)	Ah 0-19
Sand 600 μm-2 mm %	2
200-600 μm %	7
60-200 μm %	12
Silt 2-60 μm %	53
Clay <2 μm %	26
Organic carbon %	3.5
pH in water (1:2.5)	6.3
pH in 0.01M CaCl$_2$ (1:2.5)	5.7

DENBIGH SERIES
Profile no.: SH 96/3353
Definition: Typical brown earths. Fine loamy over lithoskeletal
mudstone and sandstone or slate.
Elevation: 183 m O.D.   Slope and aspect: 3° NNW, concave.
Land use: Permanent grassland.
Horizons:
0-22 cm Ap   Dark brown (10 YR 3/3) slightly stony clay loam; medium
subangular and platy, siltstone; dry; strongly developed medium granular
with light brownish grey (2.5 Y 6/2) faces; low packing density; very
weak soil and ped strength; abundant fine fibrous roots; non-calcareous;
gradual wavy boundary.

22-40 cm Bw1   Brown to dark brown (10 YR 4/3) moderately stony clay
loam; medium subangular and platy, siltstone; dry; moderately developed
fine subangular blocky with pale brown (10 YR 6/3) faces; low packing
density; very weak soil and ped strength; many fine fibrous roots; non-
calcareous; gradual wavy boundary.

40-70 cm Bw2   Brown to dark brown (10 YR 4/3) moderately stony clay
loam; medium subangular and platy, siltstone; dry; moderately developed
fine subangular blocky with pale brown (10 YR 6/3) faces; low packing
density; very weak soil and ped strength; common fine fibrous roots; non-
calcareous; abrupt irregular boundary.

70-100 cm BCu   Yellowish brown (10 YR 5/4) very stony sandy loam with
many very fine light olive brown (2.5 Y 5/4) mottles; large subangular
and platy, siltstone; massive; high packing density; very firm soil
strength; few fine fibrous roots; non-calcareous.

Horizon: Depth (cm)	Ap 0-22	Bw1 22-40	Bw2 40-70	BCu 70-100
Sand 600 μm-2 mm %	11	11	15	21
200-600 μm %	9	9	11	28
60-200 μm %	5	4	5	13
Silt 2-60 μm %	52	51	45	25
Clay <2 μm %	23	25	24	13
<0.2 μm %	6	5	5	2
Organic carbon %	4.4	2.4	1.9	0.5
pH in water (1:2.5)	5.6	5.0	5.0	5.1
pH in 0.01M CaCl$_2$ (1:2.5)	4.9	4.3	4.4	4.6
Pyrophosphate ext.				
Fe %	0.4	0.4	0.5	0.2
Al %	0.2	0.2	0.3	0.2
C %	1.2	0.7	1.0	0.3
Residual dithionite ext.				
Fe %	1.1	1.1	1.1	1.1
Bulk density g cm^{-3}	1.15	1.25	1.20	
Available water capacity				
% vol. <2 bar	13	11	11	
<15 bar	20	17	18	
Air capacity % vol.	12	19	24	
Retained water capacity				
% vol.	42	33	30	
Illuvial clay %	0	0	<1	

DAVIDSTOW SERIES
Profile no.: SX 18/6962
Definition: Typical brown podzolic soils. Coarse loamy over
lithoskeletal basic crystalline rock.
Elevation: 240 m O.D.   Slope and aspect: 5° S, convex.
Land use: Permanent grassland.
Horizons:
0-19 cm Ap   Brown to dark brown (10 YR 4/3) slightly stony silt
loam with few fine mottles along root channels; very small to medium
angular platy, tabular and blocky, lava and tuff; moist; moderately
developed medium blocky breaking to fine angular and subangular;
moderately weak soil strength; gradual boundary.

19-50 cm Bs   Dark yellowish brown (10 YR 4/4) moderately stony coarse
sandy silt loam; very small to medium angular platy, tabular and angular,
lava and tuff; moist; moderately developed fine and very fine blocky and
granular; moderately weak soil strength; gradual boundary.

50-65 cm Cu   Olive brown (2.5 Y 4/4) very stony, becoming extremely
stony at base, fine sandy loam; very small to large angular platy and
tabular, lava and tuff; moist; apedal.

Horizon: Depth (cm)	Bs 19-50
Sand 600 μm-2 mm %	13
200-600 μm %	9
60-200 μm %	10
Silt 2-60 μm %	58
Clay <2 μm %	10
Organic carbon %	1.9
pH in water (1:2.5)	6.3
pH in 0.01M CaCl$_2$ (1:2.5)	5.9
Pyrophosphate ext.	
Fe %	0.7
Al %	0.3
C %	0.9
Residual dithionite ext.	
Fe %	1.4

DUNWELL SERIES
Profile no.: NT 91/2213
Definition: Brown rankers. Loamy, lithoskeletal basic crystalline rock.
Elevation: 335 m O.D.   Slope and aspect: 30° S, straight.
Land use: Rough grazing.
Horizons:
0-38 cm Ah   Dark brown (7.5 YR 3/3) moderately stony coarse sandy
loam; small to large angular, andesite; moist; weakly developed fine
granular; low packing density; moderately weak soil strength; abundant
very fine fibrous roots; non-calcareous; sharp wavy boundary.

At 38 cm   Cr   Very stony scree with andesite stones.

Horizon: Depth (cm)	Ah 0-38
Sand 600 μm-2 mm %	47
200-600 μm %	13
60-200 μm %	6
Silt 2-60 μm %	21
Clay <2 μm %	13
Organic carbon %	3.5
pH in water (1:2.5)	4.9
pH in 0.01M CaCl$_2$ (1:2.5)	4.2
Pyrophosphate ext.	
Fe %	0.3
Al %	0.3
C %	1.2
Residual dithionite ext.	
Fe %	2.2

EAST KESWICK SERIES
Profile no.: SJ 21/6286
Definition: Typical brown earths. Fine loamy; drift with siliceous stones.
Elevation: 73 m O.D.   Slope and aspect: Level.
Land use and vegetation: Ley grassland, with Rye Grass and Clover sp.
Horizons:

0-18 cm Ap Dark brown (10 YR 3/3) slightly stony clay loam; medium subrounded and tabular, siltstone, sandstone and quartzite; slightly moist; moderately developed medium subangular blocky; medium packing density; moderately strong soil strength; very firm ped strength; abundant very fine fibrous roots; non-calcareous; clear smooth boundary.

18-41 cm Bw1 Olive brown (2.5 Y 4/4) slightly stony clay loam; medium subrounded and tabular, siltstone, sandstone and quartzite; slightly moist; moderately developed very coarse subangular blocky with dark greyish brown (10 YR 4/2) faces; very firm ped strength; many very fine fibrous roots; non-calcareous; many organic coats; abrupt wavy boundary.

41-63 cm Bw2 Yellowish brown (10 YR 5/6) very stony clay loam; small subangular, siltstone, sandstone and quartzite; moist; weakly developed, adherent fine subangular blocky with brown to dark brown (10 YR 4/3) faces; medium packing density; moderately firm soil strength; common very fine fibrous roots; non-calcareous; many organic coats; abrupt irregular boundary.

63-87 cm BCu Dark yellowish brown (10 YR 4/4) moderately stony clay loam; very small rounded and tabular, siltstone, sandstone and mudstone; slightly moist; massive; high packing density; moderately firm soil strength; few very fine fibrous roots; non-calcareous; abrupt irregular boundary.

87-98 cm 2BCu Olive brown to dark greyish brown (2.5 Y 4/3) moderately stony sandy silt loam; small subangular, siltstone, sandstone and quartzite; moist; massive; medium packing density; moderately firm soil strength; non-calcareous; common clay coats; abrupt smooth boundary.

98-120 cm 3BCx Brown to dark brown (7.5 YR 4/3) moderately stony clay loam; very small subrounded and platy, siltstone; moist; massive; high packing density; very firm soil strength; non-calcareous; many clay coats.

Horizon: Depth (cm)	Ap 0-18	Bw1 18-41	Bw2 41-63	BCu 63-87	2BCu 87-98	3BCx 98-120
Sand 600 µm-2 mm %	8	8	12	11	9	18
200-600 µm %	5	5	9	10	12	10
60-200 µm %	7	8	8	13	26	12
Silt 2-60 µm %	53	54	42	44	44	40
Clay <2 µm %	27	25	29	22	9	20
<0.2 µm %	2	7	5	3	2	4
Organic carbon %	2.9		1.1			0.3
pH in water (1:2.5)	7.2	7.5	7.5	7.5	7.5	7.4
pH in 0.01M CaCl₂ (1:2.5)	6.8	7.0	7.0	7.0	6.9	6.8
Pyrophosphate ext.						
Fe %	0.1	0.1	0.1	0.1		
Al %	0.1	0.1	0.1	0.1		
C %	0.3	0.2	0.2	0.1		
Residual dithionite ext.						
Fe %	1.4	1.7	0.9	0.9		

ELLERBECK SERIES
Profile no.: SJ 35/1010
Definition: Typical brown earths. Loamy-gravelly, very hard siliceous stones.
Elevation: 120 m O.D.   Slope and aspect: Level.
Land use: Permanent grassland.
Horizons:

0-22 cm Ap Very dark greyish brown (10 YR 3/2) moderately stony sandy loam; very small to medium subrounded and angular, quartzite, rhyolite and sandstone; moist; moderately developed medium subangular blocky; low packing density; moderately weak soil and ped strength; many fine fibrous roots; non-calcareous; abrupt wavy boundary.

22-53 cm Bw Dark yellowish brown (10 YR 4/5) very stony sandy loam; small to large subrounded and subangular, quartzite and sandstone; moist; moderately developed medium subangular blocky; low packing density; moderately weak soil and ped strength; many very fine fibrous roots; non-calcareous; clear irregular boundary.

53-64 cm BCx Light yellowish brown (10 YR 6/4) very stony sandy loam; small to medium subrounded and subangular, quartzite and sandstone; moist; massive; high packing density; moderately firm soil strength; common very fine fibrous roots; non-calcareous; silt caps on larger stones; clear irregular boundary.

64-120 cm Cu Yellowish brown (10 YR 5/4) very stony coarse sand; very small subrounded, quartzite; moist; single grain; medium packing density; loose soil strength; few very fine fibrous roots; non-calcareous.

Horizon: Depth (cm)	Ap 0-22	Bw 22-53	BCx 53-64	Cu 64-120
Sand 600 µm-2 mm %	8	8	33	38
200-600 µm %	20	15	21	40
60-200 µm %	23	27	16	10
Silt 2-60 µm %	36	38	21	8
Clay <2 µm %	13	12	9	4
<0.2 µm %	4	3	1	1
Organic carbon %	3.3			
pH in water (1:2.5)	6.0	6.9	7.1	7.3
pH in 0.01M CaCl₂ (1:2.5)	5.2	6.0	6.0	6.3
Bulk density g cm⁻³	1.40	1.45		
Available water capacity				
% vol. <2 bar	12	14		
<15 bar	18	18		
Air capacity % vol.	15	17		
Retained water capacity				
% vol.	31	27		
Illuvial clay %	0	0		2

ESCRICK SERIES
Profile no.: SO 43/1061
Definition: Typical argillic brown earths. Reddish coarse loamy; drift with siliceous stones.
Elevation: 99 m O.D.   Slope and aspect: 1° straight.
Land use: Fallow.
Horizons:

0-28 cm Ap Reddish brown (5 YR 4/3) slightly stony sandy silt loam; medium rounded, micaceous siltstone; moist; moderately developed medium subangular blocky; low packing density; moderately firm soil strength; moderately weak ped strength; common very fine fibrous roots; slightly calcareous; abrupt smooth boundary.

28-51 cm Eb Reddish brown (5 YR 4/3) slightly stony sandy silt loam; medium rounded, micaceous siltstone; moist; moderately developed fine subangular blocky; medium packing density; moderately firm soil and ped strength; common very fine fibrous roots; slightly calcareous; sharp smooth boundary.

51-77 cm Bt Reddish brown (2.5 YR 4/4) slightly stony clay loam; medium rounded, micaceous siltstone; moist; moderately developed medium prismatic; medium packing density; moderately firm soil and ped strength; few very fine fibrous roots; very slightly calcareous; common clay coats; clear smooth boundary.

77-110 cm BCt Dark reddish brown (2.5 YR 3/4) moderately stony sandy silt loam; many medium rounded, micaceous siltstone; moist; weakly developed medium angular blocky; medium packing density; moderately firm soil strength; very slightly calcareous; common clay coats.

Horizon: Depth (cm)	Ap 0-28	Eb 28-51	Bt 51-77	BCt 77-110
Sand 600 µm-2 mm %	5	5	3	9
200-600 µm %	7	8	7	9
60-200 µm %	15	15	17	15
Silt 2-60 µm %	60	59	55	52
Clay <2 µm %	13	13	18	15
<0.2 µm %	5	5	7	7
CaCO₃ equivalent %	2	2	<1	<1
Organic carbon %	0.9	0.9		
pH in water (1:2.5)	7.7	8.0	7.8	7.9
pH in 0.01M CaCl₂ (1:2.5)	7.1	7.3	7.1	7.1
Bulk density g cm⁻³	1.25	1.55	1.65	1.65
Available water capacity				
% vol. <2 bar	7	5	8	7
<15 bar	14	15	13	13
Air capacity % vol.	29	12	10	9
Retained water capacity				
% vol.	25	29	28	28
Illuvial clay %		0	>3	>3

EVERINGHAM SERIES
Profile no.: SE 47/1592
Definition: Typical sandy gley soils. Fine sandy; stoneless drift.
Elevation: 27 m O.D.   Slope and aspect: 1° SE, straight.
Land use: Cereals.
Horizons:

0-30 cm Ap1 Dark brown (10 YR 3/3) very slightly stony fine sandy loam; small rounded, sandstone; moist; weakly developed coarse angular blocky; medium packing density; moderately weak soil strength; common very fine fibrous roots; non-calcareous; few soft ferri-manganiferous concentrations; sharp smooth boundary.

30-33 cm Ap2 Very dark grey (N 3/0) very slightly stony fine sandy loam; small rounded, sandstone; moist; weakly developed coarse angular blocky; medium packing density; moderately weak soil strength; common very fine fibrous roots; non-calcareous; few irregular soft ferri-manganiferous concentrations; sharp smooth boundary.

33-45 cm Bg1 Light brownish grey (2.5 Y 6/2) stoneless loamy fine sand with very many medium yellowish brown (10 YR 5/6) mottles; very weakly developed coarse angular blocky; medium packing density; very weak soil strength; few very fine fibrous roots; non-calcareous; common irregular ferri-manganiferous nodules; clear irregular boundary.

45-70 cm Bg2 Yellowish brown (10 YR 5/6) stoneless loamy fine sand with common medium grey (10 YR 5/1) mottles; moist; massive; medium packing density; moderately weak soil strength; non-calcareous; clear smooth boundary.

70-88 cm Cg Dark greyish brown (2.5 Y 4/2) and yellowish brown (10 YR 5/6) stoneless loamy fine sand with common medium dark grey (10 YR 4/1) mottles; moist; massive; medium packing density; very weak soil strength; non-calcareous; abrupt smooth boundary.

88-104 cm bAh Very dark greyish brown (2.5 Y 3/1) stoneless loamy medium sand with common fine olive brown (2.5 Y 4/4) mottles; very moist; single grain; medium packing density; loose soil strength; abrupt smooth boundary.

104-120 cm bCg Dark grey (N 4/0) stoneless loamy fine sand with many medium reddish brown (5 YR 4/4) mottles; very moist; single grain; low packing density; very weak soil strength; non-calcareous.

Horizon: Depth (cm)	Ap1 0-30	Bg1 33-45	Bg2 45-70	Cg 70-88	bAh 88-104	bCg 104-120
Sand 600 µm-2 mm %	4	5	0	1	4	5
200-600 µm %	17	11	3	17	48	23
60-200 µm %	59	64	83	66	33	57
Silt 2-60 µm %	9	11	5	8	8	9
Clay <2 µm %	11	9	9	8	7	6
Organic carbon %	1.4	0.5	0.3	0.2	2.3	1.6
pH in water (1:2.5)	6.9	7.0	7.7	7.5	7.3	7.4
pH in 0.01M CaCl₂ (1:2.5)	6.3	6.4	6.8	6.8	6.5	6.8
Pyrophosphate ext.						
Fe %	0.1	tr	0.4	0.4		
Al %	0.1	tr	tr	tr		
C %	0.3	0.1	0.1	0.1		
Residual dithionite ext.						
Fe %	1.7	1.7	1.8	1.7		
Bulk density g cm⁻³	1.40	1.45	1.45			
Available water capacity						
% vol. <15 bar	16	11	20			
Air capacity % vol.	21	19	18			
Retained water capacity						
% vol.	26	26	28			
Illuvial clay %		<2	<2	<2	>2	

FENTON SERIES

Profile no.: SE 47/3476
Definition: Cambic stagnogley soils. Fine loamy over clayey; stoneless drift.
Elevation: 23 m O.D.   Slope and aspect: 3° SE, straight.
Land use: Cereals.
Horizons:

0-25 cm Ap   Very dark greyish brown (10 YR 3/2) stoneless sandy clay loam; moist; moderately developed coarse subangular blocky; medium packing density; moderately firm soil strength; common very fine fibrous roots; non-calcareous; few cylindrical soft ferri-manganiferous concretions; abrupt smooth boundary.

25-47 cm Bg1   Greyish brown (2.5 Y 5/2) and dark grey (N 4/0) stoneless clay loam with many fine yellowish brown (10 YR 5/4) and strong brown (7.5 YR 5/6) mottles; moist; weakly developed coarse angular blocky; moderately high packing density; moderately firm soil strength; few very fine fibrous roots; non-calcareous; common rounded ferri-manganiferous nodules; abrupt smooth boundary.

47-60 cm Bg2   Dark grey (5 Y 4/1) moderately stony sandy clay loam with many medium strong brown (7.5 YR 5/6) mottles; small subangular, tabular, sandstone; moist; very weakly developed coarse angular blocky; medium packing density; moderately firm soil strength; few very fine fibrous roots; non-calcareous; few rounded soft ferri-manganiferous concretions; abrupt smooth boundary.

60-85 cm 2BCg   Strong brown (7.5 YR 4/6) stoneless clay with many fine grey (N 5/0) mottles; moist; strongly developed coarse prismatic; high packing density; moderately strong ped strength; few very fine fibrous roots; non-calcareous; gradual smooth boundary.

85-120 cm 2BCgk   Brown (10 YR 5/3) stoneless clay with many fine dark grey (N 4/0) mottles; moist; strongly developed coarse prismatic; high packing density; moderately strong ped strength; few very fine fibrous roots; calcareous; few irregular calcareous concretions.

Horizon: Depth (cm)	Ap 0-25	Bg1 25-47	Bg2 47-60	2BCg 60-85	2BCgk 85-120
Sand 600 μm-2 mm %	3	4	15	0	0
200-600 μm %	25	12	19	1	0
60-200 μm %	32	26	23	2	1
Silt 2-60 μm %	19	30	20	41	42
Clay <2 μm %	21	28	23	56	57
<0.2 μm %	10	12	9	16	16
CaCO$_3$ equivalent %	0	0	0	0	6
Organic carbon %	1.0	0.5	0.4	0.9	0.9
pH in water (1:2.5)	6.8	7.2	7.7	7.6	8.0
pH in 0.01M CaCl$_2$ (1:2.5)	6.1	6.5	6.8	7.0	7.6
Bulk density g cm^{-3}	1.55				
Available water capacity % vol. <15 bar	16				
Air capacity % vol.	1				
Retained water capacity % vol.	39				
Illuvial clay %	0	<1	>2	0	

FOGGATHORPE SERIES

Profile no.: SE 47/1019
Definition: Pelo-stagnogley soils. Clayey; stoneless drift.
Elevation: 17 m O.D.   Slope and aspect: Level.
Land use: Fallow.
Horizons:

0-20 cm Ap   Very dark greyish brown (2.5 YR 3/2) stoneless clay with few very fine dark greyish brown (10 YR 4/2) mottles; moist; moderately developed medium angular blocky; medium packing density; moderately firm soil strength; common fine fleshy roots; non-calcareous; sharp wavy boundary.

20-52 cm Bg   Dark grey (N 4/0) stoneless clay with many medium strong brown (7.5 YR 5/8) mottles; moist; moderately developed coarse prismatic with dark greyish brown (2.5 YR 4/2) faces; medium packing density; very firm ped strength; few very fine fibrous roots; non-calcareous; few irregular soft ferri-manganiferous concentrations; clear smooth boundary.

52-65 cm BCg1   Dark grey (5 Y 4/1) stoneless clay with many medium light olive brown (2.5 YR 5/4) mottles; very moist; moderately developed coarse angular blocky; medium packing density; moderately strong ped strength; non-calcareous; clear smooth boundary.

65-90 cm BCg2   Dark grey (N 4/0) stoneless clay with common medium yellowish brown (10 YR 5/8) mottles; very moist; strongly developed very coarse prismatic with grey (N 5/0) faces; high packing density; moderately strong ped strength; very slightly calcareous; clear smooth boundary.

90-120 cm BCg   Dark grey (10 YR 4/1) stoneless silty clay with many coarse dark yellowish brown (10 YR 4/4) mottles; wet; strongly developed very coarse prismatic; high packing density; very firm soil strength; slightly calcareous.

Horizon: Depth (cm)	Ap 0-20	Bg 20-52	BCg1 52-65	BCg2 65-90	BCgk 90-120
Sand 600 μm-2 mm %	0	0	0	0	0
200-600 μm %	2	2	2	1	1
60-200 μm %	5	7	8	5	8
Silt 2-60 μm %	45	45	44	42	55
Clay <2 μm %	48	46	46	52	36
<0.2 μm %	19	21	18	23	15
CaCO$_3$ equivalent %	<1	<1	<1	<1	2
Organic carbon %	3.6	2.5	4.1	2.1	1.5
pH in water (1:2.5)	7.1	7.5	7.2	7.8	7.7
pH in 0.01M CaCl$_2$ (1:2.5)	6.7	6.9	6.9	7.3	7.7
Bulk density g cm^{-3}	1.05	1.25	1.25		
Available water capacity % vol. <2 bar	12	7	11		
<15 bar	21	17	17		
Air capacity % vol.	8	4	4		
Retained water capacity % vol.	51	50	48		
Illuvial clay %	0	0	0	0	0

FLADBURY SERIES

Profile no.: SE 47/2579
Definition: Pelo-alluvial gley soils. Clayey; river alluvium.
Elevation: 21 m O.D.   Slope and aspect: Level.
Land use: Fallow.
Horizons:

0-24 cm Ap   Dark brown (10 YR 3/3) stoneless clay; moist; moderately developed medium angular blocky; low packing density; moderately weak soil strength; few very fine fibrous roots; non-calcareous; sharp smooth boundary.

24-53 cm Bg1   Greyish brown (2.5 YR 5/2) stoneless clay with very many fine strong brown (7.5 YR 5/8) mottles; moist; moderately developed coarse prismatic with greyish brown (10 YR 5/2) faces; medium packing density; moderately firm ped strength; few very fine fibrous roots; non-calcareous; few irregular soft ferri-manganiferous concentrations; abrupt smooth boundary.

53-78 cm Bg2   Dark grey (10 YR 4/1) slightly stony clay with many medium yellowish brown (10 YR 5/6) mottles; very small subangular, sandstone; very moist; moderately developed coarse prismatic with dark greyish brown (10 YR 4/2) faces; medium packing density; moderately firm soil strength; few very fine fibrous roots; non-calcareous; few rounded ferri-manganiferous nodules; clear smooth boundary.

78-94 cm 2Bg1   Dark greyish brown (10 YR 4/2) stoneless clay loam with many medium reddish brown (5 YR 4/4) mottles; very moist; moderately developed medium prismatic with dark grey (10 YR 4/1) faces; medium packing density; moderately weak soil strength; common very fine fibrous roots; non-calcareous; common rounded soft ferruginous concentrations; abrupt wavy boundary.

94-120 cm 2Bg2   Light grey to grey (10 YR 6/1) stoneless clay with many fine strong brown (7.5 YR 5/6) mottles; wet; weakly developed, adherent medium angular blocky with greyish brown (10 YR 5/2) faces; medium packing density; moderately firm soil strength; few very fine fibrous roots; non-calcareous.

Horizon: Depth (cm)	Ap 0-24	Bg1 24-53	Bg2 53-78	2Bg1 78-94	2Bg2 94-120
Sand 600 μm-2 mm %	0	0	5	4	2
200-600 μm %	3	2	8	8	6
60-200 μm %	9	7	13	20	18
Silt 2-60 μm %	34	33	33	36	38
Clay <2 μm %	54	58	41	32	36
CaCO$_3$ equivalent %			0	0	0
Organic carbon %	6.3	2.2	1.3	1.1	0.9
pH in water (1:2.5)	6.2	6.2	6.8	6.9	7.0
pH in 0.01M CaCl$_2$ (1:2.5)	5.6	5.7	6.2	6.3	6.3
Dithionite ext. Fe %	2.8	3.9	3.2	2.8	3.8
Bulk density g cm^{-3}	0.80	1.15			
Available water capacity % vol. <2 bar	17	9			
<15 bar	27	16			
Air capacity % vol.	8	7			
Retained water capacity % vol.	59	51			
Illuvial clay %		0	0	0	0

FORDHAM SERIES

Profile no.: SK 00/8105
Definition: Typical humic gley soils. Coarse loamy; drift with siliceous stones.
Elevation: 120 m O.D.   Slope and aspect: 1° NNW, straight.
Land use: Permanent grassland.
Horizons:

0-31 cm Ap   Black (5 YR 2/1) slightly stony humose sandy loam; large rounded, quartzite, and subrounded and subangular, sandstone; moist; moderately developed medium subangular blocky; low packing density; moderately weak soil and ped strength; abundant fine fibrous roots; sharp smooth boundary.

31-41 cm Bgf   Yellowish brown (10 YR 5/8) moderately stony loamy sand with many medium brown to dark brown (7.5 YR 4/4) and dark reddish brown (5 YR 3/3) mottles; large rounded, quartzite; moist; very weakly developed medium subangular blocky; medium packing density; loose soil strength; common very fine fibrous roots; common ferri-manganiferous coats around many stones, ferruginous coats on sand grains; abrupt wavy boundary.

41-52 cm Bg1   Light brownish grey (10 YR 6/2) moderately stony loamy sand with common medium yellowish brown (10 YR 5/6) mottles; medium rounded, quartzite; moist; very weakly developed medium subangular blocky; high packing density; loose soil strength; very weak ped strength; common very fine fibrous roots; clear wavy boundary.

52-87 cm Bg2   Light brownish grey (10 YR 6/2) moderately stony sandy loam with very many medium reddish brown (5 YR 5/4), light grey (10 YR 7/1) and light reddish brown (5 YR 6/3) mottles; medium rounded, quartzite; moist; weakly developed very coarse prismatic with light grey to grey (10 YR 6/1) faces; high packing density; moderately firm soil strength; few very fine roots; common sand and silt coats; clear wavy boundary.

87-103 cm Cg   Pale brown (10 YR 6/3) moderately stony loamy sand with many medium reddish yellow (5 YR 6/6), strong brown (7.5 YR 5/6) and grey to light grey (10 YR 6/1) mottles; medium rounded, quartzite; moist; single grain; medium packing density; loose soil strength.

Horizon: Depth (cm)	Ap 0-31	Bgf 31-41	Bg1 41-52	Bg2 52-87	Cg 87-103
Sand 600 μm-2 mm %	7	6	4	7	6
200-600 μm %	36	47	41	33	49
60-200 μm %	31	31	33	30	32
Silt 2-60 μm %	18	11	15	17	6
Clay <2 μm %	8	5	7	13	7
<0.2 μm %	3	2	3	6	4
Organic carbon %	5.6	0.7	0.3		
pH in water (1:2.5)	6.0	6.6	6.3	6.3	6.4
pH in 0.01M CaCl$_2$ (1:2.5)	5.2	5.5	5.6	5.5	5.6
Bulk density g cm^{-3}	1.05	1.50	1.80		
Available water capacity % vol. <2 bar	11	8	5		
<15 bar	21	11	11		
Air capacity % vol.	22'	27	17		
Retained water capacity % vol.	36	16	21		
Illuvial clay %		<2			<1

GOLDSTONE SERIES
Profile no.: SK 10/4626
Definition: Humo-ferric podzols. Sandy gravelly, very hard siliceous stones.
Elevation: 146 m O.D.   Slope and aspect: 4° SE, convex.
Land use: Permanent grassland.
Horizons:
0-3 cm   H   Black (5 YR 2/1) stoneless litter; moist; common fine fibrous roots; non-calcareous; abrupt smooth boundary.

3-20 cm   Ah   Dark reddish brown (5 YR 3/3) very stony sandy loam; medium rounded, quartzite; moist; weakly developed medium subangular blocky; medium packing density; very weak ped strength; many very fine fibrous roots; non-calcareous; sharp smooth boundary.

20-45 cm   Ea   Dark reddish grey (5 YR 4/2) very stony loamy sand; medium rounded, quartzite; moist; weakly developed medium granular; low packing density; very weak ped strength; abundant fine fibrous roots; non-calcareous; sharp irregular boundary.

45-65 cm   Bh   Very dusky red (2.5 YR 2/2) very stony loamy sand; medium rounded, quartzite; moist; massive; medium packing density; many very fine fibrous roots; non-calcareous; many sesquioxidic coats.

Horizon: Depth (cm)	Ah 3-20	Ea 20-45	Bh 45-65
Sand 600 µm-2 mm %	3	1	2
200-600 µm %	42	56	63
60-200 µm %	23	22	16
Silt 2-60 µm %	28	19	15
Clay <2 µm %	4	2	4
Organic carbon %	2.7	0.6	2.3
pH in water (1:2.5)	6.4	5.8	5.2
pH in 0.01M CaCl₂ (1:2.5)	6.1	5.3	4.6
Pyrophosphate ext.			
Fe %	0.1	0.1	0.3
Al %	0.1	0.1	0.2
C %	0.7	0.3	1.0
Residual dithionite ext.			
Fe %	0.5	0.5	0.5

HAFREN SERIES
Profile no.: SD 86/0591
Definition: Ferric stagnopodzols. Loamy over lithoskeletal mudstone and sandstone or slate.
Elevation: 236 m O.D.   Slope and aspect: 4° N, convex.
Land use and vegetation: Rough grazing with Nardus stricta and Agrostis sp.
Horizons:
0-9 cm   Ah   Very dark greyish brown (10 YR 3/2) stoneless humose sandy silt loam; moist; weakly developed fine granular; many very fine fibrous roots; abrupt smooth boundary.

9-24 cm   Eag   Dark grey (10 YR 4/1) slightly stony sandy silt loam with slight ochreous staining along root channels; medium subangular and platy, silty shale; moist; weakly developed fine subangular blocky with very dark greyish brown (10 YR 3/2) faces; moderately weak soil strength; common very fine fibrous roots; abrupt wavy boundary.

24-47 cm   Bs1   Strong brown (7.5 YR 5/6) moderately stony sandy loam; medium and large subangular and platy, gritstone and silty shale; moist; weakly developed fine subangular blocky; medium packing density; moderately weak soil strength; common very fine fibrous roots; gradual wavy boundary.

47-54 cm   Bs2   Yellowish brown (10 YR 5/4) moderately stony sandy loam; medium and large subangular and platy, silty shale; moist; weakly developed medium angular blocky with yellowish brown (10 YR 5/6) faces; medium packing density; moderately weak soil strength; abrupt wavy boundary.

At 54 cm   R   Very dark greyish brown (10 YR 3/2) hard laminated silty shale.

Horizon: Depth (cm)	Ah 0-9	Eag 9-24	Bs1 24-47	Bs2 47-54
Sand 600 µm-2 mm %	8	7	11	11
200-600 µm %	10	10	10	10
60-200 µm %	22	27	31	30
Silt 2-60 µm %	46	42	39	40
Clay <2 µm %	14	14	9	9
Organic carbon %	6.3	3.6	1.3	1.1
pH in water (1:2.5)	6.5	6.6	6.4	6.5
pH in 0.01M CaCl₂ (1:2.5)	6.2	6.0	5.7	6.0
Pyrophosphate ext.				
Fe %	0.6	0.6	1.1	0.6
Al %	0.2	0.2	0.4	0.3
C %	1.3	1.1	0.9	0.5
Residual dithionite ext.				
Fe %	0.9	1.0	2.0	1.6

GREYLAND SERIES
Profile no.: SH 77/7664
Definition: Cambic stagnogley soils. Fine loamy over clayey; drift with siliceous stones.
Elevation: 64 m O.D.   Slope and aspect: Level.
Land use: Permanent grassland with rushes.
Horizons:
0-5 cm   Ahg   Greyish brown (2.5 Y 5/2) very slightly stony clay loam; small, shale; moist; small to medium angular blocky; moderately weak soil strength; abundant roots; abrupt boundary.

5-20 cm   ABg   Greyish brown (2.5 Y 5/2) very slightly stony clay loam with yellowish red (5 YR 4/6) root channels; small, shale; moist; strongly developed angular blocky to prismatic; moderately firm soil strength; abrupt boundary.

20-51 cm   Bg   Light brownish grey (2.5 Y 6/2) slightly stony clay loam with many reddish yellow (7.5 YR 6/6) mottles; large, shale; moist; prismatic; few roots; abrupt boundary.

51-68 cm   BCg   Light grey and light brownish grey (2.5 Y 7/2 and 2.5 Y 6/2) very slightly stony clay with brownish yellow (10 YR 6/6) mottles; shale stones; moist; prismatic; moderately firm soil strength.

Horizon: Depth (cm)	Ahg 0-5	ABg 5-20	Bg 20-51	BCg 51-68
Sand 50 µm-2 mm %	30	28	23	8
Silt 2-50 µm %	51	48	45	45
Clay <2 µm %	19	24	32	47
Loss on ignition %	15.5	10.5	9.5	6.8
pH in water (1:2.5)	5.8	6.0	6.8	6.7
pH in 0.01M CaCl₂ (1:2.5)	5.2	5.3	6.4	6.0

HALL SERIES
Profile no.: SO 85/4591
Definition: Typical brown earths. Coarse loamy over non-calcareous gravelly.
Elevation: 31 m O.D.   Slope and aspect: Level.
Land use and vegetation: Permanent grassland with Lolium perenne.
Horizons:
0-19 cm   Ap   Brown to dark brown (7.5 YR 4/3) very slightly stony humose sandy loam; medium to very small rounded, Bunter quartzite and hard siltstone and sandstone; slightly moist; moderately developed fine subangular blocky; low packing density; moderately weak soil strength; very weak ped strength; many very fine fibrous roots; abrupt smooth boundary.

19-44 cm   Bw1   Brown to dark brown (7.5 YR 4/4) moderately stony sandy loam; medium to very small rounded, Bunter quartzite and hard siltstone and sandstone; slightly moist; weakly developed medium angular blocky; medium packing density; moderately weak soil strength; many very fine fibrous roots; non-calcareous; clear smooth boundary.

44-75 cm   Bw2   Dark brown (7.5 YR 3/4) moderately stony sandy loam; very large to very small rounded, Bunter quartzite, vein quartz, sandstone, siltstone and conglomerate; slightly moist; very weakly developed medium and fine angular blocky; medium packing density; very weak soil strength; many very fine fibrous roots; non-calcareous; clear smooth boundary.

75-105 cm   2Cu   Brown (7.5 YR 5/3) very stony coarse sandy loam; very large to small, but mainly very small, rounded, Bunter quartzite and vein quartz, and rounded, tabular, sandstone and siltstone; very moist; single grain; low packing density; few very fine fibrous roots; non-calcareous.

Horizon: Depth (cm)	Ap 0-19	Bw1 19-44	Bw2 44-75	2Cu 75-105
Sand 600 µm-2 mm %	14	12	14	37
200-600 µm %	46	44	45	29
60-200 µm %	14	16	14	10
Silt 2-60 µm %	16	17	18	17
Clay <2 µm %	10	11	9	7
<0.2 µm %	4	4	4	3
Organic carbon %	5.3	2.1		
pH in water (1:2.5)	6.7	6.0	6.0	5.9
pH in 0.01M CaCl₂ (1:2.5)	6.0	5.0	5.1	5.1
Bulk density g cm⁻³	1.20	1.40	1.40	
Available water capacity				
% vol. <15 bar	23	13	10	
Air capacity % vol.	17	29	32	
Retained water capacity				
% vol.	38	20	16	
Illuvial clay %	0	0	0	

HAMPERLEY SERIES

Profile no.: SO 47/0057
Definition: Stagnogleyic argillic brown earths. Fine silty; drift with siliceous stones.
Elevation: 135 m O.D. Slope and aspect: 4° NNW, concave.
Land use: Permanent grassland.
Horizons:

0-18 cm Ap Brown to dark brown (10 YR 4/3) very slightly stony silty clay loam with few very fine yellowish brown (10 YR 5/6) mottles; small subrounded and tabular, siltstone; moist; strongly developed fine subangular blocky; low packing density; moderately weak soil and ped strength; abundant very fine fibrous roots; non-calcareous; abrupt smooth boundary.

18-45 cm Eb(g) Brown (10 YR 5/3) very slightly stony silt loam with few very fine yellowish brown (10 YR 5/6) mottles; small subrounded and tabular, siltstone; moist; weakly developed medium subangular blocky; low packing density; moderately weak soil and ped strength; many very fine fibrous roots; non-calcareous; smooth boundary.

45-67 cm Bg Brown (10 YR 5/3) moderately stony silt loam with common medium greyish brown (10 YR 5/2) and strong brown (7.5 YR 5/6) mottles; large subrounded and tabular, siltstone; moist; weakly developed medium angular blocky with greyish brown (10 YR 5/2) faces; medium packing density; moderately firm soil strength; moderately weak ped strength; common very fine fibrous roots; non-calcareous; abrupt smooth boundary.

67-105 cm Btg Brown to dark brown (7.5 YR 4/4) moderately stony silty clay loam with many medium strong brown (7.5 YR 5/6) and greyish brown (10 YR 5/2) mottles; large rounded and tabular, siltstone; moist; weakly developed coarse prismatic with grey brown (10 YR 5/2) faces; high packing density; moderately firm soil and ped strength; few very fine fibrous roots; non-calcareous; common clay coats.

Horizon:	Ap	Eb(g)	Bg	Btg
Depth (cm)	0-18	18-45	45-67	67-105
Sand 600 μm-2 mm %	3	4	5	3
200-600 μm %	4	3	5	6
60-200 μm %	8	9	9	7
Silt 2-60 μm %	66	67	69	65
Clay <2 μm %	19	17	12	19
<0.2 μm %	7	6	3	6
Organic carbon %	4.0	1.7		
pH in water (1:2.5)	5.5	6.0	6.3	6.3
pH in 0.01M CaCl₂ (1:2.5)	5.1	5.1	5.4	6.0
Bulk density g cm⁻³	1.05	1.20	1.60	1.65
Available water capacity				
% vol. <2 bar	14	15	14	5
<15 bar	28	22	20	9
Air capacity % vol.	8	22	11	10
Retained water capacity				
% vol.	47	32	29	27
Illuvial clay %			<0.2	>2

HIRAETHOG SERIES

Profile no.: SH 95/1597
Definition: Ironpan stagnopodzol. Loamy over lithoskeletal mudstone and sandstone or slate.
Elevation: 375 m O.D. Slope and aspect: 7° NNW, concave.

Land use and vegetation: Heather moor with Deschampsia sp., Rushes and Mat Grass.
Horizons:
0-24 cm Oh Black (5 YR 2/1) stoneless humified peat; moist; many fine fibrous roots; clear irregular boundary.

24-33 cm Eag Light brownish grey (10 YR 6/2) very slightly stony silty clay loam with few medium light grey (10 YR 7/2) mottles; medium subangular and platy, mudstone; moist; massive; low packing density; moderately weak soil strength; common organic coats; few fine fibrous roots; sharp irregular boundary.

33-34 cm Bf Thin continuous ironpan, black (2.5 YR 2/1) top with red (2.5 YR 4/6) below.

34-45 cm Bs1 Brownish yellow (10 YR 6/6) slightly stony silty clay loam; medium subangular and platy, mudstone; moist; massive; low packing density; moderately weak soil strength; few fine fibrous roots; common organic coats; abrupt irregular boundary.

45-62 cm Bs2 Strong brown (7.5 YR 5/6) slightly stony sandy silt loam; medium subangular and platy, mudstone; moist; massive; low packing density; moderately weak soil strength; few fine fibrous roots; clear wavy boundary.

62-80 cm BCu Dark yellowish brown (10 YR 4/4) slightly stony sandy silt loam; medium subangular and platy, mudstone; moist; massive; low packing density; moderately weak soil strength; clear wavy boundary.

80-120 cm Cu Dark greyish brown (2.5 Y 4/2) very stony silt loam; subangular and platy, mudstone; moist; massive; medium packing density.

Horizon:	Oh	Eag	Bf	Bs1	Bs2	BCu	Cu
Depth (cm)	0-24	24-33	33-34	34-45	45-62	62-80	80-120
Sand 600 μm-2 mm %	1	10	5	12	13	2	
200-600 μm %	2	6	3	9	12	6	
60-200 μm %	2	4	3	5	9	10	
Silt 2-60 μm %	72	55	66	58	56	73	
Clay <2 μm %	23	25	23	16	10	9	
<0.2 μm %	5	4	4	1	1	2	
Organic carbon %	48	5.0	4.3	1.2	1.4	1.0	0.5
pH in water (1:2.5)	3.5	3.9	4.3	4.3	4.5	4.8	5.3
pH in 0.01M CaCl₂ (1:2.5)	2.9	3.4	3.8	3.8	4.2	4.4	4.7
Pyrophosphate ext.							
Fe %			0.1	3.2	1.0	2.0	1.1	0.1
Al %			0.3	0.3	0.3	0.3	0.4	0.2
C %			1.6	1.1	0.5	0.7	0.6	0.3
Residual dithionite ext.							
Fe %			0.2	7.4	1.6	1.8	1.1	1.0
Bulk density g cm⁻³	0.25	0.80		0.95	0.90	1.00	1.65
Available water capacity							
% vol. <2 bar		28	11	8	16	12	
<15 bar		41	23	20	26	19	
Air capacity % vol.	14	3	13	16	13	12	
Retained water capacity							
% vol.	70	64	51	50	49	25	
Illuvial clay %	0	0	0	0	0	0	

HEXWORTHY SERIES

Profile no.: SX 67/4079
Definition: Ironpan stagnopodzols. Loamy over lithoskeletal acid crystalline rock.
Elevation: 385 m O.D. Slope and aspect: 8° WNW, straight.
Land use and vegetation: Coniferous woodland with lodgepole pine and some Sitka spruce.
Horizons:
0-8 cm F Reddish black (10 R 2/1) stoneless, laminated organic horizon of partly decomposed litter of grass, leaves and conifer needles; very moist; abundant fine woody roots; non-calcareous; sharp wavy boundary.

8-18 cm Oh Black (N 2/0) stoneless sandy peat; very moist; massive; low packing density; moderately weak soil strength; abundant very fine woody roots; non-calcareous; abrupt wavy boundary.

18-22 cm Ah Black (10 YR 2/1) moderately stony coarse sandy loam; very small subangular, granite; very moist; weakly developed fine subangular blocky; low packing density; moderately weak soil strength; abundant very fine fibrous roots; non-calcareous; abrupt wavy boundary.

22-33 cm A/Eg Dark reddish brown (5 YR 2/2) stony coarse sandy loam with common medium yellow (10 YR 7/6) mottles which become more numerous lower in horizon and are usually centred on stones many of which are soft and weathered, small subangular, granite; very moist; weakly developed fine subangular blocky; low packing density; moderately weak soil strength; many very fine fibrous roots; non-calcareous; gradual wavy boundary.

33-54 cm Eg Light brownish grey (2.5 Y 6/2) moderately stony coarse sandy loam with many coarse yellowish brown (10 YR 5/4) and brownish yellow (10 YR 6/6) mottles; very small to medium subangular, granite; very moist; weak fine subangular blocky; low packing density; moderately firm soil and ped strength; some dark reddish brown infillings; many very fine fibrous roots; sharp wavy boundary.

At 54 cm Bf Very dusky red (2.5 YR 2/2) thin ironpan; sharp wavy boundary.

54-66 cm Bs Strong brown (7.5 YR 5/6) very stony coarse sandy loam; very small to medium subangular and tabular, granite; very moist; moderately developed fine subangular blocky; low packing density; moderately weak soil strength; common medium woody roots; non-calcareous; common discontinuous ferri-manganiferous coats on stones and peds; clear wavy boundary.

66-87 cm BCu Yellowish brown (10 YR 5/6) very stony loamy coarse sand; very small to very large subangular and tabular, granite; moist; weakly developed fine subangular blocky; low packing density; moderately weak soil strength; common fine woody roots; non-calcareous; clear wavy boundary.

87-120 cm Cu Brown (10 YR 5/3) very stony loamy coarse sand; very small to large subangular and tabular, granite; moist; single grain; medium packing density; moderately weak soil strength; common very fine woody roots; non-calcareous; many sand and silt coats on stones, the larger having silt caps and often underlain by very small stones; gradual wavy boundary.

Horizon:	Oh	Ah	A/Eg	Eg	Bf	Bs	BCu
Depth (cm)	8-18	18-22	22-33	33-54	At 54	54-66	66-87
Sand 600 μm-2 mm %	30	30	33	36	32	40	49
200-600 μm %	27	20	19	21	16	18	18
60-200 μm %	10	12	9	15	11	8	10
Silt 2-60 μm %	25	26	32	22	33	29	18
Clay <2 μm %	8	12	7	6	8	5	5
Organic carbon %	14	2.6	1.4	0.4	1.7	0.6	0.3
pH in water (1:2.5)	4.1	4.3	4.7	4.7	4.6	4.7	4.9
pH in 0.01M CaCl₂ (1:2.5)	3.2	3.5	4.3	4.3	4.2	4.3	4.5
Pyrophosphate ext.							
Fe %		0.1	0.1	0.1	1.1	0.2	tr
Al %		0.3	0.3	0.3	0.4	0.2	0.2
C %		1.4	0.9	0.3	1.0	0.4	0.2
Residual dithionite ext.							
Fe %		0.4	0.6	0.4	7.0	1.0	0.3
Bulk density g cm⁻³				1.25		1.45	1.35
Available water capacity							
% vol. <2 bar				15		15	15
<15 bar				20		20	19
Air capacity % vol.				22		18	22
Retained water capacity							
% vol.				29		27	28
Illuvial clay %				0		0	

HOLLINGTON SERIES

Profile no.: SO 34/1954
Definition: Typical alluvial gley soils. Reddish fine silty; river alluvium.
Elevation: 64 m O.D. Slope and aspect: Level.
Land use: Permanent grassland.
Horizons:
0-10 cm Ah Dark yellowish brown (9 YR 4/4) stoneless silty clay loam with many very fine red (2.5 YR 4/8) mottles; very moist; moderately developed medium subangular blocky; low packing density; abundant very fine fibrous roots; non-calcareous; clear smooth boundary.

10-28 cm Bg1 Reddish brown (5 YR 5/3) stoneless silty clay loam with many medium yellowish red (5 YR 5/8) mottles; very moist; moderately developed medium prismatic with reddish grey (5 YR 5/2) faces; low packing density; abundant very fine fibrous roots; non-calcareous; clear smooth boundary.

28-100 cm Bg2 Reddish brown (2.5 YR 5/3) stoneless silty clay loam with few extremely fine red (2.5 YR 5/6) mottles; wet; weakly developed coarse prismatic with reddish grey (5 YR 5/2) faces; medium packing density; common medium fleshy roots; non-calcareous.

Horizon:	Ah	Bg1	Bg2
Depth (cm)	0-10	10-28	28-100
Sand 200-2 mm %	2	2	1
60-200 μm %	4	5	4
Silt 2-60 μm %	65	65	75
Clay <2 μm %	29	28	20
Organic carbon %	1.8		
pH in water (1:2.5)	5.3	5.6	6.6
pH in 0.01M CaCl₂ (1:2.5)	5.8	4.8	5.5
Bulk density g cm⁻³	0.65	1.10	1.40
Available water capacity			
% vol. <2 bar	28	16	9
<15 bar	35	26	20
Air capacity % vol.	7	7	8
Retained water capacity			
% vol.	64	50	37

HUNTWORTH SERIES

Profile no.: ST 22/1062
Definition: Typical argillic brown earths. Reddish fine loamy over non-calcareous gravelly.
Elevation: 110 m O.D.   Slope and aspect: 1° SSE, convex.
Land use: Ley grassland.
Horizons:

0-20 cm Ap Reddish brown (5 YR 4/3) slightly stony clay loam; medium rounded, quartzite; moist; weakly developed coarse angular blocky; medium packing density; moderately weak soil strength; many fine fibrous roots; clear smooth boundary.

20-34 cm Eb Yellowish red (5 YR 5/6) slightly stony clay loam; medium rounded, quartzite; slightly moist; weakly developed medium subangular blocky; moderately weak soil strength; many fine fibrous roots; gradual smooth boundary.

34-47 cm Bt Reddish brown (5 YR 4/4) moderately stony clay loam; medium rounded, quartzite; slightly moist; moderately developed fine subangular blocky; medium packing density; moderately weak ped strength; many fine fibrous roots; clear smooth boundary.

47-80 cm 2BCu Yellowish red (5 YR 5/6) extremely stony coarse sandy loam; small subrounded, quartzite; dry; weakly developed fine subangular blocky with reddish brown (5 YR 4/4) faces; high packing density; moderately weak ped strength; diffuse wavy boundary.

80-120 cm 2Cu Reddish brown (5 YR 5/4) extremely stony loamy coarse sand; very small angular and platy, slate; slightly moist; single grain; high packing density.

Horizon:	Ap	Eb	Bt	2BCu	2Cu
Depth (cm)	0-20	20-34	34-47	47-80	80-110
Sand 600 μm-2 mm %	11	10	14	32	44
200-600 μm %	14	14	13	30	32
60-200 μm %	16	17	14	9	9
Silt 2-60 μm %	40	41	37	14	5
Clay <2 μm %	19	18	22	15	10
<0.2 μm %	2	2	3	0	2
Organic carbon %	2.4	1.0			
pH in water (1:2.5)	5.6	5.5	5.8	6.0	5.9
pH in 0.01M CaCl₂ (1:2.5)	5.0	4.7	5.1	5.6	5.3

LESNEAGUE SERIES

Profile no.: SW 82/0502
Definition: Typical brown earths. Fine loamy over lithoskeletal basic schist.
Elevation: 64 m O.D.   Slope and aspect: 5° S, straight.
Land use: Early potatoes.
Horizons:

0-12 cm Ap Dark brown (10 YR 3/3) slightly stony clay loam; small subangular, hornblende schist; moist; very weakly developed coarse prismatic; medium packing density; moderately weak soil strength; many very fine fibrous roots; clear smooth boundary.

12-32 cm ABw Dark yellowish brown (10 YR 3/4) slightly stony clay loam; small subangular, hornblende schist; moist; moderately developed fine subangular blocky; low packing density; moderately weak soil and ped strength; many very fine fibrous roots; non-calcareous; clear smooth boundary.

32-76 cm Bw Brown to dark brown (7.5 YR 4/4) moderately stony clay loam; medium angular, hornblende schist; moist; moderately developed fine granular; low packing density; very weak soil strength; common very fine fibrous roots; non-calcareous; gradual smooth boundary.

76-120 cm BCu Yellowish brown (10 YR 5/4) extremely stony coarse sandy silt loam with common fine light yellowish brown (10 YR 6/4) mottles; large angular, hornblende schist; moist; massive; medium packing density; moderately weak soil strength; few very fine fibrous roots; non-calcareous.

Horizon:	Ap	ABw	Bw	BCu
Depth (cm)	0-12	12-32	32-76	76-120
Sand 600 μm-2 mm %	14	19	11	29
200-600 μm %	8	7	5	11
60-200 μm %	7	6	5	8
Silt 2-60 μm %	52	48	59	40
Clay <2 μm %	19	20	20	12
CaCO₃ equivalent %		0	0	0
Organic carbon %	2.8	2.4		
pH in water (1:2.5)	6.4	6.6	6.5	6.7
pH in 0.01M CaCl₂ (1:2.5)	6.0	6.2	6.3	6.4
Pyrophosphate ext.				
Fe %	0.2	0.3	0.4	
Al %	0.2	0.2	tr	
C %	0.6	0.6	0.6	
Residual dithionite ext.				
Fe %	2.5	2.6	2.4	
Bulk density g cm⁻³			1.15	1.00
Available water capacity				
% vol. <2 bar			8	13
<15 bar			17	18
Air capacity % vol.			25	28
Retained water capacity				
% vol.			32	33

ELDER SERIES

Profile no.: SK 05/6123
Definition: Cambic stagnohumic gley soils. Loamy over clayey; drift with siliceous stones.
Elevation: 274 m O.D.   Slope and aspect: 4° SE, straight.
Land use: Permanent grassland.
Horizons:

0-18 cm Ahg Black (10 YR 2/1) slightly stony humose clay loam; medium subangular, sandstone; moist; moderately developed medium subangular blocky; low packing density; abundant very fine fibrous roots; clear smooth boundary.

18-48 cm Eg Dark grey (10 YR 4/1) slightly stony clay loam with many coarse strong brown (7.5 YR 5/8) mottles; medium subangular, sandstone; moist; moderately developed coarse prismatic with grey (10 YR 5/1) faces; medium packing density; abundant very fine fibrous roots; clear irregular boundary.

48-120 cm 2Bg Dark grey (N 4/0) slightly stony silty clay with common medium strong brown (7.5 YR 5/8) mottles; medium subangular, sandstone; very moist; weakly developed very coarse prismatic with grey (10 YR 5/1) faces; high packing density; few very fine fibrous roots along ped faces.

Horizon:	Ahg	Eg	2Bg
Depth (cm)	0-18	18-48	48-100
Sand 500 μm-2 mm %	6	3	3
200-500 μm %	9	12	2
60-200 μm* %	16	21	6
Silt 2-60 μm* %	46	46	50
Clay <2 μm %	23	12	39
Organic carbon %	7.0	0.7	
pH in water (1:2.5)	6.2	5.4	6.3
pH in 0.01M CaCl₂ (1:2.5)	5.8	4.6	5.2
Bulk density g cm⁻³	0.85	1.50	1.45
Available water capacity			
% vol. <2 bar	24	7	9
<15 bar	35	16	16
Air capacity % vol.	13	8	9
Retained water capacity			
% vol.	54	34	36
Illuvial clay %	0	0	0

LLANGENDEIRNE SERIES

Profile no.: SM 90/1690
Definition: Stagnogleyic brown earths. Reddish fine loamy; drift with siliceous stones.
Elevation: 55 m O.D.   Slope and aspect: 1° S, straight.
Land use: Ley grassland.
Horizons:

0-1 cm Ap1 Dark reddish brown (5 YR 3/3) clay loam; moderate medium granular; abundant fine roots; abrupt boundary.

1-25 cm Ap2 Dark reddish brown (5 YR 3/3) very slightly stony clay loam; medium and fine subangular, red siltstone and sandstone, subangular, flint and rounded, quartz; weakly developed medium and fine prismatic breaking to moderate medium subangular blocky; medium packing density; common fine roots; slight ferruginous staining on some weathering siltstones; wormcasts in some channels.

25-55 cm Bw(g)1 Reddish brown (2.5 YR 4/3) slightly stony clay loam with occasional black manganiferous stain on stones and structure faces and yellowish red (5 YR 4/8) mottles; medium and fine angular and subangular red siltstones and occasional subangular, sandstone; weakly developed medium platy becoming massive with depth; medium to high packing density; few fine roots.

55-91 cm Bw(g)2 Reddish brown (2.5 YR 4/3) slightly stony silty clay loam with occasional black manganiferous stain on stones and structure faces and yellowish red (5 YR 4/8) mottles; medium and fine angular and subangular red siltstones and occasional subangular, sandstone; weakly developed medium platy becoming massive with depth; medium to high packing density; few fine roots.

Horizon:	Ap2	Bw(g)1	Bw(g)2	Bw(g)2
Depth (cm)	1-25	25-55	55-75	75-91
Sand 500 μm-2 mm %	7	7	2	5
200-500 μm %	7	5	3	4
60-200 μm %	13	21	9	6
Silt 2-60 μm %	46	49	61	58
Clay <2 μm %	27	18	25	27
CaCO₃ equivalent %		<1		
Organic carbon %	2.5	0.3		
pH in water (1:2.5)	5.9	7.0	6.4	7.0
pH in 0.01M CaCl₂ (1:2.5)	5.6	6.8	5.9	6.8
Pyrophosphate ext.				
Fe %	0.2	0.1	tr	
Al %	0.1	0.1	tr	
C %	0.8	0.2	0.2	
Residual dithionite ext.				
Fe %	2.1	1.9	2.2	

318

LOXHORE SERIES

Profile no.: ST 09/4065
Definition: Typical brown podzolic soils. Fine loamy over lithoskeletal sandstone.
Elevation: 244 m O.D.   Slope and aspect: 12° NNW.
Land use and vegetation: Coniferous woodland - larch.
Horizons:

0-5 cm F&H   Root mat, becoming laminated and more compact with depth with dark brown (7.5 YR 3/2) organic matter and many plant remains, over moist, black (10 YR 2/1), well decomposed humus; abrupt boundary.

5-18 cm Bs1   Strong brown (7.5 YR 5/8) slightly stony sandy clay loam; sandstone stones; slightly moist; very weakly developed subangular blocky; moderately weak soil strength; many roots; clear boundary.

18-46 cm Bs2   Reddish yellow (7.5 YR 6/8) slightly stony sandy clay loam; sandstone stones; very weakly developed subangular blocky; moderately weak soil strength; many roots; clear boundary.

46-56 cm 2Cu   Brownish yellow (10 YR 6/6) very stony sandy loam; sandstone stones; slightly moist; single grain; few roots; merging into shattered rock.

At 56 cm 2Cr   Shattered sandstone.

Horizon: Depth (cm)	Bs1 5-18	Bs2 18-46	2Cu 46-56
Sand 50 um-2 mm %	51	60	73
Silt 2-50 um %	25	18	16
Clay <2 um %	24	22	11
pH in water (1:2.5)	4.6	4.8	4.7
pH in 0.01M CaCl$_2$ (1:2.5)	3.8	4.2	4.8

MERCASTON SERIES

Profile no.: SJ 94/6060
Definition: Typical brown earths. Reddish loamy-gravelly, very hard siliceous stones.
Elevation: 259 m O.D.   Slope and aspect: 7° SE, straight.
Land use: Ley grassland.
Horizons:

0-28 cm Ap   Dark brown (7.5 YR 3/2) moderately stony clay loam; many small rounded, quartzite; moist; moderately developed medium subangular blocky; medium packing density; moderately weak soil and ped strength; common very fine fibrous roots; few soft ferri-manganiferous concentrations; abrupt smooth boundary.

28-45 cm Bw   Dark reddish brown (5 YR 3/4) moderately stony sandy loam; medium rounded, quartzite; moist; weakly developed medium granular; medium packing density; moderately weak soil and ped strength; common very fine fibrous roots; non-calcareous; clear smooth boundary.

45-70 cm BCu   Dark reddish brown (2.5 YR 3/4) very stony sandy loam; small rounded, quartzite; moist; weakly developed fine granular; high packing density; very weak soil strength; few very fine fibrous roots; non-calcareous.

Horizon: Depth (cm)	Ap 0-28	Bw 28-45	BCu 45-70
Sand 600 um-2 mm %	5	12	6
200-600 um %	23	35	47
60-200 um %	15	18	18
Silt 2-60 um %	38	24	20
Clay <2 um %	19	11	9
Organic carbon %	2.6		
pH in water (1:2.5)	5.8	6.1	6.4
pH in 0.01M CaCl$_2$ (1:2.5)	5.4	5.7	5.9

MARTOCK SERIES

Profile no.: SP 78/0992
Definition: Typical stagnogley soils. Fine loamy or fine silty over clayey passing to silty shale or siltstone.
Elevation: 114 m O.D.   Slope and aspect: Level.
Land use: Arable, oilseed rape.
Horizons:

0-24 cm Ap   Dark brown (10 YR 3/3) stoneless silty clay loam; moist; moderately developed coarse subangular blocky; medium packing density; moderately firm soil and ped strength; common fine fibrous roots; non-calcareous; abrupt smooth boundary.

24-42 cm EBg1   Dark greyish brown (10 YR 4/2) stoneless silty clay loam with many very fine dark yellowish brown (10 YR 3/4) mottles; moist; strongly developed fine subangular blocky; medium packing density; moderately firm soil and ped strength; common fine fibrous roots; non-calcareous; few irregular ferri-manganiferous nodules; abrupt smooth boundary.

42-63 cm EBg2   Greyish brown (10 YR 5/2) stoneless silty clay with many coarse reddish brown (5 YR 4/4) mottles; moist; weakly developed coarse angular blocky; medium packing density; moderately firm soil strength; non-calcareous; few irregular ferri-manganiferous nodules; sharp smooth boundary.

63-105 cm Btg   Yellowish brown (10 YR 5/6) stoneless silty clay with many coarse light brownish grey (2.5 Y 6/2) mottles; moist; very weakly developed very coarse prismatic; high packing density; moderately firm soil strength; few fine fibrous roots; non-calcareous.

Horizon: Depth (cm)	Ap 0-24	EBg1 24-42	EBg2 42-63	Btg 63-105
Sand 600 um-2 mm %	3	1	2	<1
200-600 um %	6	4	2	<1
60-200 um %	6	4	4	3
Silt 2-60 um %	54	57	54	57
Clay <2 um %	31	34	38	40
<0.2 um %	13	15	16	17
CaCO$_3$ equivalent %	<1			
Organic carbon %	2.9	3.4		
pH in water (1:2.5)	6.8	6.3	5.5	5.5
pH in 0.01M CaCl$_2$ (1:2.5)	6.6	5.7	5.1	5.1
Bulk density g cm^{-3}	1.15	1.10	1.20	1.40
Available water capacity % vol. <2 bar	13	12	9	7
<15 bar	21	19	15	14
Air capacity % vol.	9	11	12	3
Retained water capacity % vol.	44	47	43	44
Illuvial clay %		0	0	>2

MIDDLETON SERIES

Profile no.: SO 32/6010
Definition: Stagnogleyic argillic brown earths. Reddish fine silty passing to silty shale or siltstone.
Elevation: 186 m O.D.   Slope and aspect: 12° W, straight.
Land use: Permanent grassland.
Horizons:

0-21 cm Ap   Reddish brown (5 YR 4/3) stoneless silty clay loam with few fine yellowish red (5 YR 5/6) mottles; moist; strongly developed fine subangular blocky; low packing density; moderately weak soil and ped strength; many very fine fibrous roots; non-calcareous; abrupt smooth boundary.

21-48 cm Eb(g)   Reddish brown (5 YR 5/4) stoneless silty clay loam with few very fine yellowish red (5 YR 5/6) mottles; moist; moderately developed medium angular blocky with reddish brown (5 YR 5/3) faces; medium packing density; moderately firm soil strength; moderately weak ped strength; common very fine fibrous roots; non-calcareous; abrupt smooth boundary.

48-70 cm Bt(g)   Reddish brown (2.5 YR 4/4) stoneless silty clay loam with few fine yellowish red (5 YR 5/6) mottles; moist; moderately developed medium subangular blocky with reddish brown (5 YR 5/3) faces; high packing density; moderately firm soil and ped strength; few very fine fibrous roots; non-calcareous; abrupt smooth boundary.

70-100 cm BCt   Weak red to reddish brown (2.5 YR 4/3) stoneless silty clay loam; moist; strongly developed fine angular blocky; high packing density; moderately firm soil and ped strength.

Horizon: Depth (cm)	Ap 0-21	Eb(g) 21-48	Bt(g) 48-70	BCt 70-100
Sand 600 um-2 mm %	2	4	1	0
200-600 um %	2	2	1	1
60-200 um %	13	10	3	3
Silt 2-60 um %	63	63	62	67
Clay <2 um %	20	21	33	29
<0.2 um %	8	8	9	7
Organic carbon %	2.4	0.9		
pH in water (1:2.5)	4.9	5.8	5.7	6.0
pH in 0.01M CaCl$_2$ (1:2.5)	4.5	5.1	4.8	5.1
Bulk density g cm^{-3}	1.10	1.30	1.55	1.80
Available water capacity % vol. <2 bar	12	10	6	5
<15 bar	21	20	12	9
Air capacity % vol.	17	14	7	2
Retained water capacity % vol.	41	36	36	30
Illuvial clay %		<2	>2	>2

MOOR GATE SERIES

Profile no.: SX 68/9932
Definition: Humic brown podzolic soils. Coarse loamy over lithoskeletal acid crystalline rock.
Elevation: 350 m O.D.   Slope: 10⁰.
Land use and vegetation: Unenclosed acid grassland with Bent sp., Fescue sp. and bracken.
Horizons:

0-7 cm Ah Black (N 2/0) very slightly stony humose coarse sandy loam; very small subangular, granite; strongly developed fine granular; low packing density; moderately weak soil strength; abundant very fine fibrous roots; abrupt smooth boundary.

7-20 cm ABh Dark brown (7.5 YR 3/2) slightly stony humose coarse sandy loam; small to medium subangular, granite; weakly developed medium subangular blocky; low packing density; moderate firm soil and ped strength; many very fine fibrous roots; abrupt irregular boundary, basal 2 to 3 cm is dark reddish brown (5 YR 2/2) across three quarters of section.

20-30 cm Bs Reddish brown (5 YR 4/4) moderately stony coarse sandy loam; small to medium subangular, granite; weakly developed medium subangular blocky, peds break readily into weakly developed medium granular; low packing density; moderately firm soil strength; moderately weak ped strength; many very fine fibrous roots; clear irregular boundary.

30-44 cm BCu Brown to dark brown (7.5 YR 4/4) moderately stony coarse sandy loam; small subangular, granite; weakly developed coarse angular blocky; moderate packing density; moderately weak soil and ped strength; many very fine fibrous roots; clear smooth boundary.

44-100 cm BCx Brown to dark brown (10 YR 4/3) very stony coarse sandy loam; granite stones; strongly developed medium platy; high packing density; moderately firm soil and ped strength; common roots; gradual smooth boundary.

100-120 cm Cu Dark greyish brown (10 YR 4/2) extremely stony coarse sandy loam; granite stones; single grain; loose soil strength; many coats.

Horizon:	Ah	ABh	Bs	BCu	BCx	Cu
Depth (cm)	0-7	7-20	20-30	30-44	44-100	100-120
Sand 600 μm-2 mm %	23	31	36	31	39	28
200-600 μm %	20	21	17	18	16	28
60-200 μm %	7	10	9	10	10	19
Silt 2-60 μm %	35	27	26	31	28	16
Clay <2 μm %	15	11	12	10	7	9
Organic carbon %	11	5.1	1.9	1.5	0.6	0.6
pH in water (1:2.5)	4.5	4.8	5.0	5.0	5.3	5.6
pH in 0.01M CaCl₂ (1:2.5)	3.8	4.2	4.4	4.6	5.0	5.1
Pyrophosphate ext.						
Fe %	0.6	0.7	0.5	0.2	0.1	tr.
Al %	0.2	0.3	0.4	0.4	0.3	0.3
C %	2.7	2.0	1.0	0.8	0.3	0.3
Residual dithionite ext.						
Fe %	0.8	0.9	0.8	1.0	0.9	0.8

NEATH SERIES

Profile no.: SK 19/6010
Definition: Typical brown earth. Fine loamy over lithoskeletal sandstone.
Elevation: 310 m O.D.   Slope and aspect: 20⁰ SSW, straight.
Land use: Coniferous woodland.
Horizons:

0-3 cm Oh Black (10 YR 2/1) stoneless peat; dry; weakly developed medium subangular blocky; low packing density; common coarse woody roots; non-calcareous; abrupt smooth boundary.

3-23 cm Bw1 Dark yellowish brown (10 YR 4/4) slightly stony silty clay loam; medium subangular and tabular, sandstone; moist; moderately developed fine subangular blocky; medium packing density; common fine woody roots; non-calcareous; clear smooth boundary.

23-46 cm Bw2 Dark yellowish brown (10 YR 4/6) moderately stony silty clay loam; large subangular and tabular, sandstone; moist; weakly developed fine subangular blocky; medium packing density; common fine woody roots; non-calcareous; clear smooth boundary.

46-80 cm BCu Olive brown (2.5 Y 4/4) very stony sandy loam; large subangular and tabular, sandstone; moist; massive; medium packing density; common very fine fibrous roots; non-calcareous.

At 80 cm Cu Soliflucted sandstone rubble.

Horizon:	Oh	Bw1	Bw2	BCu
Depth (cm)	0-3	3-23	23-46	46-80
Sand 600 μm-2 mm %		4	7	20
200-600 μm %		10	12	20
60-200 μm %		28	28	19
Silt 2-60 μm %		29	27	25
Clay <2 μm %		29	26	16
Organic carbon %	24	3.5	1.7	1.0
pH in water (1:2.5)	3.9	3.9	4.2	4.2
pH in 0.01M CaCl₂ (1:2.5)	2.9	3.3	3.7	3.9
Pyrophosphate ext.				
Fe %	0.5	0.6	0.4	0.3
Al %	0.2	0.3	0.3	0.3
C %	7.0	1.2	0.5	0.4
Residual dithionite ext.				
Fe %	0.7	1.2	1.3	1.4

MUNSLOW SERIES

Profile no.: SO 46/7383
Definition: Typical brown earth. Coarse silty over lithoskeletal siltstone.
Elevation: 152 m O.D.   Slope and aspect: 23⁰ NE, straight.
Land use: Deciduous woodland.
Horizons:

1-0 cm L Leaf and twig litter.

0-14 cm Ah Very dark greyish brown (2.5 Y 3/2) slightly stony silt loam; small angular and platy, siltstone; moist; strongly developed fine granular; low packing density; very weak soil and ped strength; abundant very fine fibrous roots; non-calcareous; abrupt smooth boundary.

14-38 cm Bw1 Light olive brown (2.5 Y 5/4) moderately stony silt loam; medium angular and platy, siltstone; moist; moderately developed medium subangular blocky; medium packing density; very weak soil and ped strength; many very fine fibrous roots; non-calcareous; clear smooth boundary.

38-79 cm Bw2 Yellowish brown (10 YR 5/5) moderately stony silt loam; large angular and platy, siltstone; moist; weakly developed medium angular blocky breaking to fine subangular blocky and fine granular; medium packing density; moderately weak soil strength; very weak ped strength; common very fine fibrous roots; non-calcareous; sharp smooth boundary.

At 79 cm R Dark olive (5 Y 3/3) hard thinly bedded micaceous siltstone.

Horizon:	Ah	Bw1	Bw2
Depth (cm)	0-14	14-38	38-79
Sand 600 μm-2 mm %	1	1	2
200-600 μm %	1	1	1
60-200 μm %	2	2	3
Silt 2-60 μm %	80	79	79
Clay <2 μm %	16	17	15
<0.2 μm %	5	5	9
Organic carbon %	3.2	1.8	
pH in water (1:2.5)	4.1	4.4	4.9
pH in 0.01M CaCl₂ (1:2.5)	3.7	3.8	4.1

NETCHWOOD SERIES

Profile no.: SO 56/8761
Definition: Cambic stagnogley soils. Reddish fine silty passing to silty shale or siltstone.
Elevation: 96 m O.D.   Slope and aspect: Level.
Land use: Permanent grassland.
Horizons:

0-26 cm Ap Brown to dark brown (7.5 YR 4/2) stoneless silty clay loam with few very fine dark reddish brown (5 YR 3/4) mottles; moist; moderately developed medium subangular blocky; low packing density; moderately firm soil and ped strength; many fine fibrous roots; non-calcareous; few irregular ferri-manganiferous nodules; abrupt wavy boundary.

26-44 cm Eg Light brown (7.5 YR 6/4) very slightly stony silty clay loam with many medium reddish brown (5 YR 5/4) and strong brown (7.5 YR 5/6) mottles; large subangular and tabular, quartzitic sandstone; moist; weakly developed adherent, medium angular blocky; medium packing density; moderately firm soil and ped strength; many fine fibrous roots; few irregular ferri-manganiferous nodules; common organic coats; clear wavy boundary.

44-69 cm Btg Red (2.5 YR 5/6) stoneless silty clay loam with common medium yellowish red (5 YR 5/8) and light greenish grey (5 GY 7/1) mottles; moist; weakly developed medium prismatic with reddish brown (5 YR 5/3) faces; medium packing density; moderately firm soil strength; many fine fibrous roots; non-calcareous; common irregular ferri-manganiferous nodules; common clay coats; clear wavy boundary.

69-108 cm BC Red (2.5 YR 5/6) stoneless silty clay loam with common medium red (2.5 YR 4/6) mottles; wet; moderately developed fine angular blocky with reddish brown (5 YR 5/4) faces; high packing density; moderately weak soil and ped strength; common very fine fibrous roots; non-calcareous; few rounded ferri-manganiferous concretions; common clay coats.

Horizon:	Ap	Eg	Btg	BC
Depth (cm)	0-26	26-44	44-69	69-108
Sand 600 μm-2 mm %	<1	1	1	<1
200-600 μm %	2	2	2	<1
60-200 μm %	3	2	2	<1
Silt 2-60 μm %	71	69	65	79
Clay <2 μm %	24	26	30	21
<0-2 μm %	13	15	9	8
Organic carbon %	2.5	0.7		
pH in water (1:2.5)	5.5	6.0	6.6	7.0
pH in 0.01M CaCl₂ (1:2.5)	5.3	5.7	6.0	6.3
Bulk density g cm⁻³	1.15	1.40	1.45	1.60
Available water capacity				
% vol. <2 bar	16	10	7	10
<15 bar	27	18	13	20
Air capacity % vol.	2	10	8	5
Retained water capacity				
% vol.	48.	37	37	34
Illuvial clay %	0	<0.2	>2	>2

NEWBIGGIN SERIES

Profile no.: NY 43/9402
Definition: Typical brown earths. Reddish fine loamy; drift with siliceous stones.
Elevation: 167 m O.D. Slope and aspect: 3° W, straight.
Land use: Ley grassland.
Horizons:
0-19 cm Ap Reddish brown (5 YR 4/3) slightly stony clay loam; medium subangular, igneous; moist; moderately developed medium subangular blocky; medium packing density; moderately weak soil and ped strength; many fine fibrous roots; non-calcareous; clear wavy boundary.

19-85 cm Bw Reddish brown (2.5 YR 4/4) moderately stony clay loam; medium subangular, igneous stones; moist; moderately developed medium subangular blocky; medium packing density; moderately firm soil strength; moderately weak ped strength; few very fine fibrous roots; non-calcareous; abrupt wavy boundary.

Horizon: Depth (cm)	Ap 0-19	Bw 19-85
Sand 600 µm-2 mm %	9	6
200-600 µm %	8	6
60-200 µm %	30	28
Silt 2-60 µm %	34	35
Clay <2 µm %	19	25
CaCO₃ equivalent %	0	<1
Organic carbon %	3.7	0.4
pH in water (1:2.5)	5.8	6.5
pH in 0.01M CaCl₂ (1:2.5)	5.6	5.9
Pyrophosphate ext. Fe %	0.4	0.1
Al %	0.2	0.1
C %	0.7	0.2
Residual dithionite ext. Fe %	1.2	1.7

NEWCHURCH SERIES

Profile no.: TF 39/7958
Definition: Pelo-calcareous alluvial gley soils. Clayey; marine alluvium.
Elevation: 3 m O.D. Slope and aspect: Level.
Land use: Permanent grassland.
Horizons:
0-26 cm Apg Very dark greyish brown (10 YR 3/2) stoneless silty clay with very many fine yellowish brown (10 YR 5/4) mottles; moist; moderately developed medium subangular blocky with dark grey (5 Y 4/1) faces; medium packing density; moderately firm soil strength; abundant roots; slightly calcareous; abrupt smooth boundary.

26-38 cm Bg Light grey to grey (5 Y 6/1) stoneless silty clay with many coarse dark greyish brown (2.5 Y 4/2) mottles; moist; moderately developed medium subangular blocky; medium packing density; very firm soil strength; abundant roots; slightly calcareous; gradual smooth boundary.

38-64 cm BCgk Dark greyish brown (10 YR 4/2) stoneless silty clay with common medium brown to dark brown (10 YR 4/3) mottles; moist; moderately developed coarse angular blocky and prismatic with dark grey (5 Y 4/1) faces; high packing density; very firm soil strength; common roots; slightly calcareous with secondary calcium carbonate common on ped faces; gradual boundary.

64-100 cm Cgk Dark reddish grey (5 YR 4/2) stoneless silty clay with many coarse brown to dark brown (10 YR 4/3) mottles; moist; weakly developed coarse angular blocky; high packing density; very firm soil strength; common roots; slightly calcareous with secondary calcium carbonate common on ped faces.

Horizon: Depth (cm)	Apg 0-26	Bg 26-38	BCgk 38-64	Cgk 64-100
Sand 600 µm-2 mm %	<1	<1	<1	<1
200-600 µm %	<1	<1	<1	<1
60-200 µm %	1	1	1	1
Silt 2-60 µm %	49	50	48	52
Clay <2 µm %	50	49	51	47
CaCO₃ equivalent %	3	2	5	4
Organic carbon %	2.2	2.4	1.1	1.0
pH in water (1:2.5)	7.8	7.8	8.2	8.2
pH in 0.01M CaCl₂ (1:2.5)	7.4	7.5	7.8	7.7
Bulk density g cm⁻³	1.15	1.20	1.35	1.40
Available water capacity % vol. <2 bar	12	14	9	11
<15 bar	19	19	12	16
Air capacity % vol.	6	7	3	3
Retained water capacity % vol.	51	49	47	44

NEWNHAM SERIES

Profile no.: SP 25/7271-1
Definition: Typical brown earths. Reddish coarse loamy over non-calcareous gravelly.
Elevation: 46 m O.D. Slope and aspect: Level.
Land use: Ley grassland.
Horizons:
0-31 cm Ap Dark brown (7.5 YR 3/3) slightly stony sandy loam; rounded and subrounded, Bunter pebbles and flint; slightly moist; moderately developed medium subangular blocky; medium packing density; very weak soil and ped strength; many fine fibrous roots; sharp smooth boundary.

31-61 cm Bw Dark brown (7.5 YR 3/4) moderately stony sandy loam; rounded, Bunter pebbles, and subangular, flint; slightly moist; weakly developed medium subangular blocky; medium packing density; very weak soil and ped strength; many fine fibrous roots; sharp wavy boundary.

61-90 cm 2BCu Reddish brown (5 YR 4/4) very stony loamy sand; small and medium, Bunter pebbles and flint; slightly moist; massive; low packing density; very weak soil strength; few fine fibrous roots; few small irregular ferri-manganiferous nodules; abrupt wavy boundary.

Horizon: Depth (cm)	Ap 0-31	Bw 31-61	2BCu 61-90
Sand 600 µm-2 mm %	6	6	9
200-600 µm %	44	45	68
60-200 µm %	20	22	8
Silt 2-60 µm %	17	16	4
Clay <2 µm %	13	11	11
Organic carbon %	3.0	0.7	
pH in water (1:2.5)	6.0	6.7	7.1
pH in 0.01M CaCl₂ (1:2.5)	5.1	6.1	6.4
Bulk density g cm⁻³	1.50	1.55	1.60
Available water capacity % vol. <2 bar	15	9	3
<15 bar	25	12	4
Air capacity % vol.	9	20	19
Retained water capacity % vol.	34	18	7
Illuvial clay %	0	0	8-10

NORDRACH SERIES

Profile no.: SK 15/5092
Definition: Typical paleo-argillic brown earths. Fine silty over clayey over lithoskeletal limestone.
Elevation: 292 m O.D. Slope and aspect: 1° SSW, straight.
Land use: Ley grassland.
Horizons:
0-18 cm Ap Very dark greyish brown (10 YR 3/2) very slightly stony silty clay loam; small angular and tabular, chert; moist; moderately developed fine subangular blocky; low packing density; moderately firm soil strength; many very fine fibrous roots; non-calcareous; sharp smooth boundary.

18-41 cm Eb Brown to dark brown (7.5 YR 4/4) very slightly stony silty clay loam; small angular and tabular, chert; moist; moderately developed fine subangular blocky with dark brown (7.5 YR 3/4) faces; low packing density; very weak soil strength; common very fine fibrous roots; non-calcareous; abrupt wavy boundary.

41-80 cm Bt Yellowish red (5 YR 4/6) stoneless silty clay; moist; moderately developed fine subangular blocky; medium packing density; moderately firm soil strength; common very fine fibrous roots; non-calcareous; abrupt irregular boundary from 71 to 90 cm depth.

At 80 cm R Greyish brown (10 YR 5/2) hard jointed limestone.

Horizon: Depth (cm)	Ap 0-18	Eb 18-41	Bt 41-80	Bt 80-85
Sand 600 µm-2 mm %	1	1	1	1
200-600 µm %	2	2	2	2
60-200 µm %	5	4	4	4
Silt 2-60 µm %	67	70	53	56
Clay <2 µm %	25	23	40	37
<0.2 µm %	7	6	13	15
CaCO₃ equivalent %	<1	0	0	<1
Organic carbon %	4.1	1.7	1.2	0.7
pH in water (1:2.5)	6.5	7.2	7.2	7.5
pH in 0.01M CaCl₂ (1:2.5)	6.1	6.6	6.6	7.0
Pyrophosphate ext. Fe %	0.4	0.3	0.1	0
Al %	0.2	0.2	0.1	0.1
C %	0.7	0.4	0.1	0.1
Residual dithionite ext. Fe %	1.6	1.5	3.2	3.2
Bulk density g cm⁻³	1.15	1.15	1.20	
Available water capacity % vol. <15 bar	22	18	12	
Air capacity % vol.	9	17	15	
Retained water capacity % vol.	47	40	39	

PEPPERTHORPE SERIES

Profile no.: TF 45/8299-2
Definition: Typical alluvial gley soils. Fine silty over clayey; marine alluvium.
Elevation: 2 m O.D.   Slope and aspect: Level.
Land use: Arable.
Horizons:

0-30 cm Ap   Brown to dark brown (7.5 YR 4/2) stoneless silty clay loam; moist; moderately developed medium subangular blocky; medium packing density; moderately firm soil strength; many fine fibrous roots; non-calcareous; abrupt smooth boundary.

30-52 cm Bg   Brown (7.5 YR 5/2) stoneless silty clay loam with many very fine reddish brown (5 YR 5/4) mottles; moist; moderately developed coarse subangular blocky; medium packing density; moderately firm soil strength; common fine fibrous roots; non-calcareous; abrupt smooth boundary.

52-75 cm 2Bg   Reddish grey (5 YR 5/2) stoneless silty clay with many fine strong brown (7.5 YR 5/6) mottles; moist; moderately developed coarse angular blocky with brown (7.5 YR 5/2) faces; medium packing density; moderately firm ped strength; common very fine fibrous roots; slightly calcareous; common soft ferri-manganiferous concentrations; clear smooth boundary.

75-100 cm 2BCg   Reddish grey (5 YR 5/2) stoneless silty clay with many fine reddish brown (5 YR 4/4) mottles; moist; weakly developed, adherent coarse angular blocky with grey (5 YR 5/1) faces; medium packing density; moderately firm soil strength; common very fine fibrous roots; slightly calcareous; common soft ferrimangniferous concentrations.

Horizon: Depth (cm)	Ap 0-30	Bg 30-52	2Bg 52-75	2BCg 75-100
Sand 600 μm-2 mm %	<1	<1	<1	<1
200-600 μm %	<1	<1	1	1
60-200 μm %	8	8	2	1
Silt 2-60 μm %	65	65	56	55
Clay <2 μm %	27	27	41	44
CaCO₃ equivalent %	<1	<1	2	4
Organic carbon %	2.6	1.2	1.1	0.9
pH in water (1:2.5)	6.9	6.8	7.9	8.1
pH in 0.01M CaCl₂ (1:2.5)	6.5	6.3	7.3	7.6

PRINCETOWN SERIES

Profile no.: SX 57/9651
Definition: Cambic stagnohumic gley soils. Loamy over lithoskeletal acid crystalline rock.
Elevation: 421 m O.D.   Slope and aspect: 1° S, convex.
Land use and vegetation: Moorland rough grazing; Molinia grassland.
Horizons:

0-21 cm Oh1   Black (N 2/0) stoneless humified peat; very moist; strongly developed fine subangular blocky; low packing density; moderately weak soil strength; 4 cm thick surface root mat, with abundant very fine fibrous roots; abrupt smooth boundary.

21-27 cm Oh2   Black (N 2/0) stoneless humified peat; very moist; medium subangular blocky; low packing density; moderately weak soil and ped strength; abundant very fine fibrous roots; abrupt smooth boundary.

27-41 cm Ah   Very dark grey (10 YR 3/1) moderately stony coarse sandy loam; very small subangular, granite; very moist; weakly developed fine subangular blocky; low packing density; moderately firm soil strength; many very fine fibrous roots; non-calcareous; few patchy organic coats on vertical fissures; gradual smooth boundary.

41-58 cm Bg   Light brownish grey (10 YR 6/2) moderately stony clay loam with a few very fine yellowish red (5 YR 4/8) and strong brown (7.5 YR 5/6) mottles; very small to medium subangular, granite; very moist; weakly developed coarse subangular; medium packing density; moderately firm ped strength; many very fine fibrous roots; abrupt smooth boundary.

58-83 cm 2Bg   Light brownish grey (10 YR 6/2) very stony coarse sandy loam with common extremely fine yellowish red (5 YR 4/8) mottles; very small to large subangular, granite; moist; mostly single grain but with some subsidiary weakly developed fine subangular blocky structure; moderately weak soil strength; many very fine fibrous roots; common red (2.5 YR 5/8) sesquioxidic coats particularly concentrated on stones; clear smooth boundary.

83-120 cm 2Cu   Brown (7.5 YR 5/4) very stony loamy coarse sand; very small to large subangular, granite; moist; single grain; loose soil strength; common very fine fibrous roots; many reddish sesquioxidic coats on stones.

Horizon: Depth (cm)	Oh1 0-21	Oh2 21-27	Ah 27-41	Bg 41-58	2BCg 58-83	2Cu 83-120
Sand 600 μm-2 mm %			33	22	43	50
200-600 μm %			15	12	21	22
60-200 μm %			11	8	12	12
Silt 2-60 μm %			32	33	14	11
Clay <2 μm %			9	25	10	5
Organic carbon %	43	20	3.3	0.9	0.4	0.2
pH in water (1:2.5)	4.1	4.2	4.6	4.9	4.9	5.1
pH in 0.01M CaCl₂ (1:2.5)	3.1	3.2	3.8	4.1	4.3	4.6
Pyrophosphate ext.						
Fe %			0.1	0.1	0.1	tr
Al %			0.3	0.2	0.2	0.1
C %			1.9	0.4	0.2	0.1
Residual dithionite ext.						
Fe %			0.2	0.1	0.2	0.1
Bulk density g cm⁻³				1.20	1.25	
Available water capacity						
% vol. <2 bar			15	11		
<15 bar			22	21		
Air capacity % vol.			21	16		
Retained water capacity						
% vol.			33	36		

PROLLEYMOOR SERIES

Profile no.: SJ 30/6161
Definition: Typical stagnogley soils. Fine silty over clayey; drift with siliceous stones.
Elevation: 101 m O.D.   Slope and aspect: 1° NW, straight.
Land use: Permanent grassland.
Horizons:

0-19 cm Ap   Dark greyish brown (10 YR 4/2) slightly stony silty clay loam with common fine strong brown (7.5 YR 5/6) mottles; small subrounded and tabular, siltstone; moist; moderately developed medium subangular blocky; medium packing density; moderately weak soil strength; moderately firm ped strength; many fine fibrous roots; non-calcareous; abrupt smooth boundary.

19-35 cm Eg   Dark greyish brown (10 YR 4/2) slightly stony silty clay loam with many fine yellowish brown (10 YR 5/6) mottles; small subrounded and tabular, siltstone; moist; moderately developed medium subangular blocky; medium packing density; moderately firm soil and ped strength; many very fine fibrous roots; non-calcareous; abrupt smooth boundary.

35-68 cm Btg   Greyish brown (10 YR 5/2) moderately stony silty clay with very many medium yellowish brown (10 YR 5/6) mottles; large subrounded and platy, siltstone; slightly moist; moderately developed medium prismatic with light grey to grey (5 Y 6/1) faces; high packing density; moderately firm soil and ped strength; common very fine fibrous roots; common soft ferri-manganiferous concentrations; common clay coats; clear smooth boundary.

68-105 cm BCtg   Brown (10 YR 5/3) slightly stony silty clay loam with very many medium strong brown (7.5 YR 5/8) mottles; common large rounded and tabular, siltstone; moist; weakly developed coarse prismatic with light grey to grey (5 Y 6/1) faces; high packing density; very firm soil and ped strength; few very fine fibrous roots; non-calcareous; common irregular soft ferri-manganiferous concentrations; many clay coats.

Horizon: Depth (cm)	Ap 0-19	Eg 19-35	Btg 35-68	BCtg 68-105
Sand 600 μm-2 mm %	5	3	6	9
200-600 μm %	3	3	3	4
60-200 μm %	6	4	2	2
Silt 2-60 μm %	58	58	51	53
Clay <2 μm %	28	32	38	32
<0.2 μm %	11	12	15	11
Organic carbon %	2.6	1.6		
pH in water (1:2.5)	6.2	6.6	7.1	5.7
pH in 0.01M CaCl₂ (1:2.5)	5.4	5.7	6.1	4.8
Bulk density g cm⁻³	1.20	1.30	1.50	1.55
Available water capacity				
% vol. <2 vol	9	8	5	6
<15 bar	18	15	13	12
Air capacity % vol.	14	12	6	6
Retained water capacity				
% vol.	39	38	36	35
Illuvial clay %		0	2-4	2-4

QUORNDON SERIES

Profile no.: SK 10/7984
Definition: Typical cambic gley soils. Coarse loamy; drift with siliceous stones.
Elevation: 57 m O.D.   Slope and aspect: Level.
Land use: Ley grassland.
Horizons:

0-31 cm Ap   Dark brown (10 YR 3/3) very slightly stony sandy loam; medium rounded, quartzite; moist; moderately developed medium subangular blocky; medium packing density; firm soil strength; moderately firm ped strength; many fine fibrous roots; non-calcareous; abrupt smooth boundary.

31-52 cm Bg1   Yellowish brown (10 YR 5/6) very slightly stony sandy silt loam with very many medium light grey (10 YR 7/2) mottles; medium rounded, quartzite; moist; weakly developed, adherent medium angular blocky with dark greyish brown (10 YR 4/2) faces; medium packing density; moderately weak soil and ped strength; many fine fibrous roots; non-calcareous; few clay coats; abrupt smooth boundary.

52-66 cm Bg2   Yellowish brown (10 YR 5/6) slightly stony sandy loam with many medium pale brown (10 YR 6/3) mottles; medium rounded, quartzite; moist; weakly developed medium subangular blocky with brown (7.5 YR 5/4) faces; medium packing density; moderately porous; moderately weak soil and ped strength; common very fine fibrous roots; non-calcareous; few clay coats; abrupt smooth boundary.

66-84 cm BCg   Light grey to grey (10 YR 6/1) moderately stony sandy loam with common medium strong brown (7.5 YR 5/6) mottles; medium rounded, quartzite; very moist; single grain; medium packing density; non-calcareous; few irregular ferri-manganiferous nodules; abrupt smooth boundary.

84-107 cm 2Cg   Grey (5 YR 5/1) very stony loamy sand with common medium strong brown (7.5 YR 5/6) mottles; small rounded, quartzite; wet; single grain; medium packing density; non-calcareous.

Horizon: Depth (cm)	Ap 0-31	Bg1 31-52	Bg2 52-66	BCg 66-84	2Cg 84-107
Sand 600 μm-2 mm %	4	2	7	7	11
200-600 μm %	28	16	31	27	43
60-200 μm %	20	21	25	31	30
Silt 2-60 μm %	29	44	26	27	13
Clay <2 μm %	19	12	11	8	3
<0.2 μm %	10	9	6	5	2
CaCO₃ equivalent %		0	0	0	0
Organic carbon %	1.2	0.3	0.2		
pH in water (1:2.5)	5.7	6.9	7.2	7.2	7.4
pH in 0.01M CaCl₂ (1:2.5)	5.0	6.4	6.7	6.7	6.8
Bulk density g cm⁻³	1.40	1.50	1.45	1.65	
Available water capacity					
% vol. <2 bar	8	6	7	10	
<15 bar	16	12	17	14	
Air capacity % vol.	12	11	10	7	
Retained water capacity					
% vol.	35	31	35	30	
Illuvial clay %		<2	<2	<2	

RHEIDOL SERIES

Profile no.: SJ 24/0638
Definition: Typical brown earths. Fine loamy over non-calcareous gravelly.
Elevation: 100 m O.D.   Slope and aspect: 4° SW, straight.
Land use: Permanent grassland.
Horizons:

0-17 cm   Ap   Very dark greyish brown (10 YR 3/2) slightly stony clay loam; small rounded and subangular, shale and sandstone; moist; moderately developed medium and fine subangular blocky with medium and fine granular; abundant fine fibrous roots; clear smooth boundary.

17-34 cm   ABw   Dark yellowish brown (10 YR 3/4) slightly stony clay loam; small rounded and subangular, shale and sandstone; moist; moderately developed medium and fine subangular blocky with medium and fine granular; abundant fine fibrous roots; abrupt smooth boundary.

34-66 cm   Bw   Strong brown (7.5 YR 4/6) moderately stony clay loam; small rounded and subangular, shale and sandstone; moist; moderately developed medium and fine subangular blocky with medium and fine granular; abundant fine fibrous roots becoming few with depth; non-calcareous; non-calcareous; clear smooth boundary.

66-90 cm   2Cu   Greyish brown (10 YR 5/2) and brown (10 YR 5/3) very stony loamy gravel; very small and medium rounded, shale and sandstone; slightly moist; single grain.

Horizon: Depth (cm)	Ap 0-17	ABw 17-34	Bw 34-66
Sand 600 µm-2 mm %	23	31	21
200-600 µm %	8	9	10
60-200 µm %	5	5	5
Silt 2-60 µm %	43	37	46
Clay <2 µm %	21	18	18
CaCO₃ equivalent %			0
Organic carbon %	5.0	2.1	
pH in water (1:2.5)	5.4	6.2	6.6
pH in 0.01M CaCl₂ (1:2.5)	5.1	5.7	6.2
Pyrophosphate ext.			
Fe %			0.3
Al %			0.3
Residual dithionite ext.			
Fe %			1.1

RHONDDA SERIES

Profile no.: SN 60/4360
Definition: Cambic stagnohumic gley soils. Loamy over lithoskeletal sandstone.
Elevation: 260 m O.D.   Slope and aspect: 2° W, straight.
Land use and vegetation: Rough grazing with Mat Grass and Wavy Hair Grass.
Horizons:

0-7 cm   Oh   Dark reddish brown (5 YR 2/2) stoneless humified peat; moist; moderately developed fine granular; abundant very fine fibrous roots; abrupt smooth boundary.

7-19 cm   Ah   Black (N 2/0) stoneless humose sandy loam; moist; strongly developed coarse angular blocky; many fine fibrous roots; abrupt wavy boundary.

19-30 cm   Eg   Greyish brown (10 YR 5/2) slightly stony sandy loam with many coarse yellowish brown (10 YR 5/6) mottles; many medium subrounded, sandstone; moist; weakly developed, adherent coarse prismatic with grey (10 YR 5/1) faces; medium packing density; very weak soil strength; many fine fibrous roots; clear smooth boundary.

30-54 cm   Bg1   Grey (10 YR 5/1) moderately stony clay loam with very many coarse yellowish brown (10 YR 5/6) mottles; many medium subrounded, sandstone; moist; weakly developed, adherent coarse prismatic with grey (10 YR 5/1) faces; low packing density; very weak soil strength; many fine fibrous roots; clear smooth boundary.

54-78 cm   Bg2   Prominently mottled grey (10 YR 5/1) and brown to dark brown (7.5 YR 4/4) moderately stony sandy silt loam; medium subangular and tabular, sandstone; moist; massive; medium packing density; moderately weak soil strength; common fine fibrous roots; clear irregular boundary.

78-105 cm   Cu   Greyish brown (10 YR 5/2) very stony sandy silt loam; medium subangular and tabular, sandstone; moist; massive; few fine fibrous roots.

Horizon: Depth (cm)	Ah 7-19	Eg 19-30	Bg1 30-54	Bg2 54-78	Cu 78-105
Sand 600 µm-2 mm %	0	1	3	3	2
200-600 µm %	18	23	10	12	14
60-200 µm %	34	34	25	17	21
Silt 2-60 µm %	41	35	44	47	42
Clay <2 µm %	7	7	18	11	11
<0.2 µm %		4	7	3	3
Organic carbon %	11	2.2	1.5	1.1	0.9
pH in water (1:2.5)	4.1	4.3	4.4	4.7	4.9
pH in 0.01M CaCl₂ (1:2.5)	3.3	3.5	3.9	4.2	4.3
Pyrophosphate ext.					
Fe %		0.1	0.9	0.5	0.2
Al %		0.1	0.3	0.2	0.2
C %		1.2	0.3	0.6	0.2
Residual dithionite ext.					
Fe %		0.2	2.0	1.2	0.9
Bulk density g cm⁻³			1.50	1.10	1.60
Available water capacity					
% vol. <2 bar			7	12	12
<15 bar			24	21	17
Air capacity % vol.			10	16	9
Retained water capacity					
% vol.			34	41	30
Illuvial clay %		0	0	0	

ROUGH TOR SERIES

Profile no.: SX 67/4078
Definition: Ferric stagnopodzols. Loamy over lithoskeletal acid crystalline rock.
Elevation: 390 m O.D.   Slope and aspect: 8° WNW, convex.
Land use and vegetation: Coniferous woodland with Sitka spruce and Lodgepole pine.
Horizons:

0-15 cm   Op   Black (N 2/0) stoneless humified peat; slightly moist; moderately developed medium subangular blocky; low packing density; moderately weak soil strength; abundant fine to coarse fibrous and woody roots; sharp wavy boundary.

15-28 cm   Oh   Black (N 2/0) stoneless humified peat; moist; strongly developed coarse angular blocky; low packing density; moderately firm soil and ped strength; abundant fine to coarse fibrous and woody roots; sharp smooth boundary.

28-36 cm   Eag   Dark grey (10 YR 4/1) moderately stony coarse sandy loam with common extremely fine dark reddish brown (5 YR 3/2) mottles on root channels; very small to large subangular, granite, some ochreous, soft and weathered towards base of horizon; slightly moist; massive; low packing density; moderately firm soil strength; root mat on some stone surfaces; some medium and coarse pockets and channels of organic matter; many fine fibrous roots; abrupt wavy boundary.

36-45 cm   Bs   Yellowish red (5 YR 5/6) moderately stony coarse sandy silt loam; small and medium subangular and tabular, granite, some ochreous, soft and weathered at top of horizon; moist; moderately developed medium angular blocky with yellowish red (5 YR 4/6) faces; low packing density; moderately firm soil strength; moderately weak ped strength; many very fine fibrous roots; common dark reddish brown (2.5 YR 3/4) sesquioxide coats on upper stone surfaces and ped faces near top of horizon; clear smooth boundary.

45-75 cm   BCu   Brown (7.5 YR 5/4) moderately stony coarse sandy silt loam; medium and large subangular, granite; moist; massive; low packing density; moderately weak soil strength; many fine woody roots; gradual wavy boundary.

75-95 cm   BCx   Brown (7.5 YR 5/4) very stony coarse sandy loam; very small to large subangular, granite; slightly moist; strongly developed medium platy with brown to dark brown (7.5 YR 4/2) faces; medium packing density; moderately firm soil and ped strength; common very fine fibrous roots; patchy ferri-manganiferous coats on few stones; gradual wavy boundary.

95-120 cm   Cu   Dark greyish brown (10 YR 4/2) moderately stony coarse sandy loam with horizontal banding of light reddish brown (2.5 YR 6/4); very small to very large subangular, granite; moist; weakly developed coarse platy; medium packing density; moderately weak soil and ped strength; few very fine fibrous roots.

Horizon: Depth (cm)	Oh 15-28	Eag 28-36	Bs 36-45	BCu 45-75	BCx 75-95	Cu 95-120
Sand 600 µm-2 mm %	9	24	22	24	25	40
200-600 µm %	16	18	15	16	22	19
60-200 µm %	13	12	10	9	15	8
Silt 2-60 µm %	49	39	38	43	29	21
Clay <2 µm %	13	7	15	8	9	12
Organic carbon %	35	2.8	2.1	0.7	0.3	0.2
pH in water (1:2.5)	3.8	4.5	4.7	4.8	4.8	5.5
pH in 0.01M CaCl₂ (1:2.5)	3.1	3.9	4.4	4.6	4.5	4.8
Pyrophosphate ext.						
Fe %		0.1	1.2	0.1	0.1	tr
Al %		0.2	0.5	0.3	0.2	0.2
C %		1.6	1.9	0.4	0.2	0.1
Residual dithionite ext.						
Fe %		0.2	0.8	0.8	0.6	0.5
Bulk density g cm⁻³		1.20	0.80	1.30		1.60
Available water capacity						
% vol. <2 bar		19	18	14		10
<15 bar		31	24	19		15
Air capacity % vol.		10	22	21		18
Retained water capacity						
% vol.		43	47	30		21
Illuvial clay %		0	0			

ROWTON SERIES

Profile no.: SO 46/4422
Definition: Typical argillic brown earths. Silty over non-calcareous gravelly.
Elevation: 87 m O.D.   Slope and aspect: Level.
Land use: Permanent grass in apple orchard.
Horizons:

0-18 cm   Ah   Brown to dark brown (10 YR 4/3) slightly stony silt loam with few very fine yellowish brown (10 YR 5/6) mottles associated with roots; very small subrounded and tabular, siltstone; moist; strongly developed fine subangular blocky; low packing density; very weak soil and ped strength; abundant very fine fibrous roots; non-calcareous; abrupt smooth boundary.

18-43 cm   Eb   Yellowish brown (10 YR 5/4) slightly stony silt loam with small subrounded and tabular, siltstone; moist; moderately developed medium subangular blocky; medium packing density; moderately firm soil and ped strength; many very fine fibrous roots; non-calcareous; abrupt smooth boundary.

43-65 cm   Bt1   Dark yellowish brown (10 YR 4/4) slightly stony silty clay loam with medium subrounded and tabular, siltstone; moist; strongly developed medium prismatic with brown to dark brown (10 YR 4/3) faces; medium packing density; moderately firm soil and ped strength; common very fine fibrous roots; non-calcareous; common clay coats; gradual smooth boundary.

65-79 cm   Bt2   Dark yellowish brown (10 YR 4/4) moderately stony silty clay loam; medium subrounded and tabular, siltstone; moist; moderately developed medium prismatic; medium packing density; moderately firm soil and ped strength; few very fine fibrous roots; non-calcareous; common clay coats; sharp wavy boundary.

At 79 cm   Cu   Medium to very small, non-calcareous, bedded siltstone gravel.

Horizon: Depth (cm)	Ah 0-18	Eb 18-43	Bt1 43-65	Bt2 65-79
Sand 600 µm-2 mm %	2	2	1	2
200-600 µm %	1	1	1	2
60-200 µm %	2	2	1	1
Silt 2-60 µm %	79	82	65	61
Clay <2 µm %	16	13	32	34
<0.2 µm %	7	6	13	14
Organic carbon %	2.3	0.6		
pH in water (1:2.5)	5.9	5.0	5.6	5.9
pH in 0.01M CaCl₂ (1:2.5)	5.2	4.4	4.8	5.0
Illuvial clay %			>2	>2

SANDWICH SERIES

Profile no.: SN 69/0840
Definition: Typical sand pararendzinas. Sandy; stoneless drift.
Elevation: 5 m O.D.   • Slope and aspect: 10° W, convex.
Land use and vegetation: Stabilized sand dune grassland with red fescue, rest harrow, buttercup, moss, lichen, ribwort and _Viola_ spp (100 per cent cover).
Horizons:
  0-7 cm   Ah   Dark brown (7.5 YR 3/2) stoneless fine sand; dry; very weakly developed fine subangular blocky; low packing density; very weak soil and ped strength; non-calcareous; clear smooth boundary.

  7-120 cm Cu   Light brownish grey (2.5 Y 6/2) stoneless fine sand; slightly moist; single grain; low packing density; loose soil strength; few fine fibrous roots; slightly calcareous.

Horizon: Depth (cm)	Ah 0-7	Cu 7-50	Cu 50-90
Sand 600 µm-2 mm %	0	0	<1
200-600 µm %	9	9	21
60-200 µm %	88	91	79
Silt 2-60 µm %	2	<1	<1
Clay <2 µm %	1	<1	0
CaCO₃ equivalent %	<1	3	3
Organic carbon %	3.9	0.1	0.1
pH in water (1:2.5)	7.1	7.9	8.2
pH in 0.01M CaCl₂ (1:2.5)	6.7	7.4	7.6

SNARGATE SERIES

Profile no.: TF 45/4551
Definition: Gleyic brown alluvial soils. Coarse silty; marine alluvium.
Elevation: 1 m O.D.   Slope and aspect: Level.
Land use: Fallow.
Horizons:
  0-30 cm   Ap   Brown to dark brown (7.5 YR 4/2) stoneless silt loam; moist; moderately developed coarse subangular blocky; medium packing density; moderately weak soil strength; many very fine fibrous roots; non-calcareous; abrupt smooth boundary.

  30-60 cm Bw(g)   Brown (7.5 YR 5/4) stoneless silt loam with common very fine yellowish red (5 YR 4/8) mottles; moist; weakly developed coarse subangular blocky; medium packing density; moderately weak soil strength; common very fine fibrous roots; non-calcareous; common rounded soft ferri-manganiferous concentrations; clear smooth boundary.

  60-80 cm Bg   Brown (7.5 YR 5/4) stoneless silt loam with common very fine yellowish red (5 YR 5/6) mottles; moist; weakly developed medium angular blocky with brown (7.5 YR 5/2) faces; medium packing density; moderately weak soil strength; common very fine fibrous roots; calcareous; common rounded soft ferri-manganiferous concentrations; clear smooth boundary.

  80-120 cm BCg   Brown (7.5 YR 5/4) stoneless silty clay loam with many very fine pinkish grey (5 YR 7/2) mottles; moist; weakly developed coarse angular blocky with light grey to grey (N 6/0) faces; medium packing density; moderately weak soil strength; few very fine fibrous roots; very calcareous.

Horizon: Depth (cm)	Ap 0-30	Bw(g) 30-60	Bg 60-80	BCg 80-120
Sand 600 µm-2 mm %	0	0	0	0
200-600 µm %	0	0	1	0
60-200 µm %	18	20	14	6
Silt 2-60 µm %	64	64	72	71
Clay <2 µm %	18	16	13	23
CaCO₃ equivalent %	0	0	8	12
Organic carbon %	1.7	0.9	0.5	0.6
pH in water (1:2.5)	6.0	5.9	8.2	8.2
pH in 0.01M CaCl₂ (1:2.5)	5.4	5.5	7.7	7.7

SKIDDAW SERIES

Profile no.: SD 58/9677
Definition: Humic rankers. Loamy or peaty; lithoskeletal mudstone and sandstone or slate.
Elevation: 186 m O.D.   Slope and aspect: Level.
Land use: Rough grazing.
Horizons:
  1-0 cm   LF   Root mat of moss, sheep's fescue, heather and lichens; abrupt smooth boundary.

  0-9 cm   Oh   Black (5 YR 2/1) humified peat; weak medium granular; abundant fine fibrous roots; abrupt irregular boundary.

  At 9 cm   R   Greyish brown (10 YR 5/2) weathered Silurian greywacke - siltstone.

Horizon: Depth (cm)	Oh 0-9
Loss on ignition %	74
N %	1.5
C %	37.5
C:N %	25
pH in water (1:2.5)	4.3
pH in 0.01M CaCl₂ (1:2.5)	3.2

STANWAY SERIES

Profile no.: SO 47/6217
Definition: Typical stagnogley soils. Fine silty passing to silty shale or siltstone.
Elevation: 165 m O.D.   Slope and aspect: 3° NW, straight.
Land use: Permanent grassland.
Horizons:
  0-17 cm   Apg   Dark greyish brown (2.5 Y 4/2) stoneless silty clay loam with many fine yellowish brown (10 YR 5/6) mottles; moist; strongly developed medium subangular blocky; low packing density; moderately weak soil and ped strength; abundant very fine fibrous roots; non-calcareous; abrupt smooth boundary.

  17-37 cm Eg   Greyish brown (2.5 Y 5/2) stoneless silty clay loam with common very fine light olive brown (2.5 Y 5/4) mottles; moist; moderately developed coarse subangular blocky with light brownish grey (2.5 Y 6/2) faces; medium packing density; moderately weak soil strength; moderately firm ped strength; many very fine fibrous roots; non-calcareous; abrupt smooth boundary.

  37-65 cm Btg1   Light brownish grey (2.5 Y 6/2) stoneless silty clay loam with common very fine yellowish brown (10 YR 5/6) mottles; moist; strongly developed medium prismatic; medium packing density; moderately firm soil and ped strength; many very fine fibrous roots; non-calcareous; abrupt smooth boundary.

  65-90 cm Btg2   Light olive brown (2.5 Y 5/4) slightly stony silty clay loam with very many medium light olive grey (5 Y 6/2) and common fine light olive brown (2.5 Y 5/6) mottles; small angular and platy, siltstone; moist; moderately developed medium prismatic with light olive grey (5 Y 6/2) faces; high packing density; moderately firm soil and ped strength; common very fine fibrous roots; non-calcareous; common irregular soft ferri-manganiferous concentrations; common clay coats; gradual smooth boundary.

  90-110 cm BCtg   Olive (5 Y 5/3) slightly stony silty clay loam with many medium light grey to grey (5 Y 6/1) mottles; small angular and platy, siltstone; moist; weakly developed, adherent with light grey to grey (5 Y 6/1) faces, also medium platy structure associated with weathering stones; high packing density; moderately firm soil strength; few very fine fibrous roots; common irregular soft ferri-manganiferous concentrations associated with weathering siltstone; few patchy clay coats around stones.

Horizon: Depth (cm)	Apg 0-17	Eg 17-37	Btg1 37-65	Btg2 65-90	BCtg 90-110
Sand 600 µm-2 mm %	1	3	2	2	1
200-600 µm %	3	3	3	8	2
60-200 µm %	5	4	3	7	4
Silt 2-60 µm %	68	69	67	52	58
Clay <2 µm %	23	21	25	31	35
<0.2 µm %	9	8	7	11	11
CaCO₃ equivalent %					0
Organic carbon %	3.5	1.3			
pH in water (1:2.5)	5.0	5.8	6.1	6.4	6.5
pH in 0.01M CaCl₂ (1:2.5)	4.4	5.0	5.3	5.4	5.7
Bulk density g cm⁻³	1.05	1.25	1.45	1.60	1.60
Available water capacity % vol. <2 bar	15	15	9	6	6
<15 bar	30	25	15	10	11
Air capacity % vol.	3	13	8	4	1
Retained water capacity % vol.	52	39	37	36	40

TAMVATS SERIES
Profile no.: TF 45/8299
Definition: Typical alluvial gley soils. Fine silty; marine alluvium.
Elevation: 2 m O.D.   Slope and aspect: Level.
Land use: Horticultural crops.
Horizons:

0-30 cm Ap   Brown to dark brown (7.5 YR 4/2) stoneless silty clay loam; moist; moderately developed medium subangular blocky; medium packing density; moderately firm soil strength; many fine fibrous roots; non-calcareous; abrupt smooth boundary.

30-60 cm Bg1   Brown (7.5 YR 5/2) stoneless silty clay loam with many very fine yellowish brown (5 YR 5/6) mottles; moist; moderately developed coarse subangular blocky with reddish grey (5 YR 5/2) faces; medium packing density; moderately firm ped strength; common fine fibrous roots; non-calcareous; clear smooth boundary.

60-80 cm Bg2   Brown (7.5 YR 5/4) stoneless silty clay loam with many fine yellowish red (5 YR 5/6) mottles; moist; moderately developed coarse angular blocky with reddish brown (7.5 YR 5/2) faces; medium packing density; moderately firm ped strength; common fine fibrous roots; non-calcareous; clear smooth boundary.

80-100 cm BCg   Reddish brown (5 YR 4/4) stoneless silty clay with many fine yellowish red (5 YR 5/6) mottles; moist; weakly developed medium angular blocky with light grey to grey (N 6/0) faces; medium packing density; moderately firm soil strength; few fine fibrous roots; slightly calcareous.

Horizon: Depth (cm)	Ap 0-30	Bg1 30-60	Bg2 60-80	BCg 80-100
Sand 600 μm-2 mm %	0	0	0	0
200-600 μm %	0	0	1	0
60-200 μm %	8	8	6	3
Silt 2-60 μm %	63	63	64	61
Clay <2 μm %	29	29	29	36
CaCO₃ equivalent %	<1	0	0	4
Organic carbon %	2.7	1.3	0.9	0.9
pH in water (1:2.5)	7.5	7.6	8.0	8.4
pH in 0.01M CaCl₂ (1:2.5)	7.1	7.0	7.5	7.8

TRUSHAM SERIES
Profile no.: SX 65/6613
Definition: Typical brown earths. Fine loamy over lithoskeletal basic crystalline rock.
Elevation: 76 m O.D.   Slope and aspect: 10° SSE, straight.
Land use: Permanent grassland.
Horizons:

0-28 cm Ah   Brown to dark brown (7.5 YR 4/3) moderately stony clay loam; medium subangular and platy, tuff; moist; strongly developed fine subangular blocky; medium packing density; moderately weak soil strength; moderately firm ped strength; many very fine fibrous roots; non-calcareous; gradual smooth boundary.

28-45 cm Bw   Brown to dark brown (7.5 YR 4/4) moderately stony clay loam; large subangular and platy, tuff; moist; strongly developed fine subangular blocky; medium packing density; moderately weak soil and ped strength; many very fine fibrous roots; non-calcareous; clear irregular boundary.

45-120 cm BCu   Strong brown (7.5 YR 5/6) very stony clay loam; large subangular and platy, tuff; moist; weakly developed medium angular blocky; medium packing density; moderately firm soil strength; common very fine fibrous roots; non-calcareous.

Horizon: Depth (cm)	Ah 0-28	Bw 28-45	BCu 45-120
Sand 600 μm-2 mm %	14	14	16
200-600 μm %	11	9	14
60-200 μm %	9	7	10
Silt 2-60 μm %	38	39	39
Clay <2 μm %	28	31	21
Organic carbon %	3.2	1.8	
pH in water (1:2.5)	5.5	6.2	6.3
pH in 0.01M CaCl₂ (1:2.5)	5.0	5.5	5.8
Pyrophosphate ext.			
Fe %	0.7	0.5	0.3
Al %	0.3	0.2	0.6
C %	0.6	0.5	0.2
Residual dithionite ext.			
Fe %	5.4	6.4	9.7

TRINK SERIES
Profile no.: SW 43/9496
Definition: Humus-ironpan stagnopodzols. Loamy over lithoskeletal acid crystalline rock.
Elevation: 137 m O.D.   Slope and aspect: 3° N straight.
Land use: Lowland heath.
Horizons:

0-13 cm Ah   Black (N 2/0) slightly stony humose coarse sandy loam; small angular, bleached quartzite; moist; weakly developed coarse angular blocky; low packing density; moderately weak soil strength; many fine fibrous roots; clear smooth boundary.

13-30 cm Ea(g)   Dark greyish brown (10 YR 4/2) moderately stony sandy loam with common coarse reddish yellow (7.5 YR 6/6) mottles; medium subangular, granite; moist; weakly developed medium angular blocky; medium packing density; moderately weak soil strength; common fine fibrous roots; clear wavy boundary.

30-37 cm Bh   Black (5 YR 2/1) stoneless humose coarse sandy loam; medium subangular, granite, and some boulders; moist; strongly developed fine subangular blocky; low packing density; moderately weak soil and ped strength; common fine fibrous roots; common organic coats; abrupt broken boundary.

37-40 cm Eag   Pinkish grey (7.5 YR 6/2) moderately stony sandy loam with many fine strong brown (7.5 YR 5/8) mottles; many small subangular, granite; moist; massive; medium packing density; moderately firm soil strength; common fine fibrous dead roots concentrated on underlying pan; sharp irregular boundary.

40-41 cm Bf   Thin ironpan of very many plate-like ferruginous nodules.

41-57 cm Bs   Reddish yellow (7.5 YR 6/6) moderately stony coarse sandy loam with few extremely fine reddish yellow (5 YR 6/8) mottles; small subangular, granite; moist; weakly developed medium angular blocky; medium packing density; moderately weak soil strength; few sand or silt coats; gradual smooth boundary.

57-120 cm BCu   Very pale brown to yellow (10 YR 7/5) very stony coarse sandy loam; small subangular, granite; moist; weakly developed coarse platy; moderately firm soil strength; many sand or silt coats on upper stone and structure surface.

Horizon: Depth (cm)	Ah 0-13	Ea(g) 13-30	Bh 30-37	Bs 41-57	BCu 57-120
Sand 600 μm-2 mm %	29	17	22	28	31
200-600 μm %	23	23	21	21	28
60-200 μm %	13	16	14	12	13
Silt 2-60 μm %	27	37	29	32	25
Clay <2 μm %	8	7	14	7	3
Organic carbon %	7.9	1.9	7.6		
pH in water (1:2.5)	4.9	4.9	4.6	5.1	5.1
pH in 0.01M CaCl₂ (1:2.5)	3.9	3.9	4.1	4.5	4.7
Pyrophosphate ext.					
Fe %	0.4	0.1	0.5	1.1	0.2
Al %	0.3	0.2	0.3	0.5	0.2
C %	3.0	1.4	5.1	1.7	0.3
Residual dithionite ext.					
Fe %	0.2	0.1	0.3	0.1	0.5

UNRIPENED GLEY SOILS
Profile no.: TM 22/3829
Definition: Unripened gley soils.
Elevation: 1 m O.D.   Slope and aspect: Level.
Land use: Saltmarsh.
Horizons:

0-17 cm Ahg   Very dark grey (10 YR 3/1) stoneless humose silty clay with common very fine yellowish red (5 YR 4/6) mottles; wet; low packing density; very weak soil strength; slightly fluid when very moist or wet; abundant very fine fibrous roots; non-calcareous.

17-29 cm Cg   Greyish brown (2.5 Y 5/2) stoneless humose silty clay with common fine reddish brown (5 YR 4/4) mottles; wet; low packing density; very weak soil strength; moderately fluid when very moist or wet; many very fine fibrous roots; non-calcareous.

29-46 cm 2Cg1   Greenish grey (5 GY 5/1) stoneless fine sandy silt loam with a few fine yellowish red (5 YR 4/6) mottles; wet; medium packing density; moderately weak soil strength; slightly fluid when very moist or wet; common very fine fibrous roots; non-calcareous.

46-76 cm 2Cg2   Greyish brown (2.5 Y 5/2) stoneless silt loam with very many medium strong brown (7.5 YR 5/8) mottles; wet; moderately firm soil strength; deformable when very moist or wet; common very fine fibrous roots; non-calcareous.

Horizon: Depth (cm)	Ahg 0-17	Cg 17-29	2Cg1 29-46	2Cg2 46-76
Sand 600 μm-2 mm %	0	0	0	1
200-600 μm %	1	2	6	5
60-200 μm %	2	2.	17	12
Silt 2-60 μm %	56	53	65	69
Clay <2 μm %	41	43	12	13
CaCO₃ equivalent %			0	0
Organic carbon %	16	9.6	1.2	0.8
pH in water (1:2.5)	6.0	6.1	7.4	6.7
pH in 0.01M CaCl₂ (1:2.5)	6.0	6.1	7.2	6.5
Conductivity siemens m⁻¹	52	63	31	29
Bulk density g cm⁻³	0.25	0.40	1.60	

VERNOLDS SERIES

Profile no.: SO 34/1984
Definition: Typical stagnogley soils. Reddish fine silty; drift with siliceous stones.
Elevation: 70 m O.D.   Slope and aspect: 1° ENE, straight.
Land use: Permanent grassland under cider orchard.
Horizons:

0-10 cm Ah1 Brown to dark brown (7.5 YR 4/2) stoneless silty clay loam with few extremely fine yellowish brown (10 YR 5/6) mottles along root channels; moist; strongly developed fine subangular blocky; medium packing density; moderately weak soil and ped strength; abundant very fine fibrous roots; non-calcareous; abrupt smooth boundary.

10-22 cm Ah2 Brown to dark brown (10 YR 4/3) very slightly stony silty clay loam; small subrounded and platy, siltstone; moist; moderately developed fine subangular blocky; medium packing density; moderately weak soil and ped strength; many very fine fibrous roots; non-calcareous; abrupt smooth boundary.

22-53 cm Eg Brown (7.5 YR 5/3) very slightly stony silty clay loam with many fine yellowish brown (10 YR 5/6) mottles; small subrounded and platy, siltstone; moist; moderately developed medium prismatic breaking to medium and fine angular blocky; medium packing density; moderately weak soil and ped strength; common very fine fibrous roots; non-calcareous; sharp wavy boundary.

53-77 cm Btg1 Reddish brown (5 YR 4/4) very slightly stony silty clay loam with many medium strong brown (7.5 YR 5/6) mottles; small subrounded and tabular, siltstone; moist; moderately developed coarse prismatic with brown (7.5 YR 5/3) faces; medium packing density; moderately firm ped strength; common very fine fibrous roots; non-calcareous; few clay coats around pores; clear smooth boundary.

77-110 cm Btg2 Reddish brown (2.5 YR 4/4) slightly stony silty clay loam with few very fine brown (7.5 YR 5/3) mottles; small rounded and tabular, siltstone; moist; weakly developed medium prismatic with reddish brown (5 YR 5/3) faces becoming platy towards the base; high packing density; very firm ped strength; few very fine fibrous roots; non-calcareous; common irregular soft ferri-manganiferous concentrations; few clay coats around stones.

Horizon:	Ah1	Ah2	Eg	Btg1	Btg2
Depth (cm)	0-10	10-22	22-53	53-77	77-110
Sand 600 μm-2 mm %	1	6	1	2	3
200-600 μm %	4	4	5	5	8
60-200 μm %	7	8	6	6	3
Silt 2-60 μm %	66	63	65	60	59
Clay <2 μm %	20	19	23	27	27
<0.2 μm %	7	6	6	6	6
Organic carbon %	3.8	1.8			
pH in water (1:2.5)	5.3	6.0	6.6	6.8	6.7
pH in 0.01M CaCl2 (1:2.5)	5.0	5.8	6.2	6.4	6.3

WALLASEA SERIES

Profile no.: TF 45/5338
Definition: Pelo-alluvial gley soils. Clayey; marine aluvium.
Slope and aspect: Level.
Land use: Fallow.
Horizons:

0-30 cm Apg Dark greyish brown (10 YR 4/2) stoneless silty clay with common very fine strong brown (7.5 YR 5/6) mottles; moist; moderately developed coarse subangular blocky; high packing density; very firm soil strength; many very fine fibrous roots; non-calcareous; abrupt smooth boundary.

30-50 cm Bg1 Grey (10 YR 5/1) stoneless clay with many fine yellowish red (5 YR 5/6) mottles; moist; moderately developed medium angular blocky; medium packing density; very firm ped strength; common very fine fibrous roots; non-calcareous; abrupt smooth boundary.

50-55 cm Bg2 Grey (5 Y 5/1) stoneless silty clay with common fine strong brown (7.5 YR 5/6) mottles; moist; moderately developed medium angular blocky; medium packing density; moderately firm ped strength; common very fine fibrous roots; non-calcareous; clear smooth boundary.

55-70 cm Bg3 Grey (5 Y 5/0) stoneless silty clay with many fine strong brown (7.5 YR 5/6) mottles; moist; moderately developed medium angular blocky; medium packing density; moderately firm ped strength; common very fine fibrous roots; non-calcareous; abrupt smooth boundary.

70-110 cm BCg Reddish grey (5 YR 5/2) stoneless silty clay with many medium yellowish red (5 YR 4/6) mottles; very moist; moderately developed coarse prismatic with brown to dark brown (7.5 YR 4/2) faces; medium packing density; moderately firm ped strength; common very fine fibrous roots; non-calcareous.

Horizon:	Apg	Bg1	Bg2	Bg3	BCg
Depth (cm)	0-30	30-50	50-55	55-70	70-110
Sand 600 μm-2 mm %	0	0	0	0	0
200-600 μm %	0	0	0	0	0
60-200 μm %	2	1	0	0	0
Silt 2-60 μm %	50	42	50	50	45
Clay <2 μm %	48	57	50	50	55
<0.2 μm %	19	17	14	15	20
CaCO3 equivalent %	0	0	0	0	0
Organic carbon %	2.1	1.1	1.1	0.9	1.1
pH in water (1:2.5)	6.7	7.3	7.4	7.5	7.2
pH in 0.01M CaCl2 (1:2.5)	6.3	6.7	6.5	6.8	6.9

WALFORD SERIES

Profile no.: SO 53/2277
Definition: Typical brown alluvial soils. Reddish coarse loamy; river alluvium.
Elevation: 50 m O.D.   Slope and aspect: Level.
Land use: Permanent grassland.
Horizons:

0-30 cm Ap Reddish brown to yellowish red (5 YR 4/5) silt loam; moderate medium and fine subangular blocky with reddish brown (3.5 YR 4/3) faces; abundant fine and medium fibrous roots; gradual smooth boundary.

30-100 cm Bw Yellowish red (5 YR 4/6) sandy silt loam; moderate medium prismatic breaking to fine subangular blocky; abundant fine fibrous roots becoming common with depth.

Horizon:	Ap	Bw
Depth (cm)	0-30	30-100
Sand 200-2 mm %	1	1
60-200 μm %	16	24
Silt 2-60 μm %	65	61
Clay <2 μm %	18	14
Organic carbon %	0.9	0.4
pH in water (1:2.5)	6.1	6.3
pH in 0.01M CaCl2 (1:2.5)	5.4	5.6

WETTON SERIES

Profile no.: SN 71/3490
Definition: Humic rankers. Loamy or peaty, lithoskeletal limestone.
Elevation: 490 m O.D.   Slope and aspect: 1° NNE, convex.
Land use: Rough grazing.
Horizons:

0-14 cm Ah Very dark greyish brown (10 YR 3/2) stoneless humose clay loam; moist; strongly developed fine granular; low packing density; moderately weak soil strength; many very fine fibrous roots; non-calcareous; abrupt smooth boundary.

14-23 cm A/Cu Very dark greyish brown (10 YR 3/2) very stony humose clay loam; medium angular and tabular, limestone; moist; strongly developed fine granular; low packing density; moderately weak soil strength; many very fine fibrous roots; non-calcareous.

At 23 cm   Limestone.

Horizon:	Ah
Depth (cm)	0-14
Sand 600 μm-2 mm %	2
200-600 μm %	11
60-200 μm %	22
Silt 2-60 μm %	45
Clay <2 μm %	20
Organic carbon %	6.3
pH in water (1:2.5)	5.4
pH in 0.01M CaCl2 (1:2.5)	4.7

WICKHAM SERIES

Profile no.: TQ 64/0792
Definition: Typical stagnogley soils. Fine loamy or fine silty drift over clayey passing to clay or mudstone.
Elevation: 42 m O.D.   Slope and aspect: 1° NNE, straight.
Land use: Orchard.
Horizons:

0-22 cm Ah Dark greyish brown (10 YR 4/2) very slightly stony silty clay loam with few fine greyish brown (10 YR 5/2) mottles; medium subangular and tabular, chert; moist; moderately developed medium subangular blocky; medium packing density; moderately firm soil strength; many very fine fibrous roots; non-calcareous; clear wavy boundary.

22-45 cm Eg Brown (10 YR 5/3) slightly stony silty clay loam with many fine strong brown (7.5 YR 5/6) mottles; medium subrounded and tabular, chert; very moist; weakly developed, adherent coarse subangular blocky with light brownish grey (2.5 Y 6/2) faces; medium packing density; moderately firm soil and ped strength; common very fine fibrous roots; non-calcareous; few rounded ferruginous concretions; gradual wavy boundary.

45-65 cm 2Btg Light grey (5 Y 7/1) slightly stony silty clay with many fine strong brown (7.5 YR 5/8) mottles; medium subangular and tabular, chert; very moist; weakly developed, adherent medium prismatic; high packing density; moderately firm soil and ped strength; few fine fibrous roots; very slightly calcareous; gradual wavy boundary.

65-110 cm 2BCg Light grey to grey (5 Y 6/1) stoneless silty clay with many fine strong brown (7.5 YR 5/8) mottles; moist; weakly developed, adherent coarse prismatic; high packing density; very firm soil and ped strength; very slightly calcareous.

Horizon: Depth (cm)	Ah 0-22	Eg 22-45	2Btg 45-65	2BCg 65-110
Sand 600 µm-2 mm %	1	4	2	0
200-600 µm %	1	3	4	0
60-200 µm %	1	5	4	0
Silt 2-60 µm %	74	65	51	56
Clay <2 µm %	23	23	41	44
<0.2 µm %	11	10	30	21
CaCO₃ equivalent %		0	0.9	0.6
Organic carbon %	1.6	0.5		0.2
pH in water (1:2.5)	6.1	7.4	7.5	7.4
pH in 0.01M CaCl₂ (1:2.5)	5.8	6.8	6.8	5.7
Bulk density g cm⁻³	1.45	1.50	1.45	1.50
Available water capacity				
% vol. <2 bar	12	5	7	7
<15 bar	16	10	14	11
Air capacity % vol.	16	9	14	11
Retained water capacity				
% vol.	38	32	43	41
Illuvial clay %	0	<1	>2	<2

WIGTON MOOR SERIES

Profile no.: SE 54/5480
Definition: Typical cambic gley soils. Fine loamy; drift with siliceous stones.
Elevation: 23 m O.D.   Slope and aspect: Level.
Land use: Field vegetables.
Horizons:

0-30 cm Ap Very dark greyish brown (10 YR 3/2) slightly stony clay loam; small stones; moist; moderately developed medium subangular blocky with very dark grey (10 YR 3/1) faces; high packing density; moderately firm soil strength; abundant fine fibrous roots; non-calcareous; abrupt irregular boundary.

30-53 cm Bg1 Strong brown (7.5 YR 5/6) slightly stony clay loam with many extremely fine brown (10 YR 5/3) mottles; very small stones; moist; coarse subangular blocky with dark greyish brown (10 YR 4/2) faces; high packing density; moderately strong soil strength; common fine fibrous roots; non-calcareous; few ferri-manganiferous nodules; clear irregular boundary.

53-70 cm Bg2 Dark grey (N 4/0) slightly stony clay loam with common very fine strong brown (7.5 YR 5/6) mottles; coarse subangular blocky with dark greyish brown (10 YR 4/2) ped faces; high packing density; moderately strong soil strength; common fine fibrous roots; non-calcareous; few ferri-manganiferous nodules; clear irregular boundary.

70-120 cm BCg Brown to dark brown (7.5 YR 4/4) slightly stony sandy clay loam with common very fine yellowish red (5 YR 5/6) mottles; very small stones; moist; weakly developed very coarse prismatic with brown to dark brown (7.5 YR 4/2) ped faces; high packing density; moderately firm soil strength; common fine fibrous roots; non-calcareous; few ferri-manganiferous nodules.

Horizon: Depth (cm)	Ap 0-30	Bg1 30-53	Bg2 53-70	BCg 70-120
Sand 600 µm-2 mm %	27	3	3	2
200-600 µm %	9	9	7	10
60-200 µm %	5	34	21	43
Silt 2-60 µm %	32	24	36	22
Clay <2 µm %	27	30	33	23
<0.2 µm %	6	10	9	10
CaCO₃ equivalent %		0	0	0
Organic carbon %	3.7	0.8		
pH in water (1:2.5)	4.3	7.2	7.4	7.5
pH in 0.01M CaCl₂ (1:2.5)	4.1	6.7	6.6	6.9
Dithionite ext. Fe %	2.4	2.6	1.9	2.0
Bulk density g cm-3	1.60	1.55		1.55
Available water capacity				
% vol. <2 bar	10	10		10
<15 bar	17	16		14
Air capacity % vol.	8	8		11
Retained water capacity				
% vol.	33	34		31
Illuvial clay %		<2	1-2	5-7

WILDERHOPE SERIES

Profile no.: SJ 17/7721
Definition: Typical argillic brown earths. Fine loamy over litho-skeletal limestone.
Elevation: 250 m O.D.   Slope and aspect: 4° W, straight.
Land use: Ley grassland.
Horizons:

0-24 cm Ap Dark brown (10 YR 3/3) very slightly stony clay loam; very small to medium tabular, mudstone and rounded, sandstone, limestone and rhyolite; moist; moderately developed medium and coarse subangular blocky; moderately weak soil strength; common to abundant fine fibrous roots; non-calcareous; abrupt wavy boundary.

24-36 cm AEb Brown to dark brown (10 YR 4/3) very slightly stony clay loam; very small to medium tabular, mudstone and rounded, sandstone and limestone; moist; weakly to moderately developed coarse subangular blocky; moderately weak soil strength; common fine fibrous roots; non-calcareous; gradual irregular boundary.

36-47 cm Eb Strong brown (7.5 YR 5/8 and 5/6) very slightly stony clay loam; very small and small tabular, mudstone and rounded, limestone; moderately developed fine subangular blocky; moderately weak soil strength; few fine fibrous roots; non-calcareous; clear undulating boundary.

47-74 cm Bt Brown (7.5 YR 5/4) stoneless clay loam; moist; moderately developed medium and coarse subangular blocky to prismatic becoming medium subangular and angular blocky above limestone; moderately weak soil strength; few fine fibrous roots; very slightly calcareous; abrupt irregular boundary.

At 74 cm R Jointed weathered limestone with brown to dark brown (10 YR 4/4) manganiferous staining in hollows.

Horizon: Depth (cm)	Ap 0-24	AEb 24-36	Eb 36-47	Bt 47-74
Sand 600 µm-2 mm %	4	4	3	1
200-600 µm %	20	19	16	7
60-200 µm %	23	23	21	12
Silt 2-60 µm %	35	36	41	56
Clay <2 µm %	18	18	19	24
CaCO₃ equivalent %		0	0	<1
Organic carbon %	5.5			
pH in water (1:2.5)	6.1	6.8	7.0	7.3
pH in 0.01M CaCl₂ (1:2.5)	5.4	6.2	6.4	6.7
Pyrophosphate ext.				
Fe %	0.3	0.3	0.2	0.1
Al %	0.2	0.2	0.2	0.1
Residual dithionite ext.				
Fe %	1.3	1.4	1.7	2.1
Illuvial clay %	0	0	0	2-3

WISBECH SERIES

Profile no.: TF 45/9843
Definition: Calcareous alluvial gley soils. Coarse silty; marine alluvium.
Elevation: 4 m O.D.   Slope and aspect: Level.
Land use: Cereals.
Horizons:

0-25 cm Ap Dark greyish brown (10 YR 4/2) stoneless silt loam with few very fine strong brown (7.5 YR 5/6) mottles; moist; moderately developed medium subangular blocky; medium packing density; moderately weak soil strength; many very fine fibrous roots; very calcareous; abrupt smooth boundary.

25-33 cm BCg Greyish brown (10 YR 5/2) stoneless silt loam with many fine and common medium reddish brown (5 YR 4/4) mottles; moist; moderately developed medium platy with superimposed weak fine angular blocky; medium packing density; very weak soil strength; common very fine fibrous roots; very calcareous; abrupt smooth boundary.

33-49 cm Cg Grey (10 YR 5/1) stoneless fine sandy silt loam with many fine and common medium reddish brown (5 YR 4/4) mottles; moist; strongly developed medium platy with pale brown (10 YR 6/3) faces; low packing density; very weak soil strength; common very fine fibrous roots; calcareous; clear smooth boundary.

49-59 cm 2Cg Grey (10 YR 5/1) stoneless fine sandy loam with many fine and common medium reddish brown (5 YR 4/4) mottles; moist; moderately developed medium platy with pale brown (10 YR 6/3) faces; medium packing density; very weak soil strength; few very fine fibrous roots; calcareous; abrupt smooth boundary.

59-88 cm 3Cg1 Brown to dark brown (10 YR 4/3) stoneless loamy fine sand with many fine dark reddish brown (5 YR 2/2) mottles; moist; strongly developed fine platy; medium packing density; very weak soil strength; few very fine fibrous roots; calcareous; abrupt smooth boundary.

88-120 cm 3Cg2 Dark brown to brown (10 YR 4/3) stoneless loamy fine sand with common fine and medium strong brown (7.5 YR 5/8) mottles; moist; strongly developed medium platy; medium packing density; very weak soil strength; calcareous.

Horizon: Depth (cm)	Ap 0-25	BCg 25-33	Cg 33-49	2Cg 49-59	3Cg1 59-88	3Cg2 88-120
Sand 600 µm-2 mm %	<1	<1	<1	0	0	0
200-600 µm %	<1	<1	<1	<1	0	0
100-200 µm %	<1	1	1	1	3	5
60-100 µm %	17	10	37	56	71	74
Silt 2-60 µm %	70	76	57	39	24	19
Clay <2 µm %	12	13	5	4	2	2
CaCO₃ equivalent %	12	11	7	7	5	5
Organic carbon %	1.6	1.6	0.8	0.4	0.2	0.2
pH in water (1:2.5)	8.2	8.3	8.5	8.5	8.6	8.4
pH in 0.01M CaCl₂ (1:2.5)	7.6	7.8	7.9	7.9	8.0	8.1
Dithionite ext. Fe %	1.1	1.3	1.0	0.8	0.7	0.4
Bulk density g cm⁻³	1.35		1.35	1.45	1.40	
Available water capacity						
% vol. <2 bar	13		24	30	27	
<15 bar	21		30	34	30	
Air capacity % vol.	12		10	5	11	
Retained water capacity						
% vol.	34		39	41	36	

WYRE SERIES
Profile no.: SP 25/3863
Definition: Pelogleyic brown alluvial soils. Clayey; river alluvium.
Elevation: 37 m O.D.   Slope and aspect: Level.
Land use: Permanent grassland.
Horizons:
  0-16 cm Ah(g)   Dark brown (7.5 YR 3/3) stoneless clay with few fine
brown to dark brown (7.5 YR 4/2) mottles; moist; moderately developed
coarse subangular blocky; medium packing density; moderately weak soil
strength; moderately firm ped strength; abundant fine fibrous roots; non-
calcareous; abrupt wavy boundary.

  16-58 cm Bw(g)   Brown to dark brown (10 YR 4/3) stoneless clay;
moist; moderately developed medium subangular blocky with dark greyish
brown (10 YR 4/2) faces; medium packing density; moderately firm soil and
ped strength; common fine fibrous roots; non-calcareous; gradual smooth
boundary.

  58-100 cm BC(g)   Brown to dark brown (10 YR 4/3) stoneless clay with
many medium greyish brown (10 YR 5/2) mottles; very moist; weakly
developed, adherent medium angular blocky; high packing density;
moderately firm soil strength; few fine fibrous roots; non-calcareous;
few irregular ferri-manganiferous nodules.

Horizon: Depth (cm)	Ah(g) 0-16	Bw(g) 16-58	BC(g) 58-100
Sand 600 µm-2 mm %	0	0	0
200-600 µm %	1	1	1
60-200 µm %	5	5	5
Silt 2-60 µm %	34	34	35
Clay <2 µm %	60	60	59
<0.2 µm %	21	29	28
$CaCO_3$ equivalent %		0	0
Organic carbon %	5.9	1.9	
pH in water (1:2.5)	6.4	6.7	6.9
pH in 0.01M $CaCl_2$ (1:2.5)	6.2	6.3	6.4
Bulk density g cm^{-3}	1.05	1.15	1.25
Available water capacity			
% vol. <2	7	6	20
<15 bar	18	15	25
Air capacity % vol.	10	7	1
Retained water capacity			
% vol.	50	50	52

INDEX

National Grid references are given in brackets after place names